AIRCRAFT WIRING & ELECTRICAL INSTALLATION

Production Staff

Lead Illustrator Amy Siever
Illustrator Dustin Blyer
Designer/Production Coordinator Roberta Byerly

© Copyright 2005 by
Avotek Information Resources, LLC.
All Rights Reserved

International Standard Book Number 1-933189-07-X
ISBN 13: 978-1-933189-07-9
Order # T-WIEL-0101

For Sale by: Avotek
A Select Aerospace Industries company

Mail to:
P.O. Box 219
Weyers Cave, Virginia 24486
USA

Ship to:
200 Packaging Drive
Weyers Cave, Virginia 24486
USA

Toll Free: 1-800-828-6835
Telephone: 1-540-234-9090
Fax: 1-540-234-9399

Third Printing
Printed in the USA

www.avotek.com

Preface

This training manual is part of the Avotek Aviation Maintenance Technician training series. It is intended for use in general training and familiarization of aircraft electrical wiring installations.

The satisfactory performance of the present-day aircraft depends greatly on the continuing reliability of its electrical system. Improperly or carelessly installed wiring can be a source of both immediate and potential danger, and many malfunctions and failures of the electrical system can be traced to this cause. The performance of the system depends on the quality of the design, plus the workmanship used in making the installation. The continued proper performance of the system dependent on the knowledge of the persons who do the inspection, repair and maintenance.

It is critical therefore that maintenance and repair operations, as well as the original installation, be made in accordance with the best available techniques in order to eliminate possible failures or at least to minimize them.

Intended Use

This manual is intended primarily for general training and familiarization. ALWAYS follow the aircraft manufacturer's FAA approved repair manuals or other FAA approved procedures. The material in this manual is believed to be accurate, but is not to be used as a substitute for FAA approved methods or data.

E-mail us at comments@avotek.com for comments or suggestions.

Avotek® Aircraft Maintenance Series:
Introduction to Aircraft Maintenance
Aircraft Structural Maintenance
Aircraft System Maintenance
Aircraft Powerplant Maintenance

Other Books by Avotek®:
Aircraft Turbine Engines
Aircraft Corrosion Control Guide
Aircraft Structural Technician
Aircraft Wiring & Electrical Installation
Aviation Maintenance Technician Reference Handbook
Avionics: Systems and Troubleshooting
Avotek Aeronautical Dictionary
Fundamentals of Modern Aviation
Light Sport Aircraft Inspection Procedures

Contents

Preface .... *iii*

Chapter 1. Wire and Cable Preparation
    Section 1. Definitions *1-1*
    Section 2. Cutting Wire and Cable *1-3*
    Section 3. Identifying Wire and Cable *1-6*
    Section 4. Wire and Cable Preparation *1-13*

Chapter 2. General Purpose Connectors
    Section 1. Description and Identification *2-1*
    Section 2. Types of AN-MS Connectors *2-4*
    Section 3. Disassembly of Connectors *2-15*
    Section 4. Assembly of Wires to Connectors *2-19*
    Section 5. Installation of MS3057 Series Connector Cable Clamps *2-25*
    Section 6. Disassembly and Reassembly of Connectors *2-28*
    Section 7. Contact Tools *2-41*

Chapter 3. RF Connectors and Cabling
    Section 1. RF Connectors *3-1*
    Section 2. BNC Series Connectors *3-5*
    Section 3. C Series Connectors *3-8*
    Section 4. HN Series Connectors *3-9*
    Section 5. N Series Connectors *3-11*
    Section 6. Pulse Series Connectors *3-15*
    Section 7. TNC Series Connectors *3-18*
    Section 8. RF Connectors *3-19*

Chapter 4. Solderless Terminations and Splices
    Section 1. Identification *4-1*
    Section 2. Terminating Small Copper Wires (Sizes No. 26 through No. 10) *4-2*
    Section 3. Terminating Large Copper Wires *4-6*
    Section 4. High Temperature Terminal Lugs *4-12*
    Section 5. Terminating Aluminum Wire *4-14*
    Section 6. Splicing *4-18*

Chapter 5. Thermocouple Wire Soldering and Insulation
    Section 1. Description and Identification *5-1*
    Section 2. Thermocouple Wire Preparation *5-4*
    Section 3. Hard Soldering Thermocouple Wire *5-5*
    Section 4. Soft Soldering Thermocouple Wire *5-8*
    Section 5. Thermocouple Wiring Installation *5-10*

Chapter 6. Bonding and Grounding
    Section 1. Description and Identification *6-1*
    Section 2. Bonding & Grounding Hardware *6-3*
    Section 3. Preparation of Bonding or Grounding Surfaces *6-4*

Section 4. Methods of Bonding or Grounding *6-5*
Section 5. Testing Bonds and Grounds *6-7*
Section 6. Refinishing *6-7*

Chapter 7. Soldering Methods
Section 1. Description and Identification *7-1*
Section 2. Heat Application Methods *7-2*
Section 3. Soldering Precautions and Procedures *7-3*
Section 4. Inspecting a Finished Solder Joint *7-6*
Section 5. Repair and Soldering of Printed Circuit Assemblies *7-6*

Chapter 8. Electric Connector Sealing
Section 1. Description and Identification *8-1*
Section 2. Preparation of Sealing Compound *8-2*

Chapter 9. Installation of Electrical Hardware
Section 1. Preparation and Installation of Bus Bars *9-1*
Section 2. Installation of Conduit *9-3*
Section 3. Installation of Junction Boxes *9-4*
Section 4. Installation of Protective Devices *9-5*
Section 5. Installation of Terminal Boards *9-7*

Chapter 10. Electrical Wiring Installation
Section 1. Description and Identification *10-1*
Section 2. Wire Groups and Bundles *10-2*
Section 3. Connections to Terminal Boards and Bus Bars *10-9*
Section 4. Installation of Wires in Conduit *10-12*
Section 5. Installation of Connectors *10-13*
Section 6. Installation of Wire in Junction Boxes *10-14*
Section 7. Lacing and Tying *10-15*
Section 8. General Precautions *10-15*
Section 9. Lacing *10-15*
Section 10. Tying *10-16*
Section 11. Wiring: Lock, Shear and Seal *10-18*
Section 12. Procedures for Lock, Shear and Seal Wiring *10-19*
Section 13. ARINC Digital Bus *10-21*

Chapter 11. Aircraft Electrical System Lamps & Fuses
Section 1. Lamps and Fuses *11-1*
Section 2. Aircraft Electrical Fuses *11-3*

Chapter 12. Emergency Repairs
Section 1. Repairing Broken or Damaged Wires *12-1*
Section 2. Repairing Damaged MS Connectors *12-4*

Appendix *A-1*

Index *I-1*

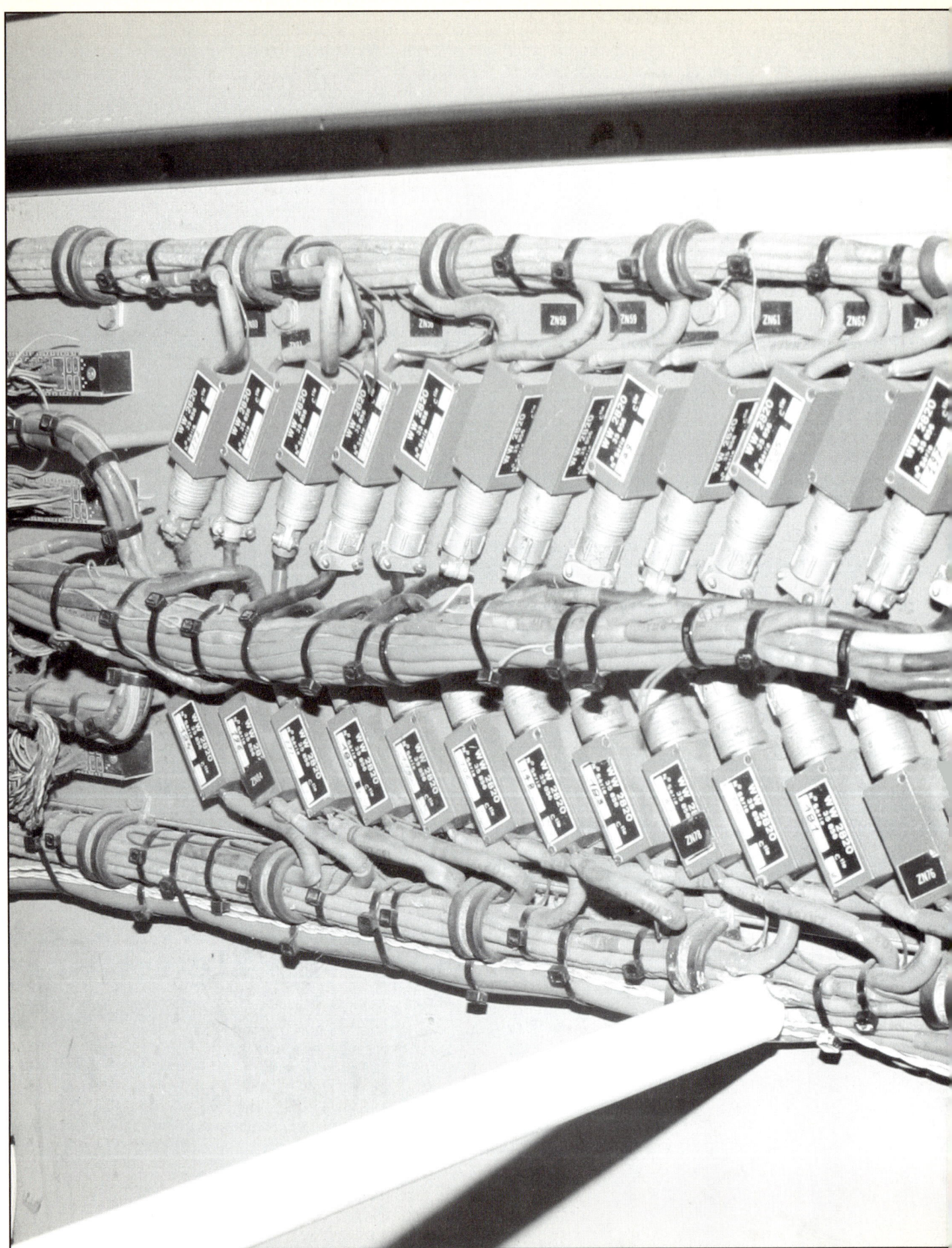

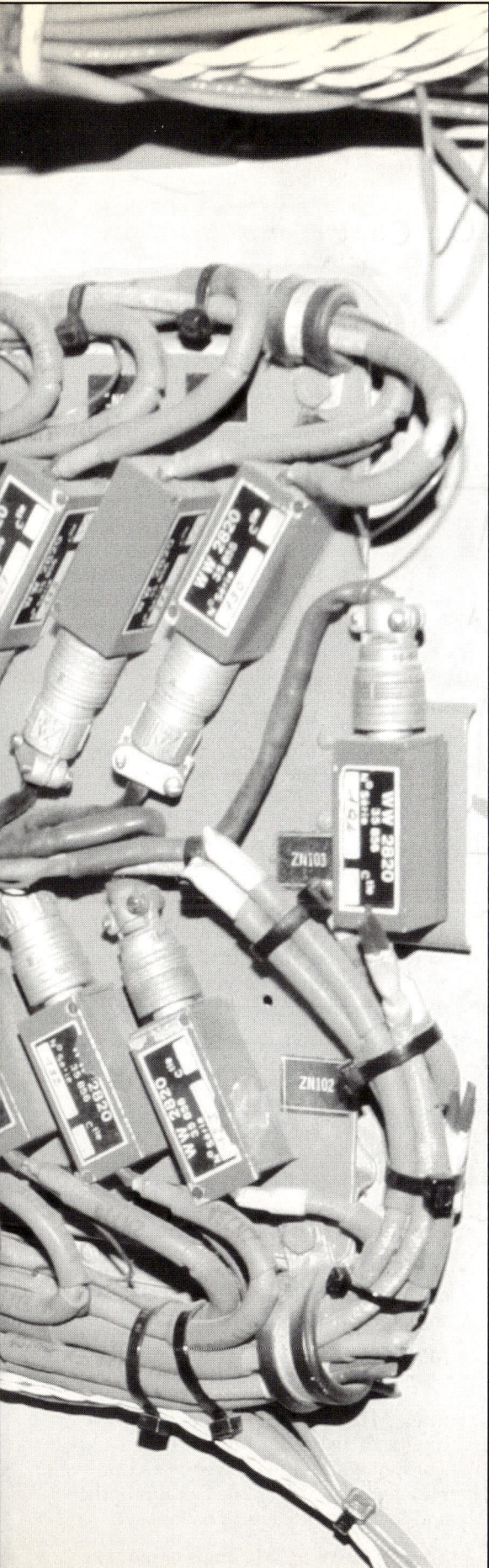

Left. To make maintenance easier, wires and cables, like the ones shown here are marked with a combination of letters and numbers which identify a wire, its size, and the circuit it belongs to.

Chapter 1
WIRE and cable preparation

Electrical installations can be complex to assemble and troubleshoot. Runs of electric wire and cable are broken at specific locations by junctions, to ease assembly and repair. These junctions can be connectors, junction boxes, buses or others. Wires and cables must be prepared prior to installation, which involves cutting to length, identification, stripping and, if necessary, tinning.

Common procedures for preparing wire and cable are described and illustrated in this chapter. Both preparation for attachment to junctions and procedures for terminating shielded cables are covered.

Learning Objectives:

- Definitions
- Wire & Cable Cutting
- Wire & Cable Identification
- Wire & Cable Preparation

Section 1
Definitions

Insulated wire. For electric and electronic installation in aircraft, insulated wire is a metal conductor covered with a dielectric or insulating material (see Figure 1-1, next page). Insulated wire is usually referred to as "wire" and will be so designated in this manual. Wires used in aircraft contain stranded conductors for flexibility. Insulation may consist of several materials and layers to provide dielectric insulation, thermal protection, abrasion resistance, moisture resistance and fluid resistance. Wires commonly used in aircraft are described in Table 1-1 on pages 1-4 and 1-5.

Cable. (See Figure 1-2 next page). The term cable, as used in aircraft electrical installations, includes the following:

1. Two or more insulated conductors, contained in a common covering, or twisted together without a common covering (multi-conductor cable).

1-2 | Wire and Cable Preparation

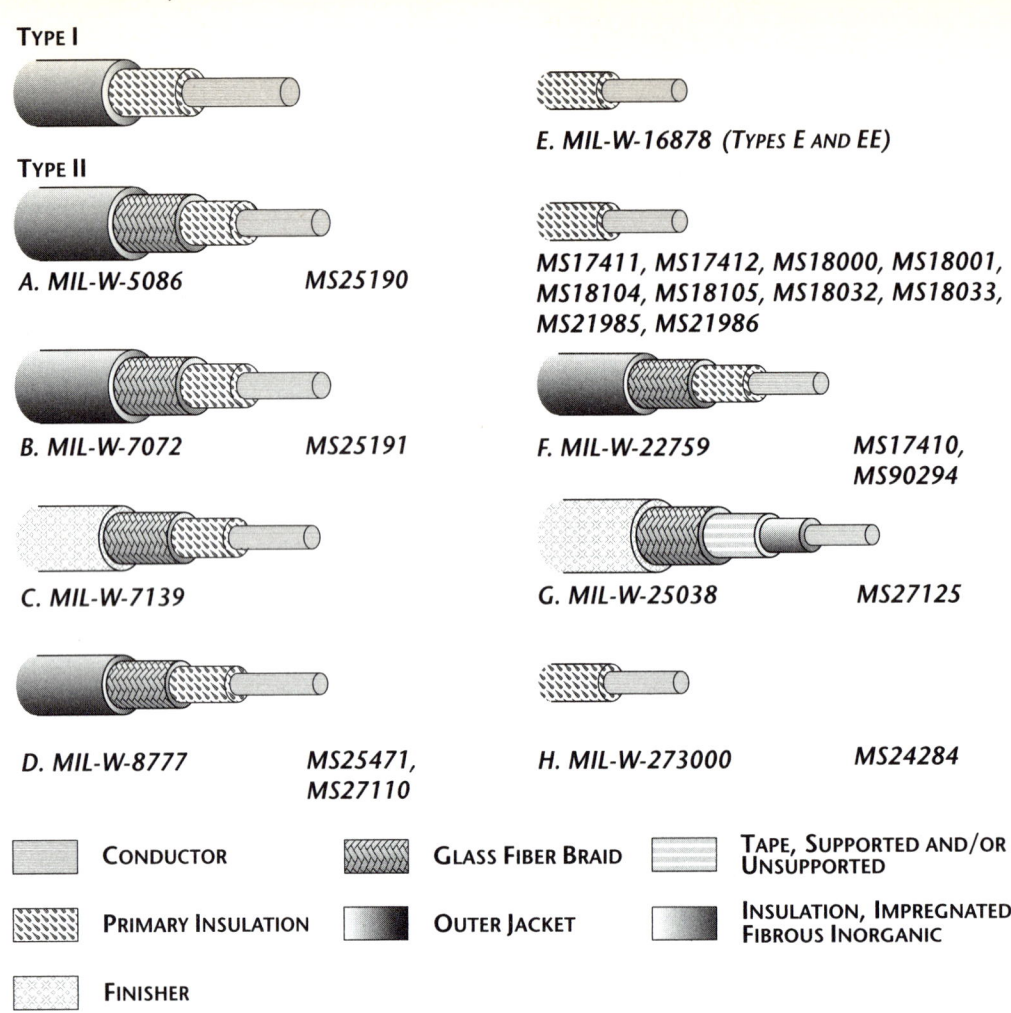

Figure 1-1. Wires commonly used in aircraft

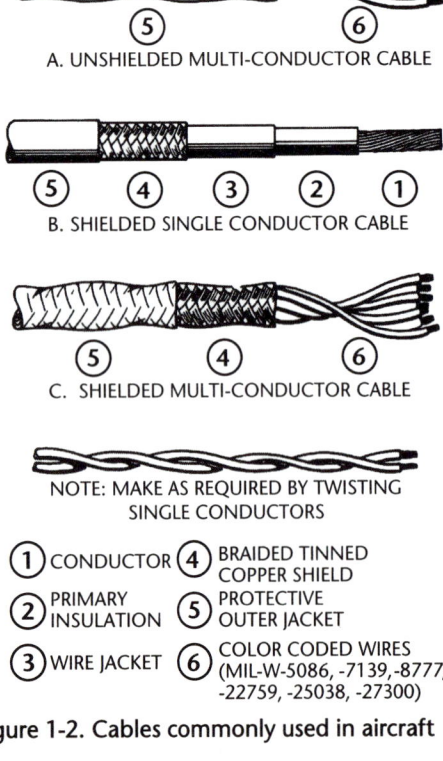

Figure 1-2. Cables commonly used in aircraft

2. A single insulated conductor, or two or more insulated conductors with an overall shield, or with an overall shield and a jacket over the shield (shielded cable).

3. Two conductors twisted together (twisted pair).

4. A single insulated center conductor with a metallic braided outer conductor (coaxial cable). The concentricity of center and outer conductor is carefully controlled during manufacture to insure that they are coaxial.

Cables commonly used in aircraft are:

- **MS25192, Spec. MIL-C-7078.** Single or multiple conductor, using MS25190 wire, shielded with tinned copper braid.

- **MS25313, Spec. MIL-C-7078.** Similar to MS25192, but covered with a nylon jacket.

- **Spec. MIL-W-22759.** Single or multiple conductor, using any wire in Table 1-1, shielded with tinned, silver-coated, or nickel-coated copper braid as appropriate and covered with appropriate jacket.

- **Spec. MIL-W-81381.** Single or multiple

Wire and Cable Preparation | 1-3

conductor, rubber insulated conductor, rubber jacket.

- **Spec. MIL-C-17.** Coaxial cable

Soft Solder. For use in aircraft electrical installations, soft solder is a mixture of 60% tin and 40% lead, as described in Federal Specification QQ-S-571. It may be in bar form to be melted for tinning, or in the form of rosin core wire solder for use with a soldering iron.

Flux. For use with soft solder, flux is water-white rosin, dissolved to paste-like consistency in denatured alcohol.

Section 2
Cutting Wire and Cable

General. Cut all wires and cables to lengths given on drawings or wiring diagrams. Cut wire and cable so that cut is clean and square and wire is not deformed. See Figure 1-3. After cutting reshape large diameter wire with pliers if necessary.

CAUTION: *Make sure that blades of cutting tools are sharp and free from nicks. A dull blade deforms and extrudes wire ends.*

Cutting copper wire and cable. To cut a large number of heavy wires or cables, use a circular saw with a cable cutting blade. A cable cutting blade is similar to a meat slicing blade (no teeth). See Figure 1-4a.

WARNING: *Do not use a circular saw without an adequate guard over the blade.*

Heavy or light copper wires can also be cut with bench shears such as shown in Figure 1-4b.

To cut a few heavy gage copper wires or cables, use a fine tooth hacksaw. A fine tooth hacksaw has 20 or more teeth per inch. See Figure 1-4c for use of hacksaw and saw vise that protects heavy wire during cutting.

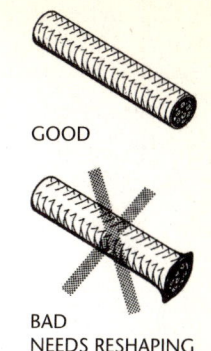

Figure 1-3. Wires after cutting

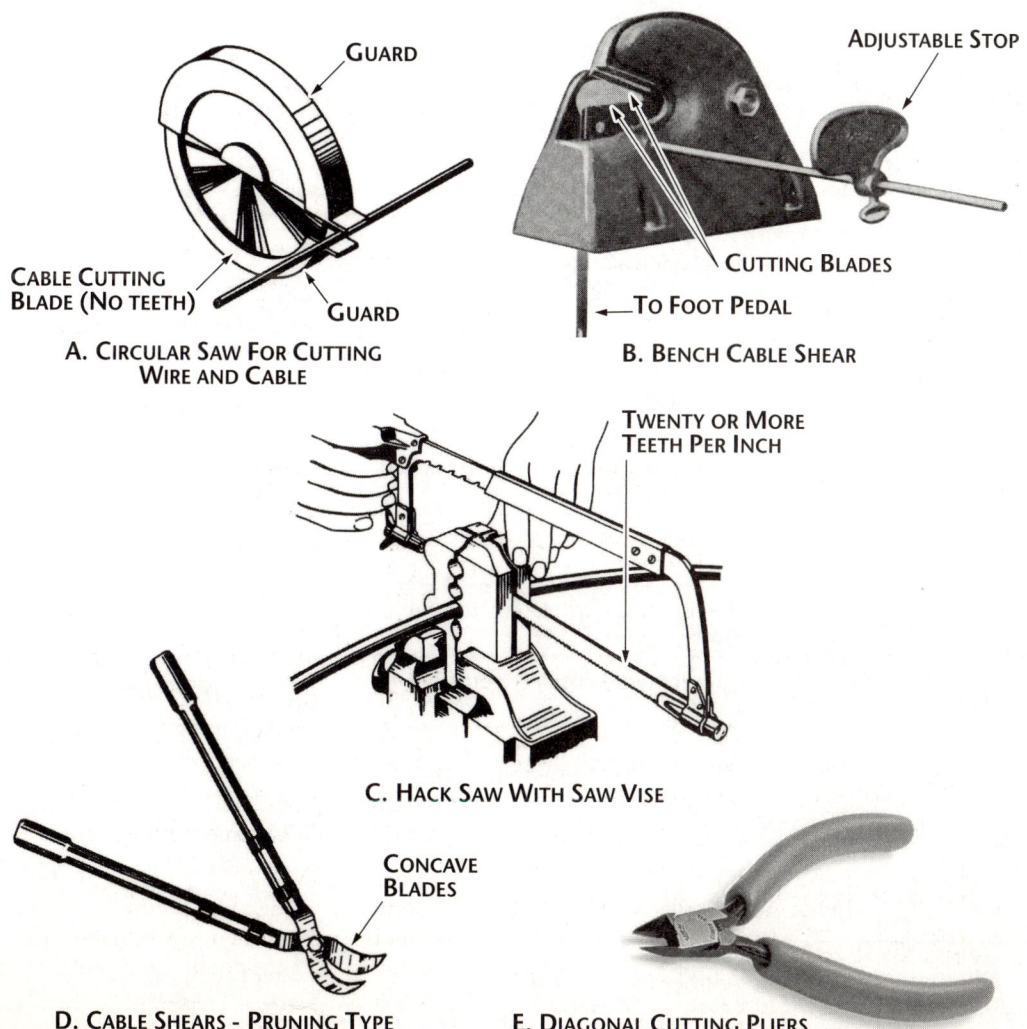

A. Circular Saw For Cutting Wire and Cable

B. Bench Cable Shear

C. Hack Saw With Saw Vise

D. Cable Shears - Pruning Type

E. Diagonal Cutting Pliers

Figure 1-4. Wire cutting tools

Document	Voltage Rating (Maximum)	Rated Wire Temperature (°C)	Insulation Type	Conductor Type
MIL-W-22759/1	600	200	Fluoropolymer insulated TFE and TFE coated glass	Silver coated copper
MIL-W-22759/2	600	260	Fluoropolymer insulated TFE and TFE coated glass	Nickel coated copper
MIL-W-22759/3	600	260	Fluoropolymer insulated TFE -glass-TFE	Nickel coated copper
MIL-W-22759/4	600	200	Fluoropolymer insulated TFE -glass-FEP	Silver coated copper
MIL-W-22759/5	600	200	Fluoropolymer insulated extruded TFE	Silver coated copper
MIL-W-22759/6	600	260	Fluoropolymer insulated extruded TFE	Nickel coated copper
MIL-W-22759/7	600	200	Fluoropolymer insulated extruded TFE	Silver coated copper
MIL-W-22759/8	600	260	Fluoropolymer insulated extruded TFE	Nickel coated copper
MIL-W-22759/9	1000	200	Fluoropolymer insulated extruded TFE	Silver coated copper
MIL-W-22759/10	1000	260	Fluoropolymer insulated extruded TFE	Nickel coated copper
MIL-W-22759/13	600	135	Fluoropolymer insulated FEP PVF2	Tin coated copper
MIL-W-22759/16	600	150	Fluoropolymer insulated extruded ETFE	Tin coated copper
MIL-W-22759/17	600	150	Fluoropolymer insulated extruded ETFE	Silver coated high strength copper alloy
MIL-W-22759/20	1000	200	Fluoropolymer insulated extruded TFE	Silver coated high strength copper alloy
MIL-W-22759/21	1000	260	Fluoropolymer insulated extruded TFE	Nickel coated high strength copper alloy
MIL-W-22759/34	600	150	Fluoropolymer insulated crosslinked modified ETFE	Tin coated copper
MIL-W-22759/35	600	200	Fluoropolymer insulated crosslinked modified ETFE	Silver coated high strength copper alloy
MIL-W-22759/41	600	200	Fluoropolymer insulated crosslinked modified ETFE	Nickel coated copper
MIL-W-22759/42	600	200	Fluoropolymer insulated crosslinked modified ETFE	Nickel coated high strength copper alloy
MIL-W-22759/43	600	200	Fluoropolymer insulated crosslinked modified ETFE	Silver coated copper
MIL-W-25038/3/2/	600	260	See specification sheet *	See specification sheet
MIL-W-81044/6	600	150	Crosslinked polyalkene	Tin coated copper
MIL-W-81044/7	600	150	Crosslinked polyalkene	Silver coated high strength copper alloy
MIL-W-81044/9	600	150	Crosslinked polyalkene	Tin coated copper
MIL-W-81044/10	600	150	Crosslinked polyalkene	Silver coated high strength copper alloy

Table 1-1. Open wiring

Document	Voltage Rating (Maximum)	Rated Wire Temperature (°C)	Insulation Type	Conductor Type
MIL-W-22759/11	600	200	Fluoropolymer insulated extruded TFE	Silver coated copper
MIL-W-22759/12	600	260	Fluoropolymer insulated extruded TFE	Nickel coated copper
MIL-W-22759/14	600	135	Fluoropolymer insulated FEP-PVF2	Tin coated copper
MIL-W-22759/15	600	135	Fluoropolymer insulated FEP-PVF2	Silver plated high strength copper alloy
MIL-W-22759/18	600	150	Fluoropolymer insulated extruded ETFE	Tin coated copper
MIL-W-22759/19	600	150	Fluoropolymer insulated extruded ETFE	Silver coated high strength copper alloy
MIL-W-22759/22	600	200	Fluoropolymer insulated extruded TFE	Silver coated high strength copper alloy
MIL-W-22759/23	600	260	Fluoropolymer insulated extruded TFE	Nickel coated high strength copper alloy
MIL-W-22759/32	600	150	Fluoropolymer insulated crosslinked modified ETFE	Tin coated copper
MIL-W-22759/33	600	200	Fluoropolymer insulated crosslinked modified ETFE	Silver coated high strength copper alloy
MIL-W-22759/44	600	200	Fluoropolymer insulated crosslinked modified ETFE	Silver coated copper
MIL-W-22759/45	600	200	Fluoropolymer insulated crosslinked modified ETFE	Nickel coated copper
MIL-W-22759/46	600	200	Fluoropolymer insulated crosslinked modified ETFE	Nickel coated high strength copper alloy
MIL-W-81044/12	600	150	Crosslinked polyalkene - PVF2	Tin coated copper
MIL-W-81044/13	600	150	Crosslinked polyalkene - PVF2	Silver coated high strength copper alloy
MIL-W-81381/17	600	200	Fluorocarbon polyimide	Silver coated copper
MIL-W-81381/18	600	200	Fluorocarbon polyimide	Nickel coated copper
MIL-W-81381/19	600	200	Fluorocarbon polyimide	Silver coated high strength copper alloy
MIL-W-81381/20	600	200	Fluorocarbon polyimide	Nickel coated high strength copper alloy
MIL-W-81381/21	600	150	Fluorocarbon polyimide	Tin coated copper

Table 1-2. Protected wiring

Wire and Cable Preparation

To cut a few light gage copper wires, use diagonal pliers. Do not attempt to cut wires larger than AN-8 with diagonal pliers.

Cutting aluminum wire. Be careful when cutting aluminum wire to avoid deforming the conductors. Aluminum is more brittle than copper, and if deformed, aluminum wire should be reshaped carefully.

> **CAUTION:** *Never cut aluminum wire with tools that have reciprocating motion, such as a hacksaw. Reciprocating cutting action "work hardens" aluminum. This will lead to broken and torn strands.*

To cut a large number of aluminum wires use a power circular saw with cable cutting blade as shown in Figure 1-4a. Do not use toothed blade for cutting aluminum wire.

> **CAUTION:** *If cutting tool has been used for other metals, wipe blades clean before cutting aluminum. Copper or steel chips will cause aluminum to corrode.*

Special cable shears with concave cutting edges such as pruning or dehorning shears may also be safely used to cut aluminum wire. A cable shear of this type is illustrated in Figure 1-4d.

Section 3
Identifying Wire and Cable

General. To make maintenance easier, each interconnecting wire and cable installed in an aircraft is marked with a combination of letters and numbers which identify the wire, the circuit it belongs to, its gage size and other information necessary to relate the wire to a wiring diagram. This marking is called the cable identification code. Details of the code are given in Mil-Spec MIL-W-5088. Wire as received from the manufacturer is printed in a light green color at intervals of one to five feet with the manufacturer's code designation: the MS number and dash number of the wire, and a one, two or three-digit number indicating the color of the basic wire insulation and the color of the stripes (if present). The color code is as follows:

Color	Code
Black	0
Brown	1
Red	2
Orange	3
Yellow	4
Green	5
Blue	6
Violet	7
Gray	8
White	9 (includes also uncolored insulations)

For example, a wire printed with number MS25190A20913 would designate a wire constructed in accordance with drawing MS25190, Type I, size 20, having white insulation (9), a first stripe of brown (1) and a second stripe of orange (3).

> **NOTE:** *When marking wire with the identification code described in the following paragraphs, it is permissible to overstamp the manufacturer's printing.*

Wire identification code-basic. See Figure 1-5a. The basic wire identification code described is derived from a Military Standard. It is presented as an illustration of one type of common wiring identification code methodology. While this stan-

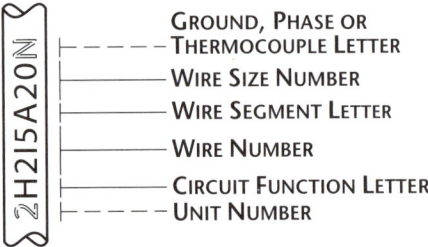

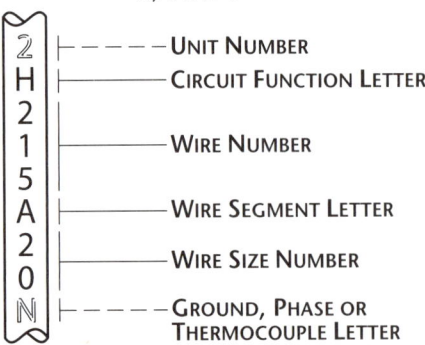

Figure 1-5. Examples of wire identification coding

dard is widely used, some aircraft use unique codes developed by the individual aircraft manufacturer. Consult the wiring diagram manual for the aircraft type you are working on before attempting to interpret codes on specific aircraft.

The basic code described is used for all circuits (except those having the circuit function letters, R, S, T or Y) is as follows, reading from left to right:

- **Unit number.** Prefixed where necessary to distinguish between wires in a circuit having identical items of equipment and identification numbers.

- **Circuit Function Letter.** Used to identify the function of the particular circuit. See Mil-Spec MIL-W-5088 for details.

- **Wire number.** Used to distinguish between wires with the same circuit function letters.

- **Wire segment letter.** Used to distinguish between conductor segments in a particular circuit.

- **Wire size number.** Used to designate AN or AL gage size of the wire. Wire size is omitted on coaxial cable. Wire size number is replaced with a dash for thermocouple wire.

- **Ground, phase or thermocouple letter.** Used to denote a wire to ground, phase of a wire in a three-phase system, or materials of a thermocouple pair.

Wire identification code (R, S and T circuits). See Figure 1-5b. The identification code for circuits R, S and T, (Radio, Radar & Special Electronic circuits), in addition to the basic numbers and letters listed in the previous paragraphs, includes another letter after the circuit function letter, called the circuit designation letter, which further identifies the circuit inside the system. See Mil-Spec MIL-W-5088 for details.

Identification methods. The identification code may be stamped on wires either horizontally, as shown in Figure 1-5a & b, or if desirable in a particular application, vertically as shown in Figure 1-5c. The preferred method of identification is to stamp the identification marking directly on the wire or cable with a hot foil stamping machine. Use this method wherever possible. If the wire insulation or outer covering will not stamp easily, lengths of insulating tubing (sleeves) are stamped with the identification marking and installed on the wire or cable. The following types of wire are usually identified by means of sleeves:

- Unjacketed shielded wire
- Thermocouple wires

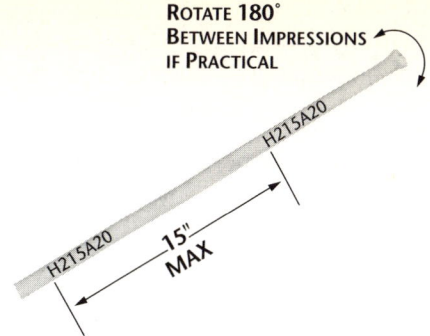

Figure 1-6. Spacing of identification stamping on wire and cable

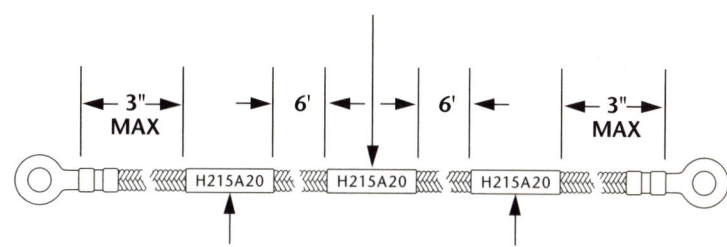

Figure 1-7. Location of identification sleeve

- Multiconductor cable
- High temperature wire with insulation difficult to mark, (such as TFE, fiberglass, etc).

  CAUTION: *Do not use metallic markers or bands for identification. Do not use any method of marking that will damage or deform the wire or cable.*

Use sleeves only if wire cannot be marked directly. With care some wires, previously thought to be unsuitable for direct marking, can be stamped with a standard marking machine using special foils.

Marking objectives. Whatever method of marking is used, be sure marking is legible and that color of stamping contrasts with the wire insulation or sleeve. Use black stamping for light colored backgrounds. Use white on dark colored backgrounds. Make sure that markings are dry so they do not smear.

Spacing of stamped marks. See Figure 1-6. Stamp wires and cables at intervals of not more than 15 inches along their entire lengths. In addition stamp wires within three inches of each junction (except permanent splices), and at each terminating point. Stamp wires which are three to seven inches long in the center. Wires less than three inches long need not be stamped.

Location of sleeve marking. See Figure 1-7. When wire or cable cannot be stamped directly, install a plastic sleeve marked with the identification number over the outer cover-

1-8 | *Wire and Cable Preparation*

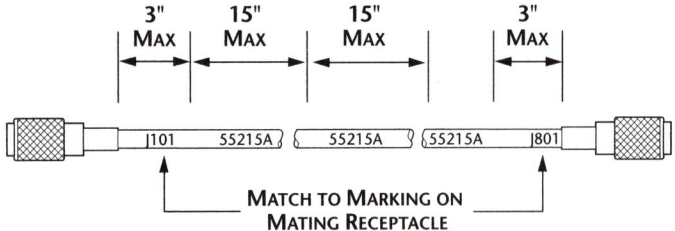

Figure 1-8. Multi-conductor cable identification

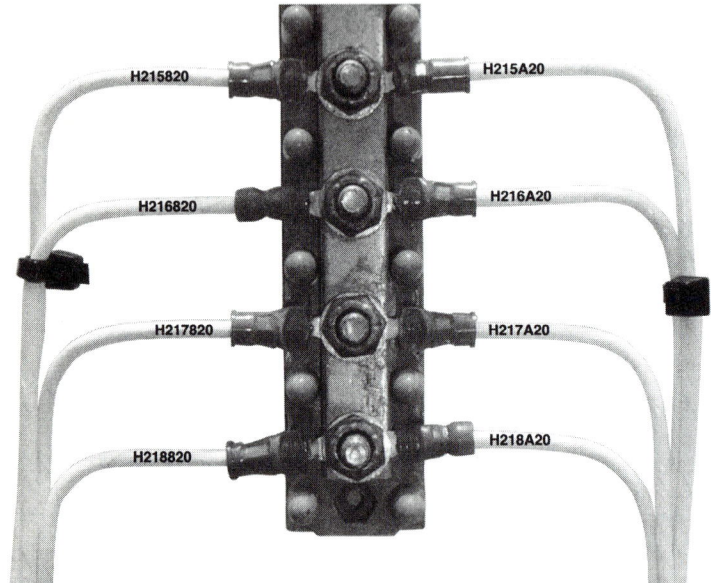

Figure 1-9. Coaxial cable identification

ing at each terminating end. Wire tracing and troubleshooting are easier when the sleeves are placed at regular intervals on the entire length of such wire or cable.

CAUTION: *Do not use sleeves to change identification of wire or cable that has already been marked, except in the case of spare wires in potted connectors.*

Multiconductor cable identification. See Figure 1-8. Identify multiconductor cables with marked sleeves installed as described above. Stamp sleeves with identification marking of each wire in the cable. Immediately following the identification code, stamp letters indicating the conductor color, using the following abbreviations:

ABBREVIATION	COLOR
BLK	Black
BLU	Blue
BRN	Brown
GY	Gray
GRN	Green
ORN	Orange
PR	Purple
RED	Red
WHT	White
YEL	Yellow

At each terminating end, strip back outer covering as far as necessary and stamp color code letters on insulation of each conductor.

Figure 1-10. Wire identification at terminal board

MIL-W-5086 TYPES I AND II AN	WIRE SIZE MIL-W-5086 TYPE III AN	MIL-W-7072 AL	SLEEVING SIZE NO.	NOMINAL ID (INCHES)
#22			11	.095
#20	#22		10	.106
#18	#20		9	.118
#16	#18		8	.133
#14	#16		7	.148
#12	#14		6	.166
#10	#12		4	.208
#8	#10		2	.263
#6	#8	#8	0	.330
#4	#6	#6	3/8	.375
#12	#4	#4	1/2	.500
#1	#2	#2	1/2	.500
#0	#1	#1	5/8	.625
#00	#0	#0	5/8	.625
#000	#00	#00	3/4	.750
#0000	#000	#000	3/4	.750
	#0000	#0000	7/8	.875

Table 1-3. Sizes of identification sleeving

Coaxial cable identification. See Figure 1-9. Identify coaxial cable by direct stamping on the cable or with sleeves. If sleeves are required, install them as previously indicated. In addition, mark coaxial cable on the end terminating in a piece of equipment to match marking on equipment terminal.

CAUTION: *When marking coaxial cable do not flatten the cable, as this may change the electrical characteristics of the cable.*

Thermocouple wire identification. Thermocouple wire that is usually duplexed (two insulated conductors laid side by side) is difficult to mark legibly. The wire size of the identification code is replaced by the full name of the material of the thermocouple conductor. Thermocouple conductor materials are: alumel, chromel, iron, constantan, copper.

Identification at terminal boards and enclosures. See Figure 1-10. If possible, mark wires attached to terminal boards and equipment terminals between termination and point where wire is brought into wire bundle. Identify wires terminating in an enclosure inside enclosure if space permits.

Selection of identification sleeving. For general purpose wiring use flexible vinyl sleeving, either clear or white opaque. For high temperature applications (over 100° C) use silicone rubber or silicone fiberglass sleeving. Where resistance to synthetic hydraulic fluids or other solvents is necessary, use nylon sleeving, either clear or white opaque. Select size of sleeving from Table 1-3. Heat-shrinkable polyethlene tubing may also be used to identify wire which cannot be marked directly.

Identification marking machines. See Figures 1-11 and 1-12 for typical marking machines. For stamping a large number of long wires use an automatic wire marking machine. In machines of this type (Figure 1-11) wire sizes No. 26 through No. 14 are fed through and stamped automatically. Wires larger than No. 14 are fed through by hand, but stamped automatically. For short wires, on repair or maintenance work, a hand-operated wire marking machine is more convenient and economical (Figure 1-12). In this type of machine wire is fed through the desired amount by hand, and stamped by operating the handle for each marking. Wire guide holders in sizes to fit wires, and slot holders to hold appropriate size type are furnished to fit the machines. Type is supplied in three sizes, to mark wire No. 26 through No. 00 as shown in Table 1-4. Marking foil is available in black or white, (and other colors if needed for special applications).

Figure 1-11. Wire marking machine – automatic

Figure 1-12. Marking machine – hand

NOTE: *Store foils at approximately 70°F and 60% relative humidity.*

Set-up of marking machines for wire stamping. After selecting the proper machine for the job, set it up for the marking procedure as follows: (Refer to Figures 1-11 and 1-12).

1. Select from Table 1-3 the correct type size for wire to be marked. Make up required identification code and insert into type holder, centering type in holder. Use spacers to prevent crowding letters and numbers.

2. Select marking foil of correct width for length of marking. Use black foil for light colored insulation, and white foil for dark insulation.

3. Select wire guide holder with wire hole to fit wire, and having a slot of same length as slot in type holder.

   CAUTION: *Use smallest guide into which wire will fit. If guide is too large, wire will not be held firmly, and will be off-centered.*

4. Install wire guide and roll of marking foil on machine. Slide type holder into slot provided for it.

Procedure for wire stamping by machine. The procedure for stamping wire by machine is as follows:

NOTE: *Good marking is obtained only by the proper combination of temperature and pressure, and is arrived at by trial.*

CAUTION: *Avoid excessive heat or pressure, as it may damage the wire insulation.*

1. Turn heat to high, then regulate downward to required temperature. See Table 1-4 for recommended temperatures.

2. Insert piece of sample or scrap wire into wire guide, and adjust pressure control until mark is sharp and clear. Impression should be just deep enough to sink slightly below surface of insulation, but should not cut into it. The pressure adjustment also controls the length of dwell in automatic machines.

3. When a satisfactory marking has been made, remove sample wire, and insert wire to be marked into wire guide, far enough so that first marking will be made about three inches from end.

4. Operate foot pedal (or hand lever) to make mark.

5. It is desirable (but not mandatory) to rotate wire 180 degrees and mark again on opposite side.

6. Mark remaining wire length. If marks are to be spaced at intervals of eight inches or less, operate machine automatically. If intervals are greater than eight inches, or if wire is larger than size No. 8 (regardless of spacing) pull wire through by hand, and operate machine at desired spacing. If practical, rotate wire back and forth at each mark through 180 degrees to mark on opposite side. During marking procedure, check permanence of mark from time to time by rubbing with a clean dry cloth. If mark smears, or becomes hard to read, adjust machine to correct condition, and remark wire.

Special instructions for marking TFE insulated wire. Because of the chemical nature of TFE, it is difficult to make a permanent marking on TFE-insulated wire. Marking machines will stamp a legible marking on TFE, but this has a tendency to rub off. The marking may be set permanently by passing the marked wire through an electrically heated oven, set at a specified speed and temperature. See Figure 1-13.

Procedure for heat-setting identification marking on TFE. The procedure for heat-setting identification marking on TFE insulated-wire is as follows:

1. Turn the temperature control to HIGH, and allow oven to heat until temperature reaches 1900°F. (Approximately 30 to 45 minutes.)

2. Turn the motor switch ON, and set speed to the desired rate; select speed from Table 1-5.

3. Insert the wire through the guide tube into the forward rollers. As the first part of the wire passes through the oven, depress the rear (exit) roller manually to allow free entry of the wire into the rear guide tube; this will prevent the wire from buckling. The rest of the wire length will pass automatically through the oven.

CAUTION: *Provide a suitable exhaust hood over the oven to carry off fumes. Make*

Wire Size	Height of Letters (inches)
#26 & #22	1/16
#20 thru #14	5/64
#12 thru #0000 & Coaxial Cable	7/64

Table 1-4. Recommended sizes of marking type

sure there is adequate ventilation in the area where the oven is used.

Set-up of marking machine for sleeve stamping. For stamping identification mark on tubing that has an OD of 1/4 inch or smaller, use the same machine that is used for stamping wire. Set up machine as follows:

1. Select type size and wire guide to suit OD of tubing.

2. Select mandrel (metal rod) of a diameter that will fit snugly inside tubing. Insert mandrel into tubing, and both into wire guide. If mandrel of proper size is not available, use piece of insulated wire of suitable diameter and length.

3. Prepare type as described on page 1-10, step 1.

Insulation Material	Recommended Temperature
Kel-F	300° to 325°F
Vinyl	400° to 425°F
*TFE	425°F
Nylon	400° to 450°F
**Silicone	400° to 450°F
Polyester	500°F
*FEP	550° to 600°F

*Using special foil
**Protective coating and short heat cure after marking recommended.

Table 1-5. Recommended marking temperatures

Figure 1-13. Electric oven for heat-setting identification marking

Wire Spec.	Wire OD	Recommended Speed (Feet per min.)
MIL-W-7139	.070-.090	15 to 20
	.090-.125	10 to 15
MIL-W-16878	.050-.070	15 to 20
	.070-.090	12 to 15
	.090-.125	10 to 15

Table 1-6. Speed settings for heat-setting electric oven wire

1-12 | Wire and Cable Preparation

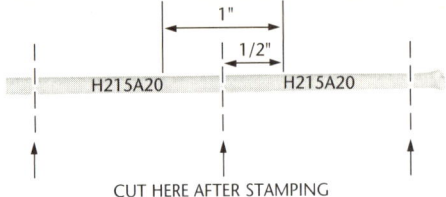

Figure 1-14. Marking on sleeves

4. Select foil, and install wire guide, foil and type holder on machine as described in steps 2, 3 and 4 in the marking machine procedure on page 1-10.

Procedure for stamping sleeves by machine. To mark tubing, follow procedure for marking wire as described in in the procedure on page 1-10, steps 1-6. After first mark, mark remaining tubing at intervals that will leave about one inch between marks (See Figure 1-14). Rotate at each marking to mark again on opposite side.

Machine stamping for large sleeving. See Figure 1-15. To mark tubing that has an OD larger than 1/4 inch, use a special machine that marks tubing flat, if it is available. Flat type rather than curved type is used on this machine. Otherwise, machine set-up and marking procedure is same as described in the previous paragraphs.

Installing identification sleeves on wiring. Cut marked tubing into lengths so that marking is approximately centered (refer to Figure 1-14). Install cut lengths of tubing over wire or cable at desired spacing, and tie at each end with clove hitch and square knot. See Chapter 10 for method of tying and knotting. When heat-shrink tubing is used, ties are not required. Before installing heat-shrink tubing on the wire, make sure that the wires are clean.

Identification of wire bundles and harnesses. See Figure 1-16. Identify wire bundles and harnesses by one of the following methods:

1. Select sleeving of proper size to fit snugly over wire bundle. Stamp with identifi-

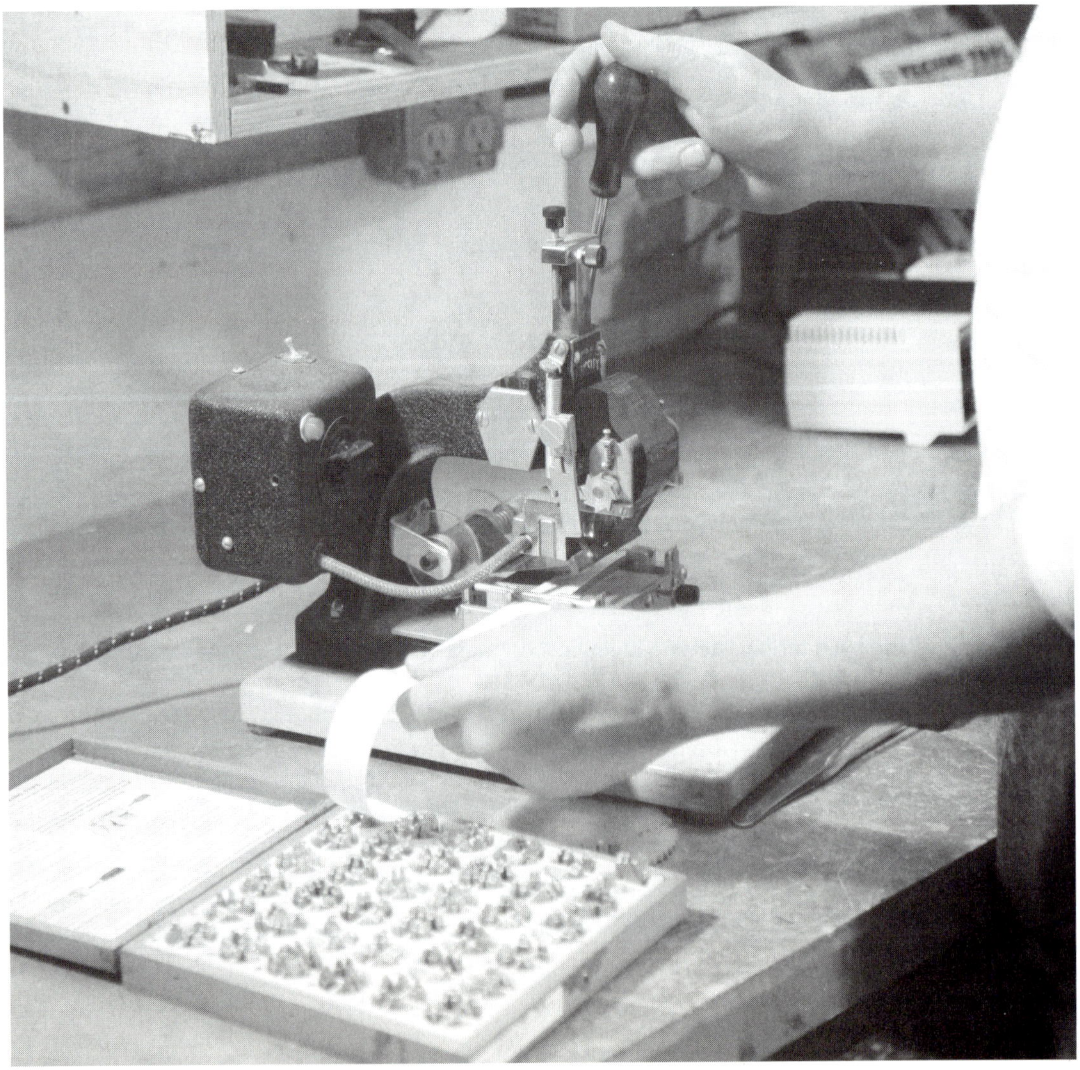

Figure 1-15. Marking machine for sleeving

cation marking and install on bundle approximately 12 inch from each terminating end. Tie securely at both ends.

NOTE: *If wires are not free at one end, sleeving must be installed on bundle before soldering wires to connectors*

Heat shrink tubing, marked with the identification code, may also be used.

2. Wire bundles up to 1 3/4 inch in diameter may be identified by means of an MS17822 cable identification strap which has a marking tab as part of it. See Figure 1-16. The procedure is as follows:

  1. Stamp the wire identification code on the marking tab.
  2. Pass the strap around the bundle with the ribbed side of the strap inside.
  3. Insert the pointed end of the strap through the eye and pull the strap snugly around the bundle.
  4. Feed the tail of the strap through MS17823 tool, and slide tool up to eye of the cable identification strap.
  5. Squeeze tool handles until strap is snug on bundle.
  6. Close tool handles all the way to cut off excess the strap.

NOTE: *Use of self-clinching adjustable plastic cable straps and installing tools is illustrated and described in Chapter 10.*

tion must be stripped from connecting ends to expose the bare conductor. For attachment to connectors enough insulation is stripped so that conductor will bottom in solder cup and leave a small gap between the top of the solder cup and cut end of insulation. Stripping dimensions for MS connectors will be found in Chapter 2, for RF connectors in Chapter 3, and for terminals in Chapter 4. See Figure 1-17 for typical tools used in wire stripping.

CAUTION: *Do not use hand strippers on wires to be installed in connectors with rubber sealing grommet.*

Stripping methods for copper wire. Copper wire may be stripped in a number of ways depending on size and insulation. See Table 1-7 for a summary of wire strippers.

Stripping methods for aluminum wire. Strip aluminum wires with a knife as described on page 1-16. Strip aluminum wire very carefully. Take extreme care not to nick aluminum wire, as strands break very easily when nicked.

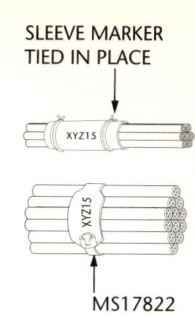

Figure 1-16. Identification of wire bundles and harnesses

Section 4
Wire and Cable Preparation
Stripping Wire and Cable

General. Before wire can be assembled to connectors, terminals, splices, etc., the insula-

Stripper	AN Gage No.	Insulations
Hot Blade	#26-#4	All except glass braid
Rotary, electric	#26-#4	All
Bench	#20-#6	All
Hand Pliers	#26-#8	All
Knife	#2-#0000	All

Table 1-7. Wire strippers for use on copper wire

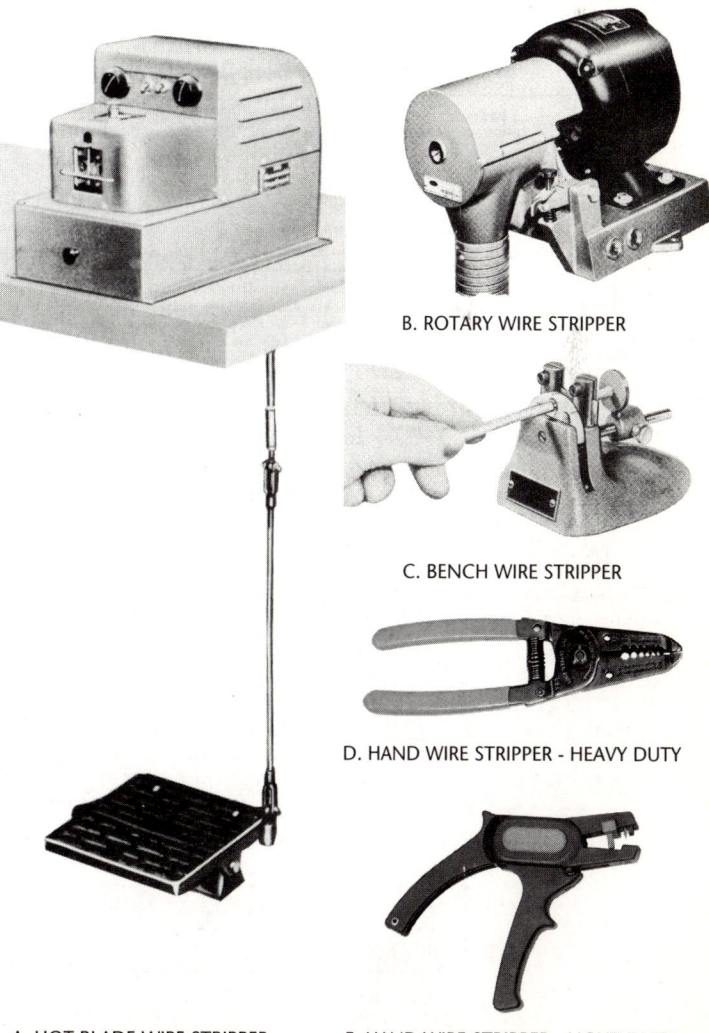

Figure 1-17. Typical wire stripping tools

Wire and Cable Preparation

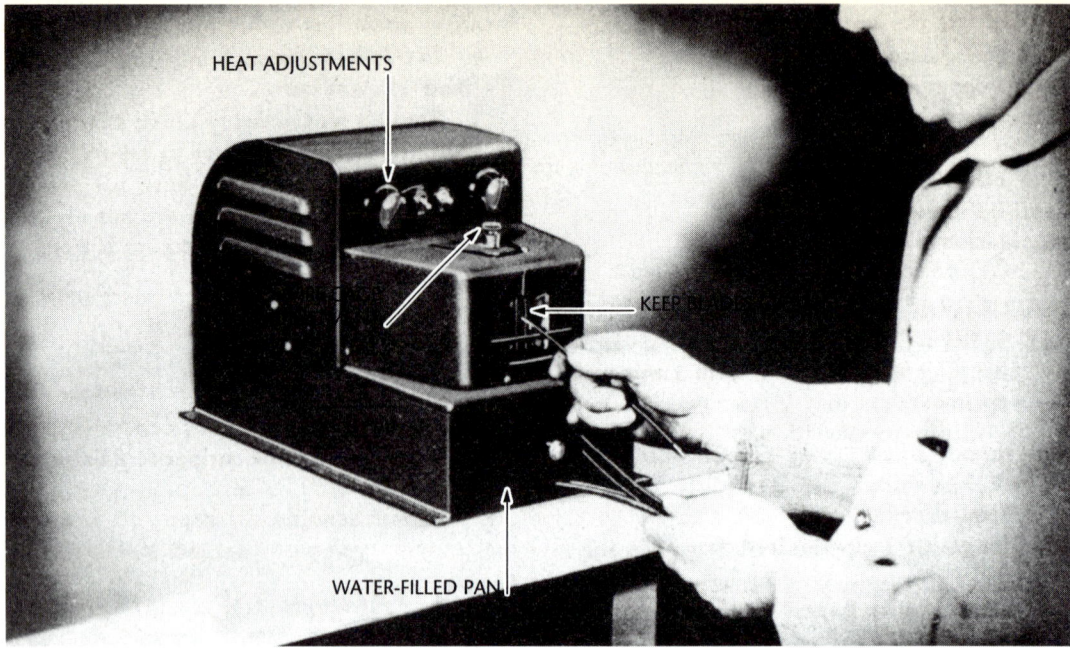

Figure 1-18. Stripping wire in a hot blade stripper

General stripping instructions. When stripping wire with any of the tools mentioned here, observe the following precautions:

- When using hot blade stripper, make sure blades are clean. Clean blades with a brass wire brush as necessary. The hot blade stripper will not strip wire with glass braid insulation.

b. Make sure all stripping blades are sharp, and free from nicks, dents, etc.

c. When using any type of wire stripper, hold wire perpendicular to cutting blades.

d. Adjust automatic stripping tools carefully; follow manufacturer's instructions, to avoid nicking, cutting, or otherwise damaging any strands. This is especially important for all aluminum wires and for copper wires smaller than No. 10. Examine stripped wires for damage, and adjust tool as necessary. Cut off and re-strip (if length is sufficient); or reject and replace any wires with more than the allowable number of nicked or broken strands given in Table 1-8.

NOTE: *Longitudinal scratches in copper wire are not considered cause for rejection or rework.*

e. Make sure insulation is clean-cut with no frayed or ragged edges. Trim if necessary.

f. Make sure all insulation is removed from stripped area. Some types of wires are supplied with a clear transparent layer

Wire	Nicked or Broken Strands
Copper	
AN #22-#12	None
#10	2
#8 -#4	4
#2 -#0	12
Aluminum	
All Sizes	None

Table 1-8. Allowable nicked or broken strands

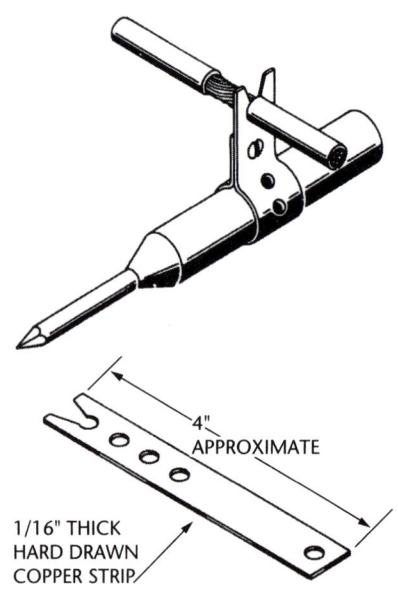

Figure 1-19. Substitute hot blade stripper

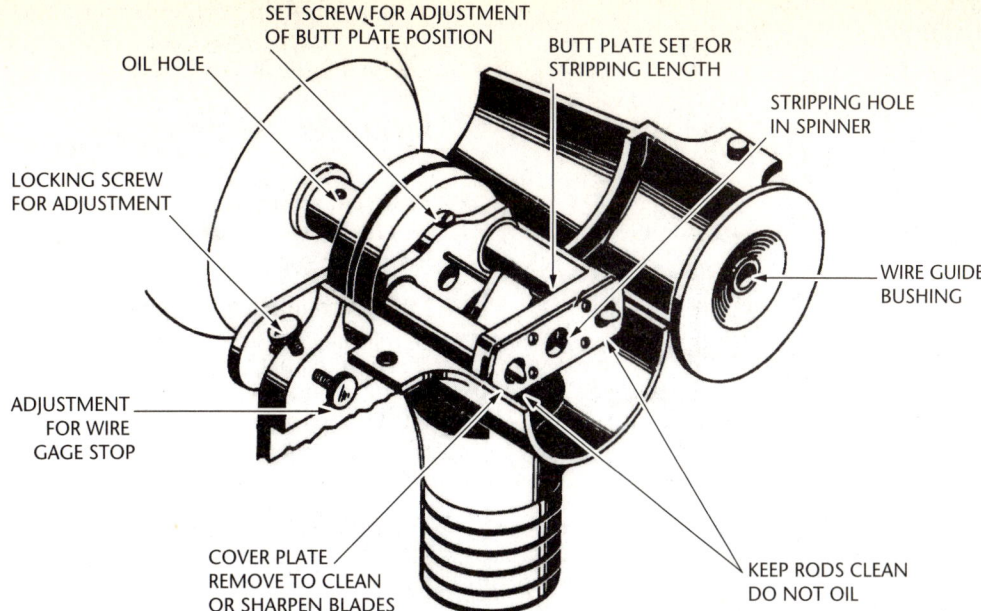

Figure 1-20. Inside view of rotary wire stripper

between conductor and primary insulation. If this is present, remove it.

g. When using hand plier strippers to remove lengths of insulation longer than 3/4 inch, it is easier to do in two or more operations.

h. Retwist copper strands by hand or with pliers if necessary to restore natural lay and tightness of strands.

Procedure for stripping wire with hot blade stripper. The procedure for stripping wire with hot-blade stripper is as follows:

1. Adjust blades to correct opening for size of wire to be stripped. See Figure 1-18.

2. Adjust stop by means of knurled brass nut on top of hood, for desired stripping length between 1/4 inch and 1 1/2 inch.

3. Adjust each blade to proper heat by trying on sample pieces of wire. Use minimum heat that will remove insulation satisfactorily without damaging strands.

4. Insert wire until it butts against stop.

5. Press foot pedal to bring heated blades against insulation.

6. Twist wire about 90 degrees and pull out.

   CAUTION: *Make sure adequate ventilation is provided when a hot-blade stripper is used to strip TFE-insulated wire.*

Substitute hot-blade stripper. See Figure 1-19. Where a hot-blade wire stripper is not available, a substitute can be made and used as follows:

1. In the end of a piece of copper strip, cut a sharp edged "V". At the bottom of the "V" make a wire slot of suitable diameter.

2. Fasten the copper strip around the heating element of an electric soldering iron as shown in Figure 1-19.

3. Lay wire or cable to be stripped in the "V"; a clear channel will be melted in the insulation.

4. Remove insulation with slight pull.

Procedure for stripping wire with power rotary stripper. Refer to Figure 1-20. The procedure for stripping wire or cable with a rotary stripper is as follows:

1. Select and install bushing of proper size for wire to be stripped. Bushings are available in 1/8 inch, 1/4 inch and 3/8 inch sizes.

2. Set butt for length of strip desired from 1/4 inch to 1 3/4 inch.

3. Make adjustment for wire gage.

4. Set switch for clockwise or counterclockwise rotation according to lay of strands.

5. Insert wire through bushing until end of wire is stopped against butt.

6. Step on foot pedal to close blades on wire.

7. Pull sharply on wires to remove insulation.

8. Examine wire to be sure all insulation is removed and also that strands are not nicked or cut. Reset wire gage adjustment (step 3) if necessary.

1-16 | Wire and Cable Preparation

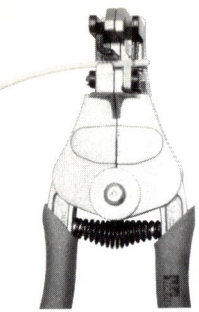

SELECT CORRECT
HOLE TO MATCH
WIRE GAGE
STEP 1

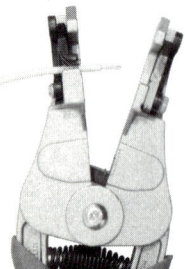

BLADES REMAIN
OPEN UNTIL WIRE
IS REMOVED
STEP 2

Figure 1-21. Stripping wire with hand stripper

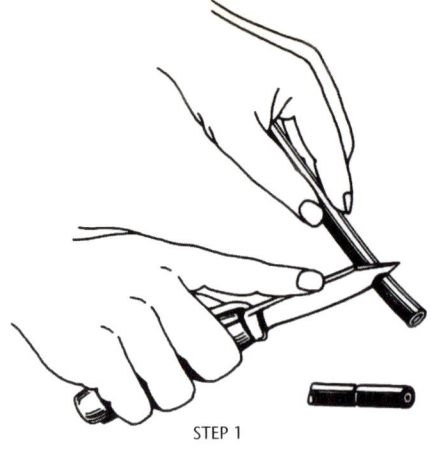

STEP 1
CUTTING AROUND INSULATION
BE CAREFUL NOT TO NICK OR CUT STRANDS

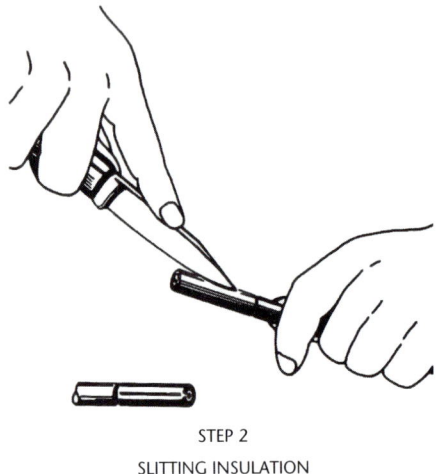

STEP 2
SLITTING INSULATION

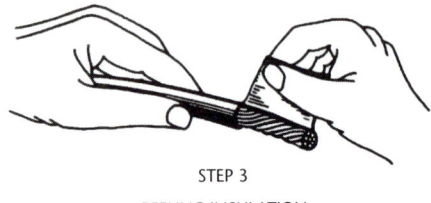

STEP 3
PEELING INSULATION

Figure 1-22. Knife stripping

Procedure for stripping wire with hand stripper. Refer to Figure 1-21. The procedure for stripping wire with plier-type hand strippers is as follows:

1. Insert wire into exact center of correct cutting for wire size to be stripped. (Each slot is marked with wire size).
2. Close handles together as far as possible.
3. Release handles, allowing wire holder to return to open position.
4. Remove stripped wire.

   NOTE: *Jaws will not snap back until wire is removed.*

Procedure for stripping wire with a knife. See Figure 1-22. The procedure for stripping wires with a knife is as follows:

   CAUTION: *Take care not to nick or cut strands.*

1. Make cut around wire at desired strip length. Do not cut completely through the insulation.
2. Make second cut lengthwise along stripping length. Do not cut completely through insulation.

   NOTE: *When a wire has two or more layers of insulation, cut through outer layers and only score innermost.*

3. Peel off insulation, following lay of strands.

Stripping dimensions for assembly to connectors. Stripped length on wires that are to be attached to connectors should be such that when stripped conductor bottoms in solder cup there will be a gap of approximately 1/32 inch between the end of the cup and the end of the insulation, for inspection purposes.

Tinning Copper Wire and Cable

General. Before copper wires are soldered to connectors the ends exposed by stripping are tinned to hold the strands solidly together. The tinning operation is considered satisfactory when the ends and sides of the wire strands are fused together with a coat of solder. Do not tin wires that are to be crimped to Class K (fireproof) connectors, wires which are to be attached to solderless terminals or splices, or wires that are to be crimped to removable crimp-style connector contacts.

Tinning methods. Copper wires are usually tinned by dipping into flux and then into a solder bath. In the field, copper wires can be tinned with a soldering iron and rosin core solder.

Extent of tinning. Tin conductor for about half its exposed length. This is enough to take advantage of closed part of solder cup. Tinning or solder on wire above cup causes wire to be stiff at point where flexing takes place. This will result in wire breakage.

Preparation of flux and solder. The flux used to tin copper wire is a mixture of denatured alcohol and freshly ground water-white rosin in the proportion of eight ounces of alcohol to one ounce of rosin, mixed together thoroughly and well shaken. During use the alcohol will evaporate and should be replaced. The solder used is a mixture of 60% tin and 40% lead. Maintain temperature of the solder pot between 450° and 500°F; this will keep solder in a liquid state. Skim surface of solder pot as necessary with a metal spoon or blade to keep solder clean and free from oxides, dirt, etc.

CAUTION: *Do not use any other flux or solder for tinning copper wires for use in aircraft electrical systems.*

Dip-tinning procedure. See Figure 1-23. Dip-tin wires smaller than No. 8 about eight or ten at a time. Dip-tin wires size No. 8 and larger individually.

CAUTION: *During tinning operation, take care not to melt, scorch or burn the insulation.*

The procedure for dip-tinning is as follows:

1. Prepare flux and solder as described above.

2. Make sure that exposed end of wire is clean and free from oil, grease and dirt. Strands should be straight and parallel. Dirty wire should be re-stripped.

3. Grasp wire(s) firmly and dip into dish of prepared flux to a depth of about 1/8 inch.

4. Remove wire and shake off excess flux.

5. Immediately dip only half of stripped conductor length into molten solder.

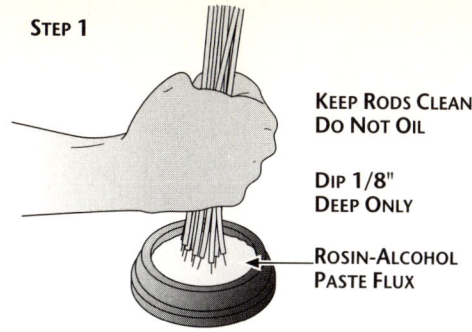

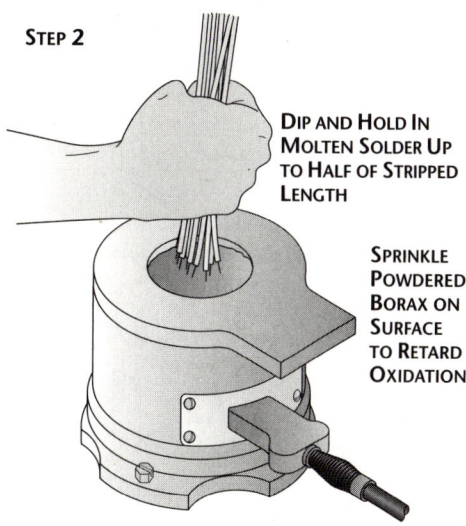

Figure 1-23. Dip tinning in solder pot

6. Manipulate wires slowly in solder bath until they are thoroughly tinned. Watch the solder fuse to wire. Do not keep wire(s) in bath longer than necessary.

7. Remove wires and shake off excess solder.

NOTE: *The thickness of the solder coat depends on the speed with which the wires are handled and shaken, and the temperature of the solder bath.*

Alternate dip-tinning procedure. See Figure 1-24. If an electrically heated solder pot is not available, a small number of wires may be tinned by means of the following procedure:

1. Cut off beveled section of tip of a discarded soldering iron tip.

2. Drill hole (to 3/8 inch diameter) in cylindrical part of tip, about two thirds through.

3. Heat up iron, and melt rosin-core solder into hole.

4. Tin wires by dipping into molten solder one at a time.

5. Keep adding fresh rosin-core solder as the flux burns away.

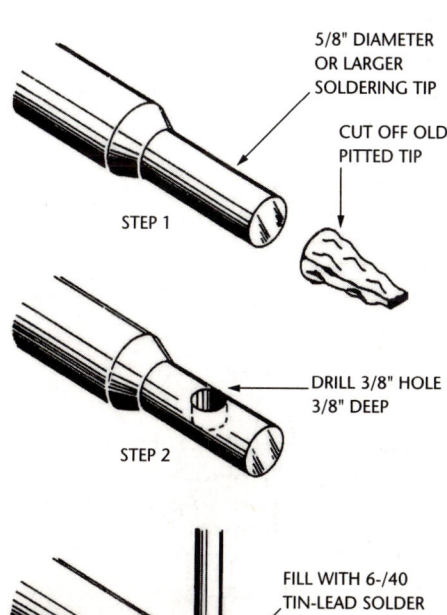

Figure 1-24. Alternate dip-tinning method

1-18 | Wire and Cable Preparation

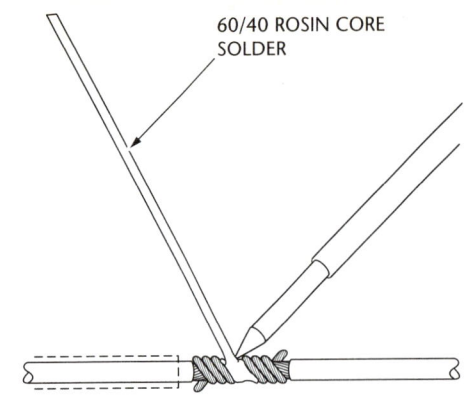

Figure 1-25. Tinning wire with a soldering iron

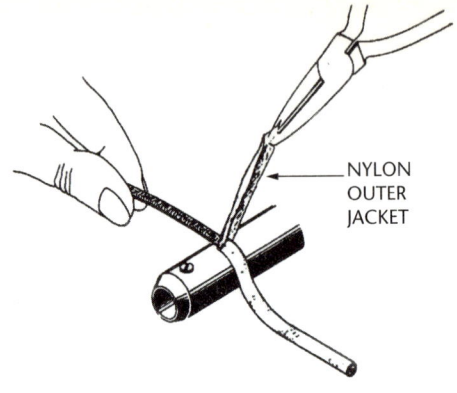

Figure 1-26. Stripping outer jacket from shielded cable

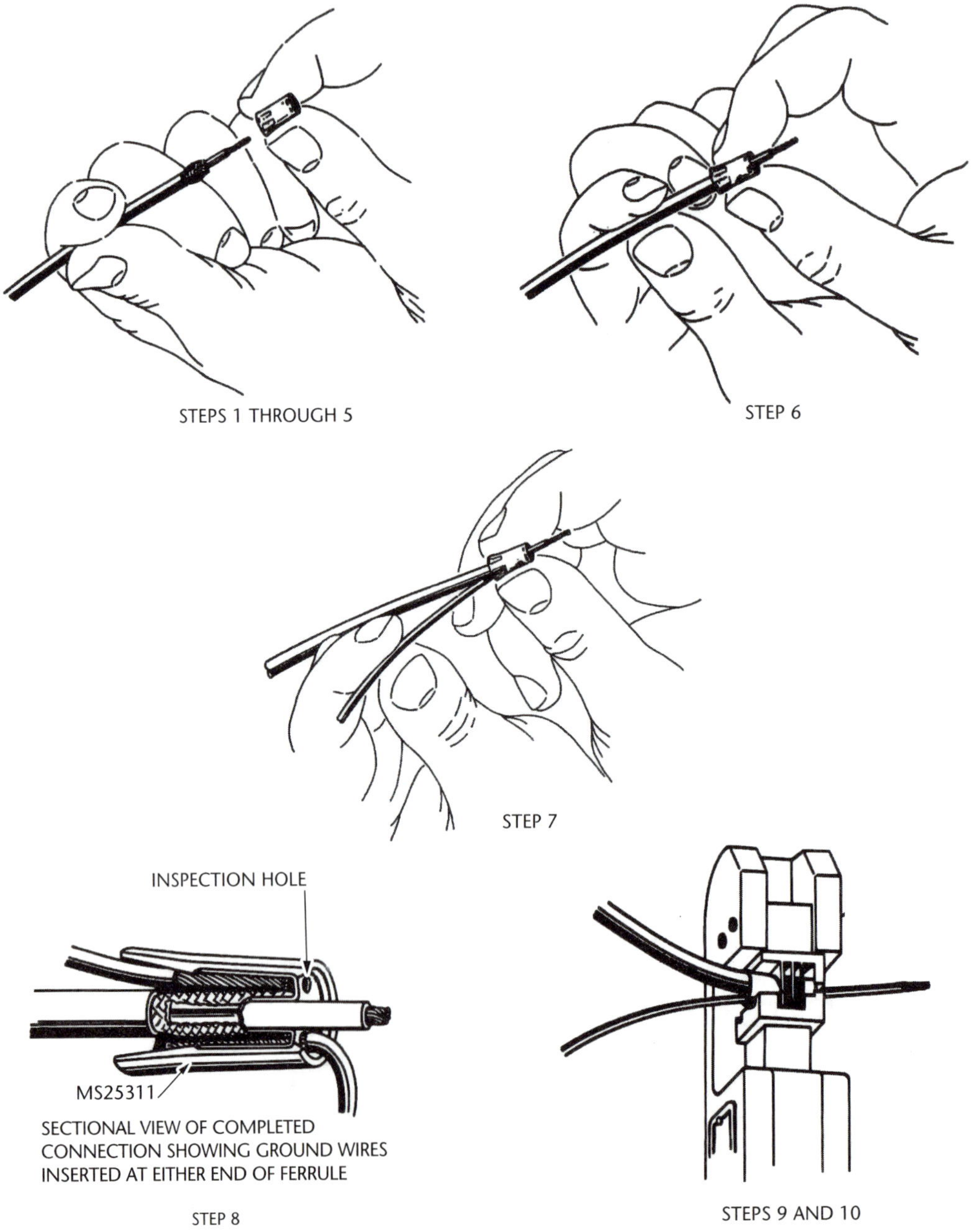

STEPS 1 THROUGH 5

STEP 6

STEP 7

INSPECTION HOLE

MS25311

SECTIONAL VIEW OF COMPLETED CONNECTION SHOWING GROUND WIRES INSERTED AT EITHER END OF FERRULE

STEP 8

STEPS 9 AND 10

Figure 1-27. One piece grounding connection or terminating shielded wire

Procedure for soldering iron tinning. See Figure 1-25. In the field wires smaller than size No. 10 may be tinned with a soldering iron and rosin core solder as follows:

1. Select a soldering iron having suitable heat capacity for wire size from Table 1-9. Make sure that iron is clean and well-tinned.
2. Prime by holding iron tip and solder together on wire until solder begins to flow.
3. Move soldering iron to opposite side of wire and tin half of the exposed length of conductor.

Terminating Shielded Cable

General. Shielded cable has a metallic braid over the insulation to provide a barrier around the conductor through which electrostatic energy (noise or interference) cannot pass. To obtain satisfactory result from shielded cable, the shield must be unbroken and must extend to a point as near the end of the conductor as practical. Shielded cable is either grounded or dead-ended at each end as required by the individual installation. The following paragraphs describe these procedures.

Stripping jacket on shielded cable. Some shielded cable has a thin extruded plastic coating over the shielding braid. Strip this off as far as necessary with a hot blade stripper, as described on page 1-15. Length of strip depends on method of shield termination and type of wire connection. Strip outer jacket back far enough for ease in working. If no hot-blade stripper is available, use plier-type hand strippers for sizes No. 22 through No. 10, and a knife for sizes larger than No. 10. Be careful not to damage shielding braid. Extruded jacket of shielded twisted wires can also be stripped by holding a soldering iron, with tip removed, against jacket, and pulling off jacket with long nose pliers as the iron melts the jacket. See Figure 1-26.

One-piece grounding ferrule method of shield termination. When the metallic braid of shielded cable can easily be flared out, the preferred method of terminating the shield is by crimping it, with or without a ground wire as required, into a one-piece insulated ferrule. Use the standard shield grounding ferrule MS25311, and the standard crimping tool MS25312. The procedure (see Figure 1-27) is as follows:

1. Determine the diameter of the insulation directly under the shield
2. Select the ferrule having the nearest larger ID from Table 1-10.
3. Strip shield braid (and outer jacket if present) 5/16 to 3/8 inch with hand strippers or scissors.

Wire Size (AN Gage)	Soldering Iron Size (Heat Capacity)
#20 - #16	65 Watts
#14 & #12	100 Watts
#10 & #8	200 Watts

Table 1-9. Approximate soldering iron sizes for tinning

4. Strip outer jacket (if present) 1/2 to 3/4 inch, being careful not to nick braid strands.
5. Fan metal braid slightly by rotating inner conductor.
6. Insert inner conductor through inner ring. Slide inner ring under the metal braid and outer ring over the metal braid.
7. If ground wire is required, use AN20 or AN22 wire, stripped 5/16 to 3/8 inch.
8. The shield braid must be visible through one of the inspection holes. The ground

MS Part Number	Tool	Insulation OD Under Shield	Ferrule ID	Tool
MS25311-90	MS25312-2	.050-.040	.058	MS25312-2
-100	-1	.055-.045	.063	-1
-110	-1	.072-.052	.080	-1
-120	-1	.093-.069	.102	-1
-130	-2	.107-.090	.115	-2
-150	-3	.126-.104	.134	-3
-160	-3	.148-.122	.156	-3
-180	-4	.171-.143	.179	-4
-200	-4	.202-.166	.210	-4

Table 1-10. Standard (MS) one piece shielded wire terminations and installing tools

1-20 | Wire and Cable Preparation

Part Number	Color Code	Insulation OD Under Shield	Inner Sleeve (Nominal) ID	Inner Sleeve (Nominal) OD
MS21981-046	Tin	.031 - .041	.046	.070
-058	Yellow	.043 - .053	.058	.083
-063	Red	.048 - .058	.063	.088
-071	Green	.056 - .066	.071	.096
-080	Blue	.065 - .075	.080	.104
-090	Orange	.075 - .085	.090	.114
-096	Purple	.081 - .091	.096	.119
-101	Yellow	.091 - .096	.101	.124
-109	Red	.096 - .104	.109	.131
-115	Tin	.104 - .110	.115	.146
-124	Green	.110 - .119	.124	.145
-128	Tin	.110 - .123	.128	.152
-134	Orange	.123 - .129	.134	.156
-149	Blue	.129 - .144	.149	.179
-156	Red	.145 - .151	.156	.192
-165	Tin	.151 - .160	.165	.194
-175	Green	.160 - .170	.175	.215
-187	Yellow	.175 - .182	.187	.227
-194	Blue	.182 - .189	.194	.225
-205	Orange	.189 - .200	.205	.245
-219	Tin	.200 - .214	.219	.248
-225	Yellow	.214 - .220	.225	.256
-232	Red	.220 - .227	.232	.263
-250	Green	—	.250	.281
-261	Blue	.227 - .255	.261	.297
-266	Tin	.261 - .271	.266	.297
-275	Orange	.255 - .270	.275	.306
-281	Yellow	.270 - .276	.281	.331
-287	Tin	.276 - .282	.287	.327
-297	Red	.282 - .292	.297	.336
-312	Purple	.292 - .307	.312	.362
-375	Blue	.370 - .380	.375	.406

Table 1-11. Shielded wire terminations inner sleeves

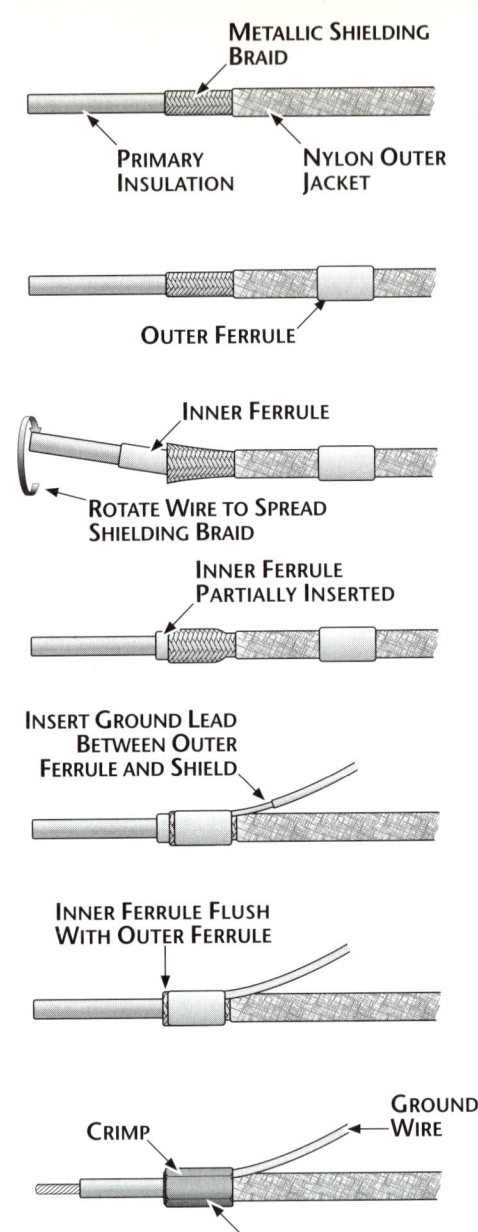

Figure 1-28. Two piece grounding connection for terminating shielded wire

lead should be visible through the inspection hole on the side where the ground lead was inserted.

9. Position the ferrule flush against the stop plate of the correct crimping tool selected from Table 1-10. Make sure the inspection holes are in the vertical position in line with direction of the crimp.

10. Close the handles of the tool all the way until the ratchet releases. Remove ferrule from tool.

Standard MS hand tool inspection. The MS25312 hand tool is checked with gages for the proper adjustment of the crimping jaws; check the tool before each series of crimping operations. The standard MS tool is checked with the standard gage MS25316. The dash number of the gage corresponds to the groove dash number of the die it will check. Check the standard tool dies with the tool fully closed; the GO gage should be able to enter between the jaws; the NO GO gage should not be able to enter.

Two-piece grounding sheath connector method of shield termination. See Figure 1-28. The metallic braid of shielded cable can also be terminated with a two-piece grounding sheath connector, by crimping it, with or without a ground wire as required, between two ferrules (or sleeves). Use the standard MS inner and outer ferrules listed in Tables 1-11, 1-12, and 1-13 and the tools listed in the tables. The procedure is as follows:

1. Strip off shielding braid (and outer jacket if present) with hand strippers or scissors. Length to be stripped is determined by

Part Number	Color Code	Sleeve ID (Inches) Nominal	Installing Tools (Thomas & Betts)
MS21980-101	Tin	.101	WT - 219
-128	Blue	.126	-200
-149	Purple	.149	-201
-156	Yellow	.156	-202
-175	Blue	.175	-203
-187	Orange	.187	-206
-194	Red	.194	-206
-199	Tin	.199	-206
-205	Yellow	.206	-208
-219	Green	.219	-208
-225	Purple	.225	-209
-232	Orange	.233	-210
-261	Yellow	.261	-211
-275	Tin	.275	-212
-281	Purple	.281	-214
-287	Blue	.287	-214
-299	Green	.299	-214
-312	Yellow	.312	-215
-327	Tin	.327	-216
-346	Orange	.346	-217
-359	Purple	.359	-221
-375	Yellow	.375	-222
-405	Red	.405	-218
-415	Blue	.415	-218
-460	Tin	.460	-220
-500	Green	.500	-223

Table 1-12. Shielded wire terminations - uninsulated outer sleeves and installing tools

Part Number	Color Code	Sleeve ID (Inches) Nominal	Installing Tools (Thomas & Betts)
MS18120-101	Tin	.101	WT-200
-128	Blue	.128	-201
-149	Purple	.149	683-51135
-156	Yellow	.156	WT-206
-175	Blue	.175	-208
-187	Orange	.187	-210
-194	Red	.194	-210
-199	Tin	.199	-210
-205	Yellow	.205	-211
-219	Green	.219	-211
-225	Purple	.225	-211
-232	Orange	.232	-212
-261	Yellow	.261	-214
-275	Tin	.275	-215
-281	Purple	.281	-217
-287	Blue	.287	-217
-297	Green	.297	-217
-312	Yellow	.312	-222
-327	Tin	.327	-222
-348	Orange	.348	683-51014
-359	Purple	.359	WT-216
-375	Yellow	.375	683-51-15
-405	Red	.405	683-51141-1
-415	Blue	.415	683-51141-1
-460	Tin	.460	683-51141-2
-500	Green	.500	683-51141-3

Table 1-13. Shielded wire terminations - insulated outer sleeves and installing tools

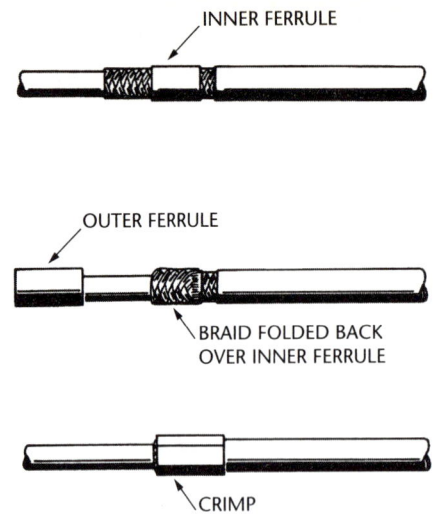

Figure 1-29. Alternate procedure for two piece grounding sheath connector

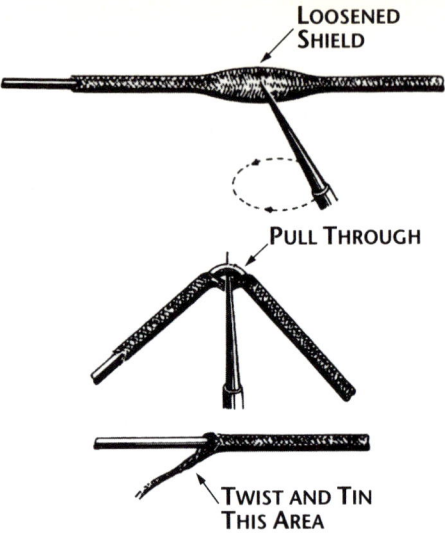

Figure 1-30. Pigtail termination for shielded wire

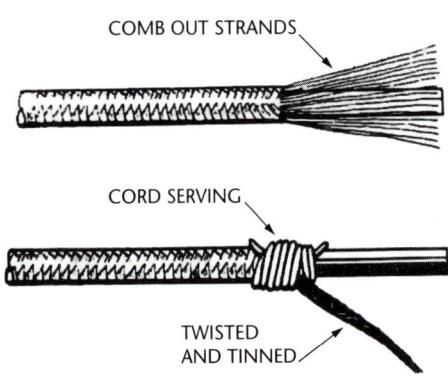

Figure 1-31. Alternate pigtail termination for shielded wire

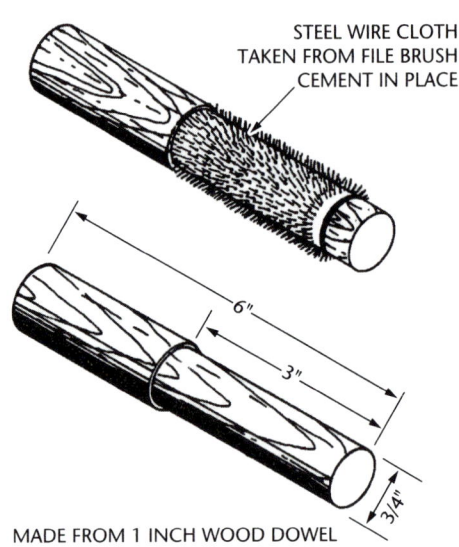

Figure 1-32. Comb for combing out shield

length of unshielded conductor necessary for making connection.

2. Strip outer jacket (if present) 1/2 to 3/4 inch.

3. Measure OD of insulation directly under shield.

4. Add .005 inch minimum to OD obtained in step 3, and select inner sleeve having the nearest larger ID from Table 1-10.

5. Note OD of inner sleeve selected in step 4, and add .025 inch minimum to it to allow for thickness of shielding braid. Add an extra .030 to .040 to allow clearance for a No. 20 or No. 18 ground wire if required. From Tables 1-12 or 1-13 select an uninsulated or insulated outer sleeve as required, with the above dimension as minimum ID.

6. Slide outer sleeve back over insulation and braid.

7. Rotate cable with circular motion to flare out braid.

8. Slip inner sleeve under braid so that about 1/16 inch of sleeve sticks out beyond braid.

9. Insert stripped ground wire under outer sleeve (if required) and slide both forward over braid and inner sleeve until only 1/32 to 1/16 inch of inner sleeve and braid protrude. See Figure 1-28. Ground wire may extend from front or back of outer sleeve as required.

CAUTION: *Examine assembly to make sure that shield braid and ground wire come through under the outer sleeve.*

10. Crimp with hand tool selected from Tables 1-12 or 1-13.

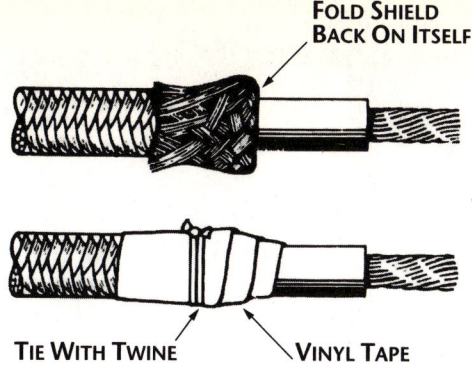

Figure 1-33. Dead ending shield with tape wrap

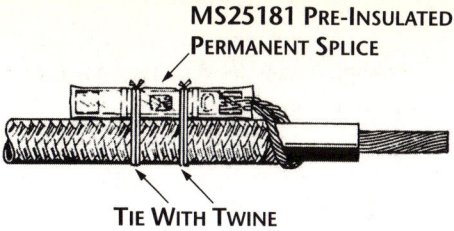

Figure 1-34. Dead ending shielding with permanent splice

The above procedure may be modified by sliding the inner ferrule over the braid, and folding the braid neatly back over the inner ferrule (See Figure 1-29). If this is to be done, add an additional .025 inch to the dimension obtained in step 4 for the extra braid thickness.

Pigtail method of shield termination. When grounding sheath connectors and tools are not available, terminate shield for grounding by making a pigtail (see Figure 1-30) as follows:

CAUTION: *Take extreme care not to damage shielding or insulated conductor while forming pigtail.*

1. Determine and mark point shielding is to terminate. This depends on the individual installation.

2. Push back shielding to form a bubble at the termination point.

3. Insert an awl, or other pointed tool into shielding braid at termination point and work an open circular area in the shield. Be careful not to cut into wire insulation.

4. Bend cable, insert tool between shielding and wire, and pull insulated conductor through hole formed by tool.

5. Pull empty part of shield taut and tin last inch to prevent fraying.

6. On unjacketed shielded cable, spot tie shielding on cable with clove hitch and square knot. This is not necessary if cable has extruded plastic jacket over shield.

Alternate pigtail method. See Figure 1-31. Some shielding braids may be too stiff for the method described in the previous paragraph. In this case, cut shielding with scissors approximately 1 1/2 inches forward of termination point. Comb out strands with comb (see Figure 1-32) or pointed bakelite rod. Twist strands into pigtail, or separate strands into three parts, twist each part, and braid together. Tie with cord as described in the previous paragraph, step 6.

Dead-ending shielded cable. When the shielding is not to be grounded, it is dead-ended so as to gather all loose shield ends together to prevent them from puncturing insulation.

Dead-ending with grounding sheath connector. When equipment is available, dead-end shielded cable with grounding sheath connector as described in the earlier sections on one or two piece grounding sheaths, omitting ground wire. Refer to Figures 1-27 or 1-28. Omit clearance allowed for ground wire when selecting outer sleeve.

Dead-ending with tape wrap. When grounding sheath connector and tools are not available, dead-end the shielding as follows (see Figure 1-33):

1. Cut shielding braid with scissors about 3/4 inch forward of termination point.

2. Loosen braid and turn back on itself 3/4 inch.

3. Wrap with two or three turns of plastic tape, making sure that all braid ends are covered.

4. Tie loose end of tape with clove hitch and square knot, or heat seal end of tape with unntinned side of soldering iron. A plastic wire strap or tie, as described in Chapter 10 may be used instead of tying cord to secure the loose end of the tape.

Alternate methods of dead-ending. As an alternate to dead-ending with grounding sheath connector make a pigtail as described in the previous pigtail or alternate pigtail methods. Trim pigtail so it is 3/4 inch long. Crimp trimmed pigtail into one end of preinsulated permanent splice of suitable size, and tieback on shielded part of cable. See Figure 1-34. If permanent splice is not available, cut tongue off solderless terminal of suitable size, and crimp pigtail into it. Protect with sleeve and tie back on wire, similar to tie used with permanent splice.

Chapter 2

GENERAL
purpose connectors

Connectors provide means of quickly connecting and disconnecting wires to simplify installation and maintenance of electric and electronic equipment.

This chapter describes and illustrates the types and classes of MS connectors and the recommended procedures for attaching wires to connector contacts. AN type connectors were formerly designated with the prefix "AN", and older connectors may still be found with this prefix. The superseding connector has the same part number except that the "AN" has been replaced by "MS".

Other connectors, commonly used in aircraft, similar to MS connectors, are also described and illustrated in this chapter. RF connectors are treated separately in Chapter 3.

Learning Objectives:

- *Description & Identification*
- *Types of AN-MS Connectors*
- *Disassembly of Connectors*
- *Assembly of Wire to Connectors*
- *Installation of Clamps*
- *Contact Tools*

Section 1

Description and Identification

AN-MS Connectors

Each complete AN-MS connector consists of two parts: a plug assembly and a receptacle assembly coupled by means of a coupling device that is part of the plug assembly. Standard AN type connectors are coupled with a threaded coupling ring except for MS3107, which has a friction coupling. Miniature MS connectors, a smaller lightweight version of the AN type connector, are coupled by means of a threaded ring, a bayonet lock or by push-pull coupling.

Left. Today's electrical connectors are designed in many different styles for use in a variety of applications and conditions.

2-2 | General Purpose Connectors

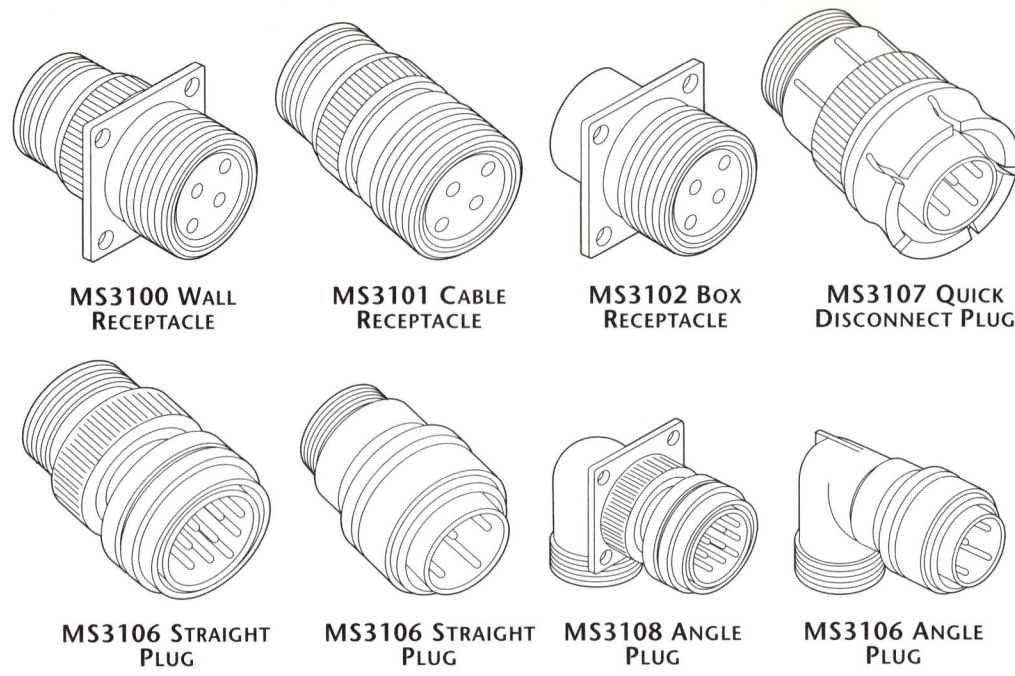

Figure 2-1. MS connectors (AN type)

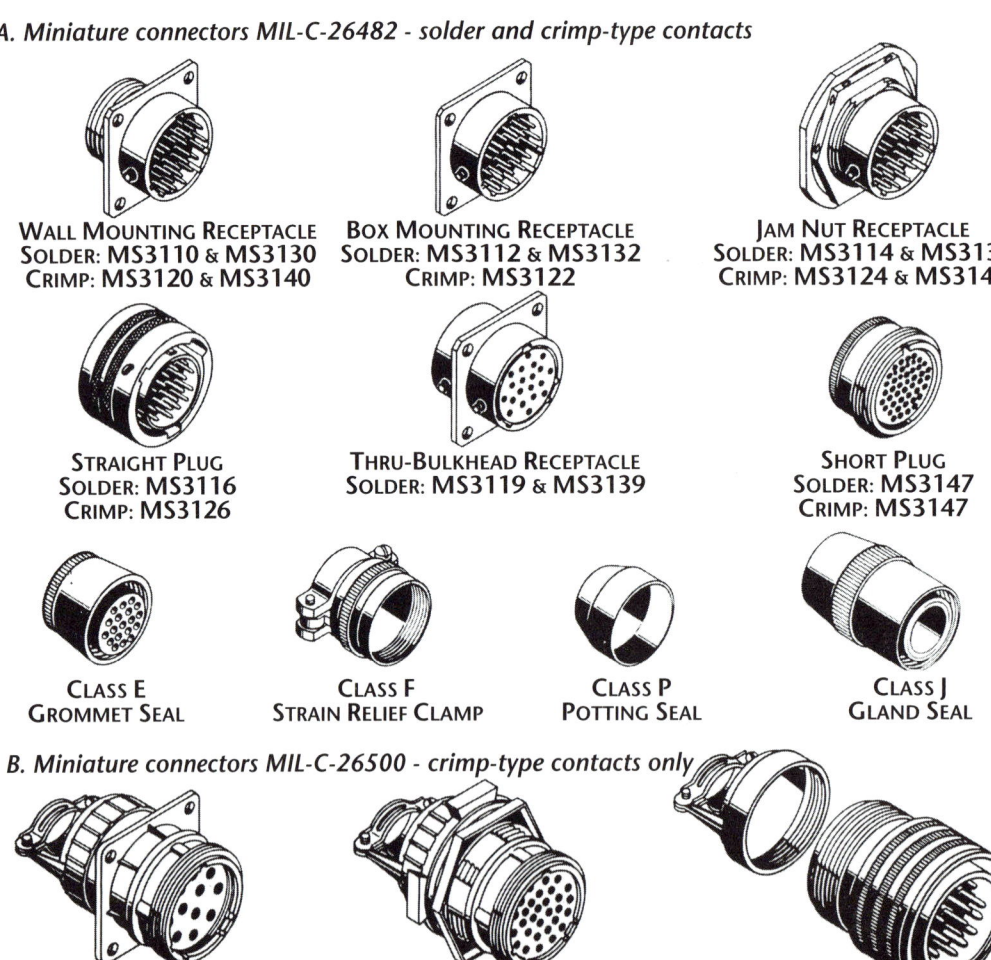

Figure 2-2. MS connectors (miniature)

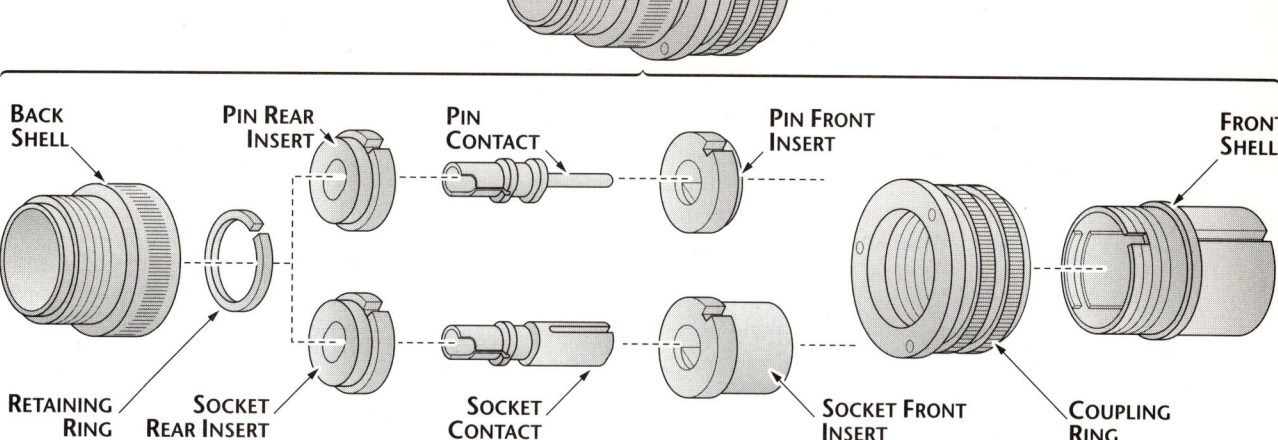

Figure 2-3. MS connector plug (exploded view)

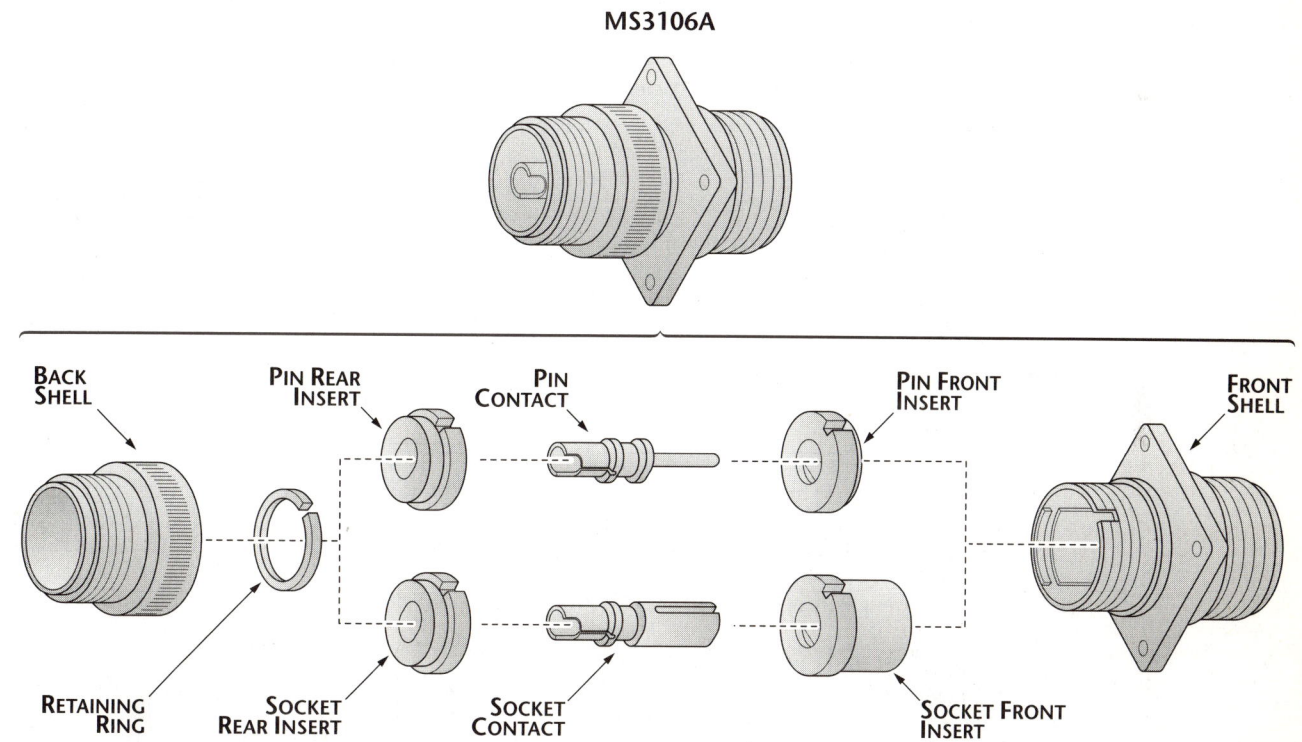

Figure 2-4. MS connector receptacle (exploded view)

See Figure 2-1 for illustrations of AN type connectors, and Figure 2-2 for illustrations of miniature MS connectors.

The receptacle is usually the "fixed" part of the connector, and is attached to a wall, bulkhead or equipment case. The plug is the removable part of the connector and includes the coupling ring. When the two parts are joined by the coupling device, the electric circuit is made by pin-and-socket contacts inside the connector. The "live" or "hot" side of the circuit usually has socket (female) contacts. Either the plug or the receptacle may contain the "live" parts of the circuit. The contacts are held in place and insulated from each other and from the shell by a dielectric insert. Insert and contacts are housed in a metal shell. See Figure 2-3 for the exploded view of a typical plug and Figure 2-4 for the exploded view of a typical receptacle.

Section 2
Types of AN-MS Connectors

AN-MS connectors are separated into types and classes, with manufacturer's variations in each type and class. These variations are in the method of meeting specification requirements, and in appearance. The variations are minor, and do not affect the ability to mate plugs and receptacles made by different manufacturers. There are six AN types of standard MS connectors as listed in Table 2-1 and shown in Figure 2-1. The connectors are further separated into the six classes listed in Table 2-2.

MS Type	Nomenclature
MS3100	Wall Mounting Receptacle
MS3101	Cable Connecting Plug
MS3102	Box Mounting Receptacle
MS3106	Straight Plug
MS3107	Quick Disconnect Plug
MS3108	90 Degree Angle Plug

Table 2-1. AN types of standard MS connectors

Air Connector Types

See Figure 2-1. The following six types of standard AN type connectors are used in aircraft:

- **MS3100.** A receptacle with flange for mounting to wall or bulkhead. Contains front shell, insert retaining ring, insert, contacts and back shell. Connectors with resilient inserts omit the insert retaining ring. See Figure 2-4 for the exploded view of a typical receptacle of this type.

- **MS3101.** A plug used at the end of a wire or wire bundle where mounting is not necessary. Similar to MS3100 except that it has no mounting flange.

- **MS3102.** A receptacle with flange for mounting to a junction box or equipment case. Similar to MS3100, except that it has no back shell.

- **MS3106.** A straight plug, used at the end of a wire or wire bundle. Consists of front shell, coupling nut, insert retaining ring, insert, contacts and back shell. Connectors with resilient inserts omit the retaining ring. See Figure 2-3 for exploded view.

- **MS3107.** A "quick-disconnect" plug, used where fast pull disconnection from the receptacle is necessary. It is similar to MS3106 except that it is coupled to an MS receptacle by means of a friction ring instead of a coupling nut.

- **MS3108.** A right-angle plug, used where wiring must make an abrupt change in direction as it leaves the plug.

MS Class	Application	Shell	Availability 3100	3101	3102	3106	3107	3108
A	General Purpose	Solid Aluminum Alloy	Yes	Yes	Yes	Yes	Yes	Yes
B	General Purpose	Split Aluminum Alloy	Yes	Yes	No	Yes	Yes	Yes
C	Pressurized Receptacle	Solid Aluminum Alloy	Yes	No	Yes	No	No	No
E	Environmental Resistant	Solid Aluminum Alloy with strain relief clamp	Yes	Yes	Yes	Yes	No	Yes
K	Fire and Flame Resistant	Solid Steel	Yes	No	Yes	Yes	No	Yes
R	Environmental Resistant	Solid Aluminum Alloy	Yes	Yes	Yes	Yes	No	Yes

Table 2-2. Classes of AN type connectors

AN-MS Connector Classes

There are six classes of AN-MS connectors. All have aluminum alloy shells except Class K which has a steel shell to achieve fire resistance.

- **Class A.** General-purpose connector with solid one piece back shell. Plugs and receptacles shown in Figure 2-1 are all Class A (solid back shells).

- **Class B.** Connector with back shell split in two, lengthwise, used where it is important to be able to get at soldered connections easily. The two halves of the back shell are held together by clamping ring, or by screws. See Figure 2-5 for exploded views of typical split shell MS connectors.

- **Class C.** Pressurized connector, used on walls and bulkheads of equipment. It looks the same as Class A receptacle, but the inside sealing arrangement is different. Inserts of Class C connectors are not removable. Mating the Class C receptacle to the other class plugs does not affect the sealing qualities of the Class C receptacle.

- **Class E.** Environment (moisture and vibration) resisting connector, used in areas where changes in temperature may cause condensation, or where there is likely to be vibration. Class E connectors have a sealing grommet in the back shell, The wires pass through tight fitting holes in the grommet and are thereby sealed against moisture. The contacts are supported in a resilient insert. For proper performance mate Class E receptacles to Class E plugs.

- **Class K.** Fireproof connector, used where it is vital that current continue to flow even though the connector may be exposed to continuous open flame. Class K connectors are longer in overall length than other classes, and have a shell made of steel instead of aluminum alloy. Inserts of Class K connectors are of special fire-resistant material, and have crimp-type contacts instead of solder-type.

- **Class R.** Lightweight environment resisting connector similar to the Class E connector, and intended to replace it where shorter length and lighter weight are required. An O-ring is provided in the MS3106 and MS3108 plugs for additional sealing.

Miniature MS Connectors

Two standards cover the miniature MS connectors most commonly used in aircraft: MIL-C-26482 and MIL-C-26500. Connectors manufactured to the requirements of MIL-C-26482 may have contacts of either the conventional solder-type, or crimp type. MIL-C-26500 connectors have crimp-type contacts only. Connectors to

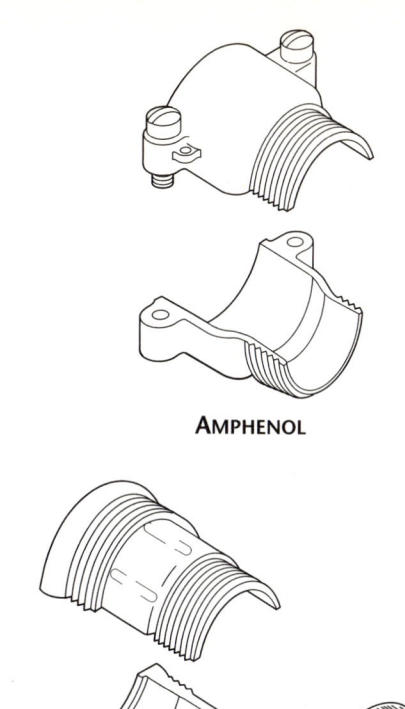

Figure 2-5. Split back shell connector

both specifications have contacts in sizes 20, 16 and 12 only. The types and classes of miniature MS connectors with solder-type contacts are listed in Table 2-3 (next page). Miniature MS connectors with crimp-type contacts are available in the types listed in Table 2-4 (next page), and are of the environment resisting classes.

Miniature MS Connector Types - Solder Contacts

The following types of miniature MS connectors with solder contacts are used in aircraft:

- **MS3110.** A receptacle with flange for mounting to a wall or bulkhead; coupled by means of a bayonet lock

- **MS3112.** A receptacle for mounting to junction box or equipment case similar to MS3110 except that it has no back shell; has bayonet lock coupling

- **MS3114.** A rear mounting receptacle, with jam nut instead of flange; has bayonet lock coupling

General Purpose Connectors

MS Type	Nomenclature	Availability				
		Class				
		E	F	P	H	J
Bayonet Coupling						
MS3110	Wall Mounting Receptacle	Yes	Yes	Yes	No	Yes
MS3112	Box Mounting Receptacle	Yes	No	Yes	Yes	No
MS3114	Rear Mounting Jam Nut Receptacle	Yes	No	Yes	Yes	No
MS3116	Straight Plug	Yes	Yes	Yes	No	Yes
MS3119	Thru-bulkhead Mounting Receptacle	Yes	No	Yes	No	No
Push-Pull Coupling						
MS3130	Wall Mounting Receptacle	Yes	No	Yes	No	Yes
MS3132	Box Mounting Receptacle	Yes	No	No	Yes	No
MS3134	Single-hole Mounting Receptacle	Yes	No	Yes	Yes	Yes
MS3137	Short Plug	Yes	No	Yes	No	Yes
MS3138	Lanyard Plug	Yes	No	Yes	No	Yes
MS3139	Thru-bulkhead Mounting Receptacle	Yes	No	No	No	No

Table 2-3. Types and classes of MIL-C-26482 miniature MS connectors with solder contacts

MS Type	Nomenclature
1. MIL-C-26482	
Bayonet Coupling:	
MS3120	Wall Mounting Receptacle
MS3122	Box Mounting Receptacle
MS3124	Rear Mounting Jam Nut Receptacle
MS3126	Straight Plug
Push-Pull Coupling:	
MS3140	Wall Mounting Receptacle
MS3144	Single-hole Mounting Receptacle
MS3147	Plug
MS3148	Lanyard Plug
2. MIL-C-26500	
MS24264	Flange Mounting Receptacle
MS24265	Single hole Mounting Receptacle
MS24266	Straight Plug

Table 2-4. Types of miniature MS connectors with crimp contacts

- **MS3116.** A straight plug for end of wire or wire bundle; has bayonet lock coupling
- **MS3119.** A thru-bulkhead mounting receptacle; has bayonet lock coupling
- **MS3130.** Similar to MS3110, except that it has push-pull (ball lock) coupling
- **MS3132.** A receptacle similar to MS3112, except that it has push-pull coupling
- **MS3134.** A single-hole mounting receptacle, similar to MS3114, except that it has push-pull coupling
- **MS3137.** A straight plug similar to MS3116 except that it has push-pull coupling
- **MS3138.** A plug with lanyard, with a push-pull coupling
- **MS3139.** A thru-bulkhead mounting receptacle, similar to MS3119 except that it has push-pull coupling

Miniature MS Connector Types - Crimp Contacts

These connectors are similar to the miniature connectors listed in above, but have removable contacts to which wires are crimped with a standard crimping tool, instead of soldered. Connectors with crimp-type contacts are available in the following types:

- **MS3120.** A receptacle with flange for mounting to a wall or bulkhead; is coupled by means of a bayonet lock

- **MS3122.** A receptacle for mounting to junction box or equipment case similar to MS3120 except that it has no back shell; bayonet lock coupling

- **MS3124.** A rear mounting receptacle, with jam nut instead of flange; bayonet lock coupling

- **MS3126.** A straight plug for use at end of wire or wire bundle; bayonet lock coupling.

- **MS3140.** A flange-mounting receptacle, similar to MS3120 except that it has push-pull coupling

- **MS3144.** A single-hole-mounting receptacle, similar to MS3124 except that it has push-pull coupling

- **MS3147.** A plug for use at end of wire or wire bundle; push-pull coupling

- **MS3148.** A plug with lanyard; push-pull coupling

- **MS24264.** A receptacle with flange for mounting to a wall or bulkhead

- **MS24265.** A receptacle with jam nut for panel mounting

- **MS24266.** A straight plug used at the end of a wire or a wire bundle

Miniature MS Connector Classes

There are five classes of miniature MS connectors with solder-type contacts. These are:

- **Class E.** An environment (moisture and vibration) resisting connector, moisture-proofed by means of a wire grommet seal and clamping nut.

- **Class F.** An environment resisting connector, similar to Class E, with addition of a strain relief clamp.

- **Class H.** Hermetic sealed receptacle that has a glass insert fused to the contacts and the shell.

- **Class J.** A connector incorporating a gland seal for sealing a jacketed cable.

- **Class P.** Connectors supplied with a plastic potting mold, so that the connectors may be sealed by the application of a potting compound.

MS connector marking. Each MS connector is marked on the shell or coupling ring with a code of letters and numbers giving all the information necessary to identify the connector. See Figure 2-6. A typical code is as follows:

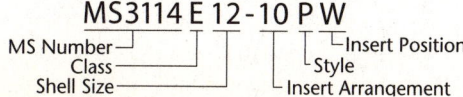

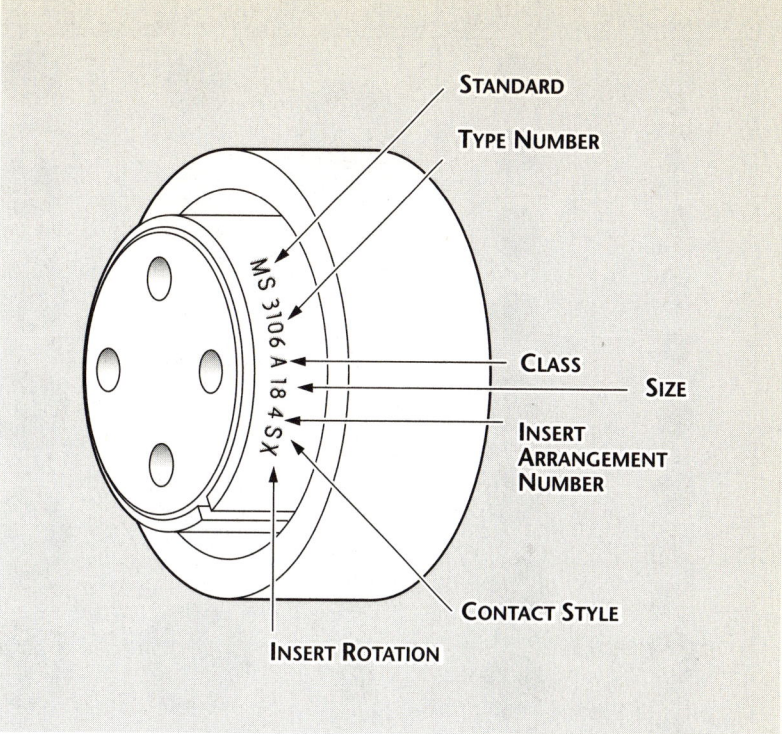

Figure 2-6. MS connector marking

- The letters "MS" indicate that the connector has been made according to Military Specifications.

- Numbers such as 3114 indicate the type of shell, and whether it is a plug or receptacle.

- Class letter indicates design of shell, and for what purpose connector is normally used.

- Numbers following class letter indicate shell size by outside diameter of mating part of receptacle in one-sixteenth inch increments, or by the diameter of the coupling thread in sixteenths of an inch. For example size 12 has an outside diameter or a coupling thread of 3/4 inches.

NOTE: *MIL-C-26500 connectors have an added letter to indicate type of coupling between the shell size and insert arrangement code numbers. These letters are "T" for thread coupling, "B" for bayonet coupling and "Q" for push-pull coupling. For example: MS24264R18B30P6, where B indicates type of coupling.*

- Numbers following hyphen indicate insert arrangement. This number does not indicate the number of contacts. MS drawings cover contact arrangements approved for service use. See Figures 2-7, 2-8 and 2-9 (following pages) for illustrations of insert arrangements.

- First letter following number indicates style of contact.

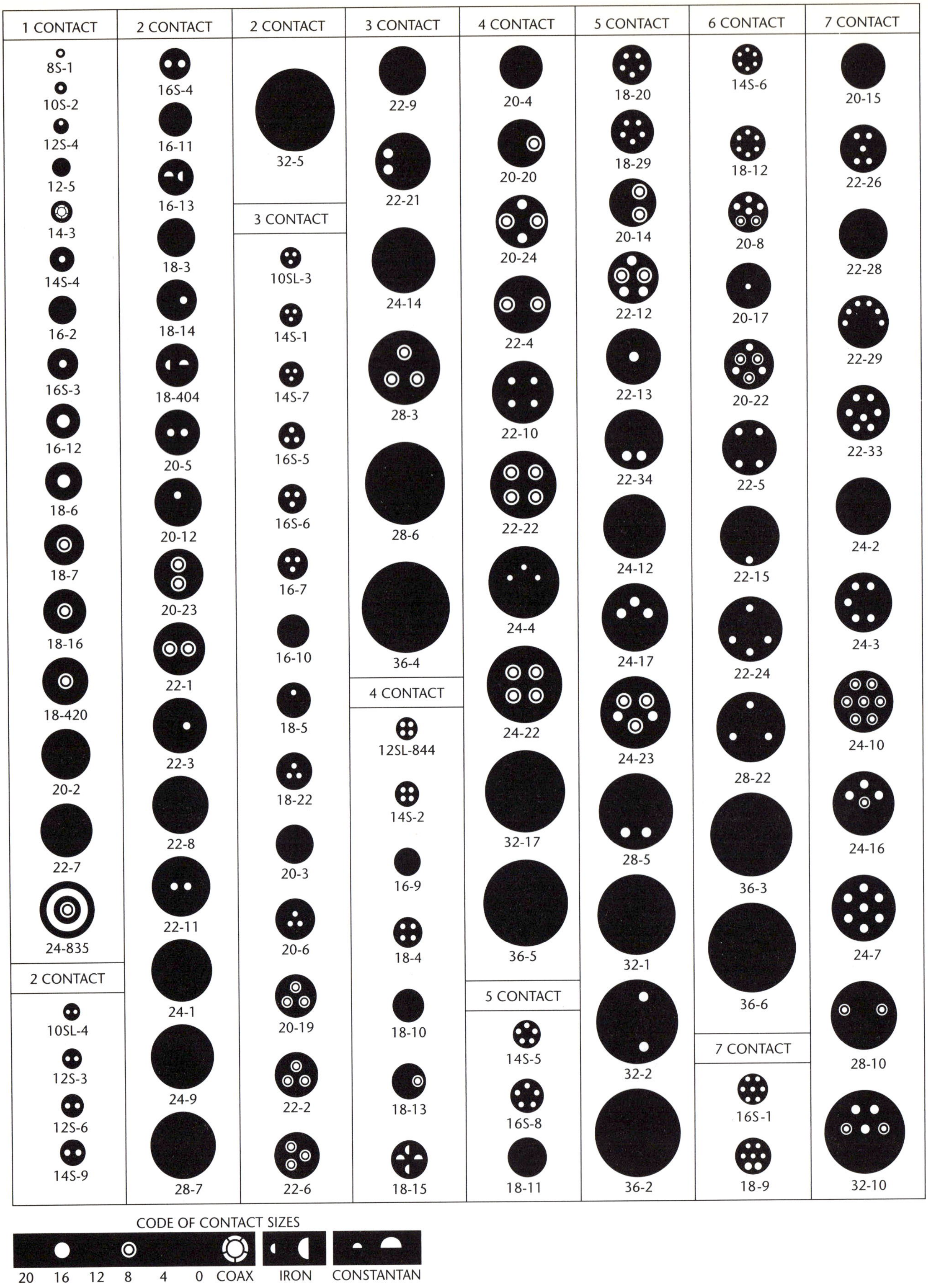

Figure 2-7. Insert arrangements - AN type connectors, MIL-C-5015

General Purpose Connectors | 2-9

8 CONTACT	9 CONTACT	11 CONTACT	14 CONTACT	16 CONTACT	23 CONTACT	30 CONTACT	47 CONTACT
18-8	22-17	24-20	28-20	24-7	32-13	40-1	36-7
20-7	22-20	28-14	32-4	36-14	40-2	**31 CONTACT**	36-8
20-9	22-27	**12 CONTACT**	32-9	**17 CONTACT**	40-3	36-9	36-8
22-18	24-11	24-19	32-102	20-29	40-4	**35 CONTACT**	40-9
22-23	28-1	28-8	**15 CONTACT**	36-13	**24 CONTACT**	28-15	**48 CONTACT**
22-36	28-4	28-9	28-17	**19 CONTACT**	24-28	32-7	32-409
22-404	32-3	28-18	32-12	22-14	**26 CONTACT**	36-15	36-10
24-6	**10 CONTACT**	32-101	36-405	**20 CONTACT**	28-12	**37 CONTACT**	**52 CONTACT**
32-15	18-1	**13 CONTACT**	36-407	28-16	**29 CONTACT**	28-21	32-414
9 CONTACT	18-19	20-11	**14 CONTACT**	**22 CONTACT**	40-10	36-406	36-403
20-16	24-21	**14 CONTACT**	20-27	28-11	**30 CONTACT**	**42z CONTACT**	36-404
20-18	28-19	20-27	22-19	36-1	32-6	44-1	
20-21	**11 CONTACT**	28-19	**16 CONTACT**	**23 CONTACT**	32-8		
22-16	20-33	28-2	24-5	32-6			

2-10 | General Purpose Connectors

SHELL SIZE	3 CONTACTS		4 CONTACTS	5 CONTACTS	6 CONTACTS	8 CONTACTS	10 CONTACTS	11 CONTACTS
Insert Arrangement Number	8-3	12-3	8-4	14-5	10-6	16-8	12-10	18-11
Number and Size of Contacts	3-20	3-16	4-20	5-16	6-20	8-16	10-20	11-16
SHELL SIZE	12 CONTACTS		15 CONTACTS	16 CONTACTS	19 CONTACTS	21 CONTACTS	26 CONTACTS	
Insert Arrangement Number	14-12		14-15	20-16	14-19	22-21	16-26	
Number and Size of Contacts	8-20 / 4-16		14-20 / 1-16	16-16	19-20	21-16	26-20	
SHELL SIZE	32 CONTACTS		39 CONTACTS		41 CONTACTS	55 CONTACTS	61 CONTACTS	
Insert Arrangement Number	18-32		20-39		20-41	22-55	24-61	
Number and Size of Contacts	32-20		37-20 / 2-16		41-20	55-20	61-20	

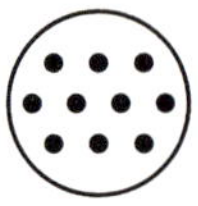

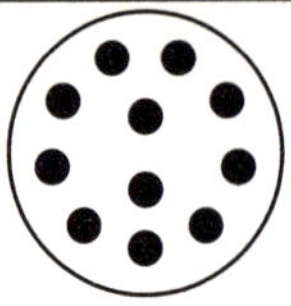

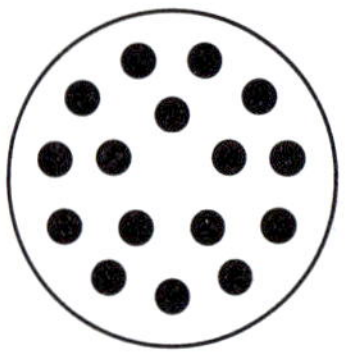

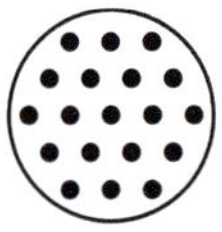

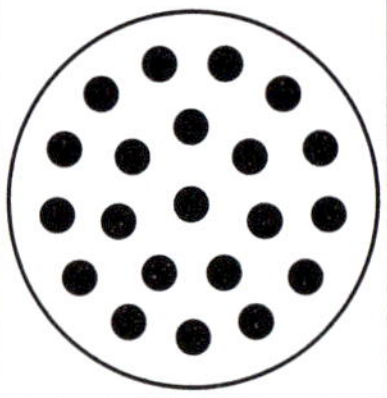

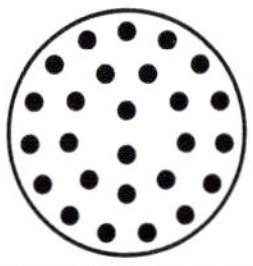

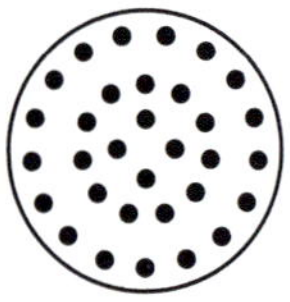

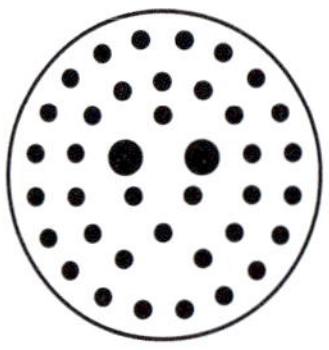

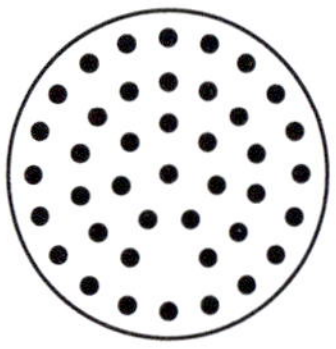

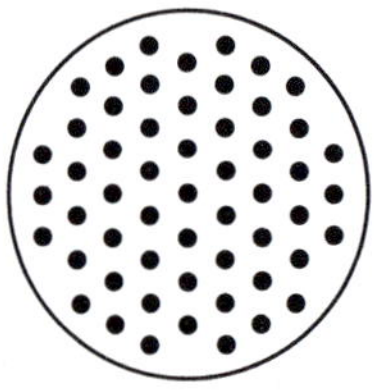

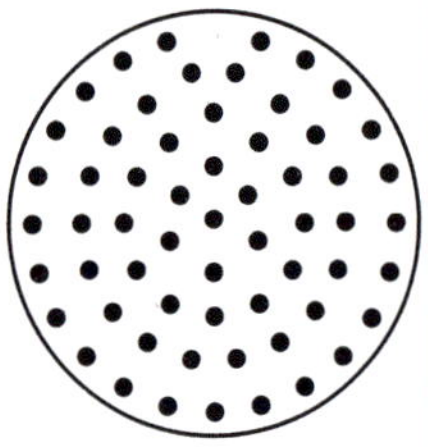

Figure 2-8A. Insert arrangements - MS miniature connectors, MIL-C-26482

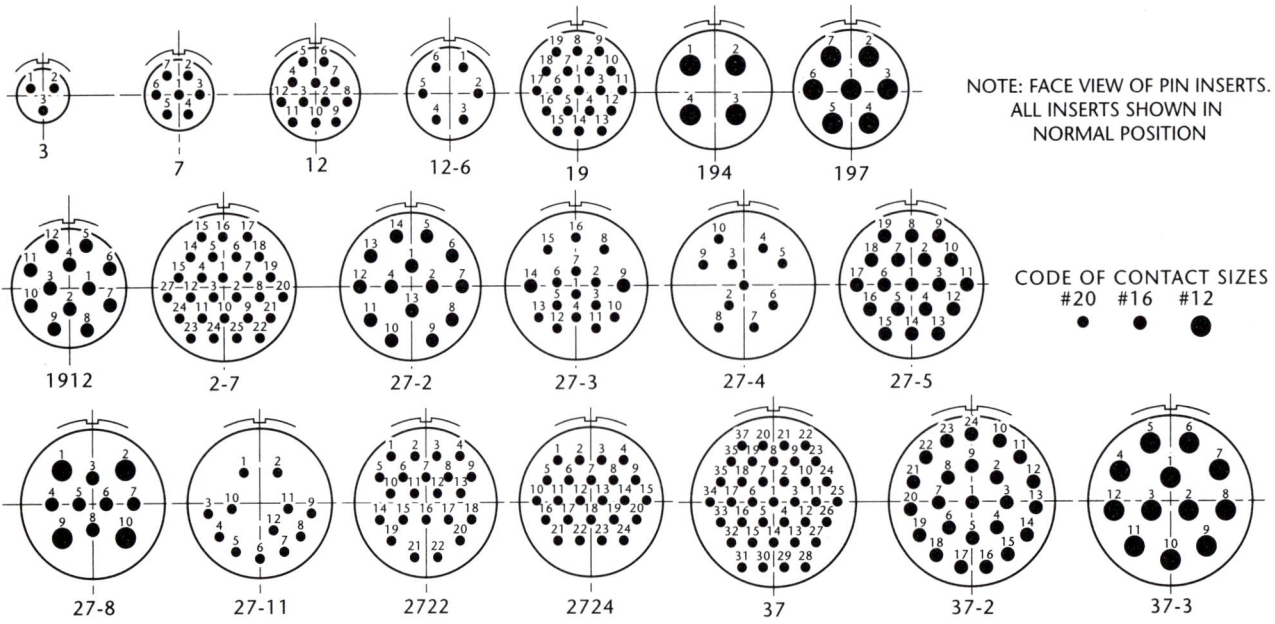

Figure 2-8B. Insert arrangements - MS miniature connectors with push-up engagement, MIL-C-26482

General Purpose Connectors | 2-11

		SHELL SIZE					
		10	12	14	16	18	22
SIZE 20 CONTACTS RATED 7.5 AMPERES FOR WIRE SIZES 24-22-20 AWG INSULATION O.D. MAX:=.090IN. MIN: = .040 IN.		●	●	●	●	●	●
		5 CONTACTS	12 CONTACTS	15 CONTACTS	24 CONTACTS	31 CONTACTS	55 CONTACTS
SIZE 16 CONTACTS RATED 20 AMPERES FOR WIRE SIZES 18-16 AWG INSULATION O.D. MAX: = .130 IN. MIN: = .064 IN.			●	●		●	●
			3 CONTACTS	7 CONTACTS		14 CONTACTS	19 CONTACTS
SIZE 12 CONTACTS RATED 35 AMPERES FOR WIRES SIZES 14-12 AWG INSULATION O.D. MAX: .170 IN MIN: = .106				●			
				4 CONTACTS			
SIZE 16 CONTACTS SEE ABOVE WIRE ACCOMMODATIONS #2 SHIELDED CONTACT RATED 7.5 AMPERES				●		●	
				3 CONTACTS		11 CONTACTS	
FOR SHIELDED CABLE SIZE 20 PER MIL-C-7078, TYPE II				2 #16 1 #2 SHIELDED		10 #16 1 #2 SHIELDED	
				Back face of pin insert as shown			

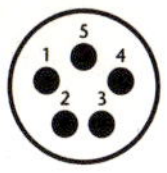

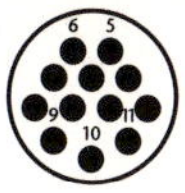

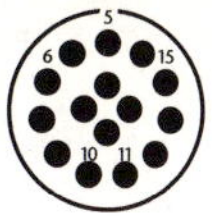

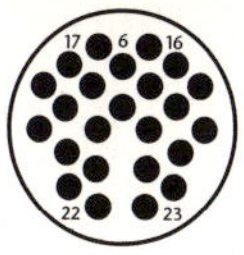

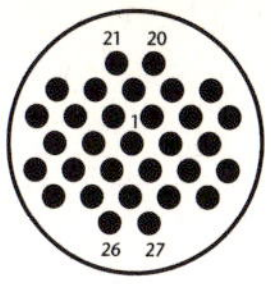

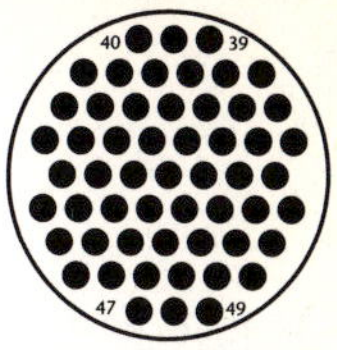

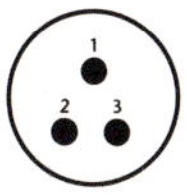

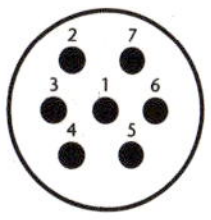

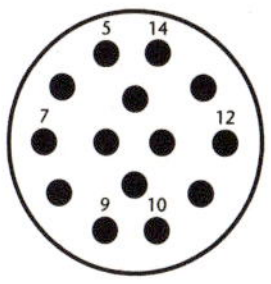

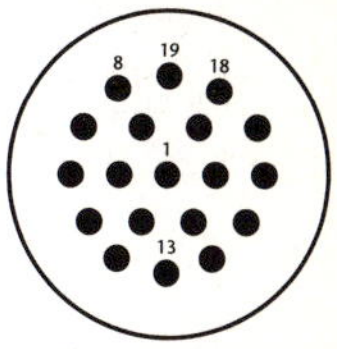

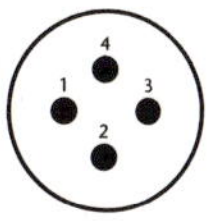

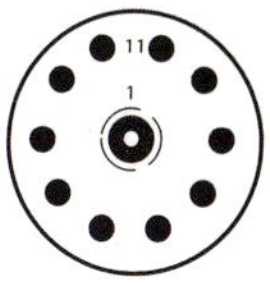

Figure 2-9. Insert arrangements - MS miniature connectors, MIL-C-26500

- Second letter indicates alternate insert position. Insert position letters W, X, Y or Z, indicate that the connector insert has been rotated with respect to the shell a specified number of degrees from the normal position. Alternate positions are specified to prevent mismating when connectors of identical size and contact arrangement are installed adjacent to each other. These alternate positions are shown on governing MS drawings. If no letter appears the insert is in the normal position. On connectors with multiple keyways the degree of rotation is measured from the widest keyway. See Figure 2-10 for typical alternate position arrangements.

NOTE: *Alternate insert positions on MIL-C-265100 connectors are indicated by numbers 6, 7, 8, 9 and 10 instead of by letters.*

NOTE: *AN-MS connectors MS3100 through MS3108 with socket contacts, have the letter "C" stamped on the connector after the code identification marking, indicating that the required prod damage test has been met.*

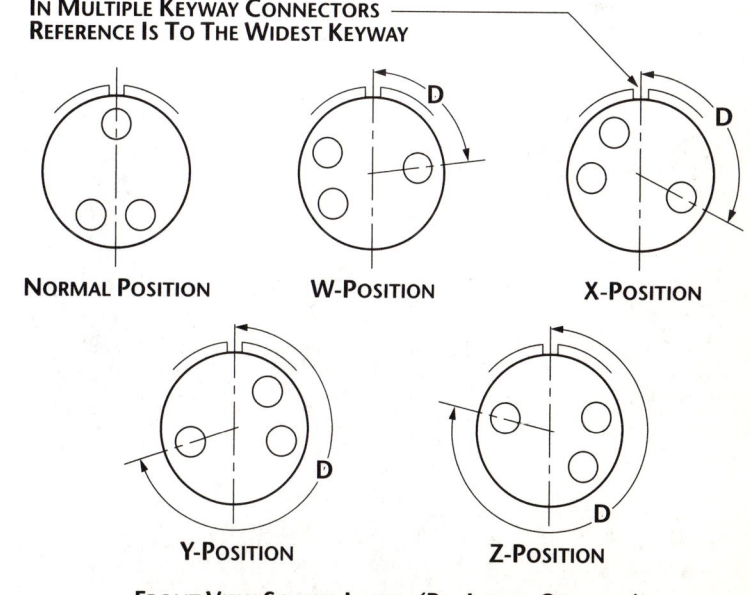

Figure 2-10. Alternate positions of connector inserts

2-12 | General Purpose Connectors

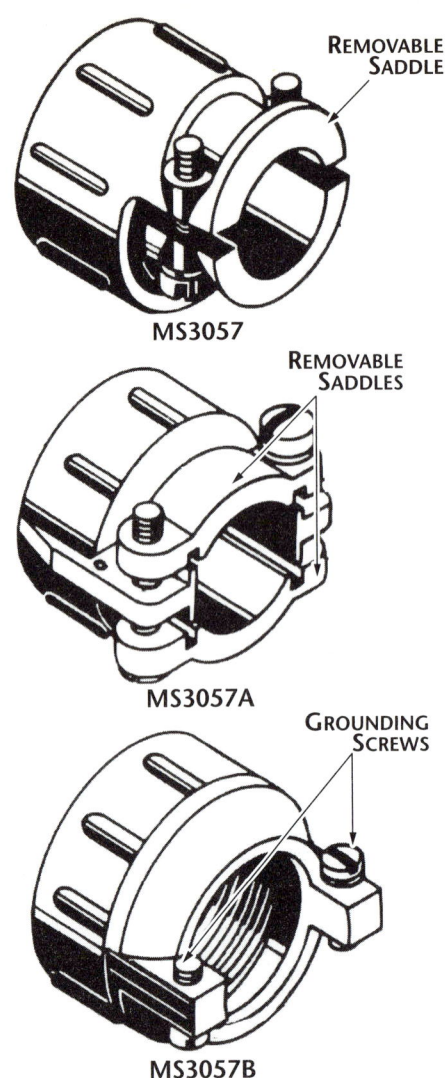

Figure 2-11. MS3057 connector cable clamp types

Contact Size	Wire Size Range
*20	*24-20
*16	*22-16
*12	*14-12
8	10-8
4	6-4
0	2-0

*Available in crimp type connectors

Table 2-5. Contacts and their wire size range

MS Connector Contacts

There are two kinds of contacts found in MS connectors: solder type, and crimp type. Crimp type contacts are removable. Contact sizes are related to AN wire sizes, but not all wire sizes have corresponding contacts. Contacts accommodate a range of wire sizes as shown in Table 2-5. It is sometimes necessary to use a wire larger than that included in the indicated range. Special instructions are given for this in a later section on reducing wire size.

CAUTION: *Use only contacts designed for use with the connector. When replacing contacts, make sure that replacement contact is identical with contact being replaced.*

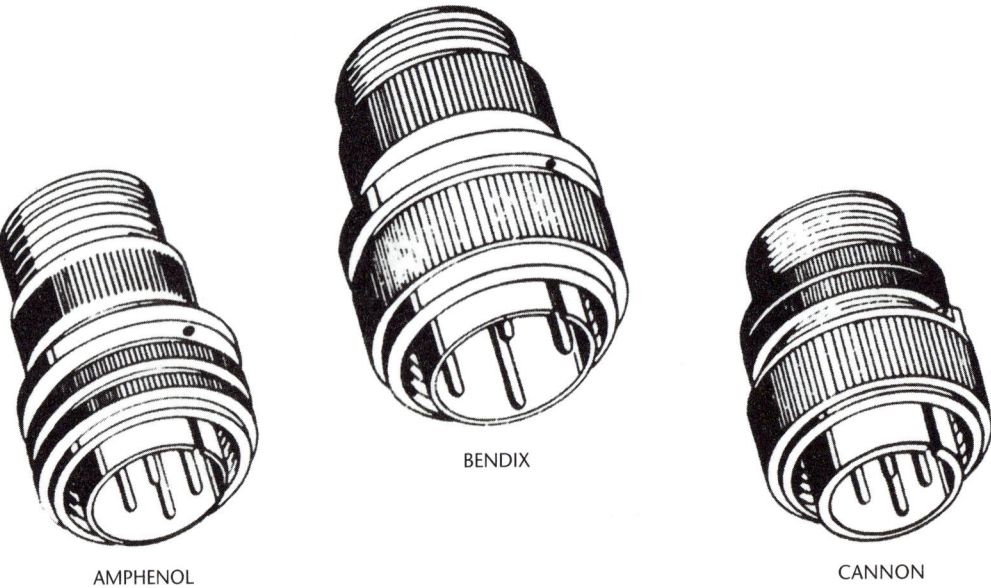

Figure 2-12. Variations in AN-MS connectors

MS Connector Cable Clamps

Connector cable clamps are used at the back end of MS connectors, except potted connectors, to support wiring, and to prevent twisting or pulling on soldered connections. There are three types of MS cable clamps as shown in Figure 2-11. These are as follows:

- **MS3057.** Consists of a clamp body, two washers, and a clamp saddle held on the clamp body by two screws and lockwashers.

- **MS3057A.** Consists of a clamp body and two saddles held on by screws and lockwashers. Used with AN3420 telescoping bushing.

- **MS3057B.** One piece clamp with no separate cap or saddles. Used with AN3420A bushings.

Manufacturers variations in MS connectors. (See Figure 2-12) Standard AN-MS plugs and receptacles made to the requirements of a Mil-Spec may show differences in appearance between one manufacturer and another. Also minor changes in disassembly and installation instructions may be required. The text and illustrations to follow will show differences in detail.

MS Potting Connectors

These connectors are used only where potting is required. They are similar to other standard types, except that they have a shorter body shell and include a potting boot. See Figure 2-13. MS potting connectors are available in the following types:

- **MS3103.** A receptacle with flange for mounting to a wall or bulkhead.

- **MS25183.** A straight plug used at the end of a wire or wire bundle.

- **MS25183A.** Similar to MS25183, with the addition of a grounding screw.

Special Purpose Connectors

In addition to connectors with MS numbers, there are some special connectors commonly used in aircraft listed below.

Connectors with crimp-type contacts. (Amphenol 69 series, Cannon Ex-A Series and Bendix 10-214000 series standard size; Bendix CE series miniature). These connectors are similar to MS standard or MS miniature solder type connectors, but with removable crimp-type contacts, are available for use where the added reliability of a crimped connection, and the greater ease of circuit change and maintenance provided by removable contacts is desirable. These special connectors will mate with the corresponding MS connectors, and are similar in appearance.

Subminiature connectors. (Cannon US series). These are wire connecting types only; they have no mounting flanges. However, they can be mounted with nut and lockwasher. Subminiature connectors are used on instruments, switches, relays, transformers, amplifiers, etc. See Figure 2-14.

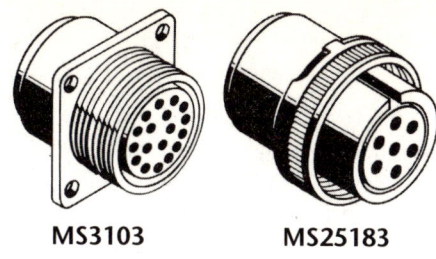

Figure 2-13. Potting connectors

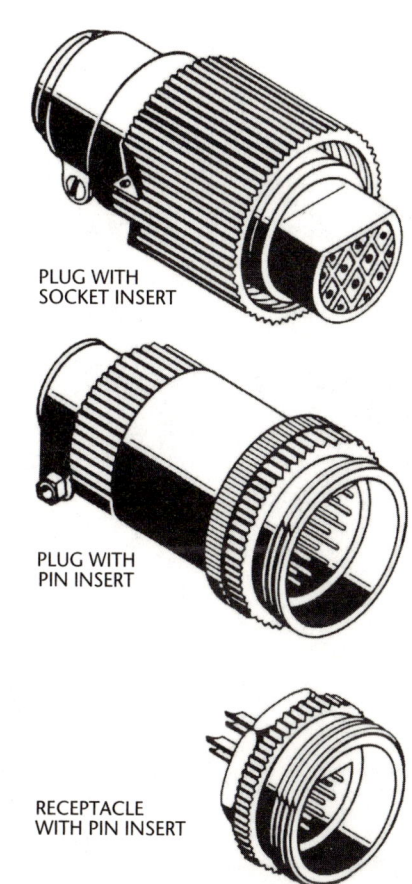

Figure 2-14. Subminiature connectors

2-14 | General Purpose Connectors

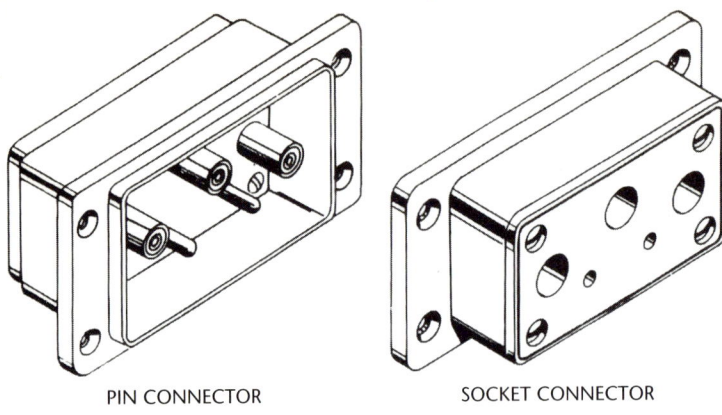

Figure 2-15. Rectangular shell connectors

Rectangular shell connectors. (Bendix SR; Cannon DPD). See Figure 2-15. These connectors are flanged for panel or equipment mounting. They consist of an aluminum alloy shell, rigid or resilient insert, and pin or socket contact. They are usually potted to protect connections against moisture at the back of the connector. The mating faces are not moisture sealed.

Miniature rectangular connectors. (Winchester A and SA series). These are rack-and-panel connectors, having alone piece molded body and pin and socket contacts. These connectors are available with either solder cup contacts or taper pin receptacles. See Figure 2-16.

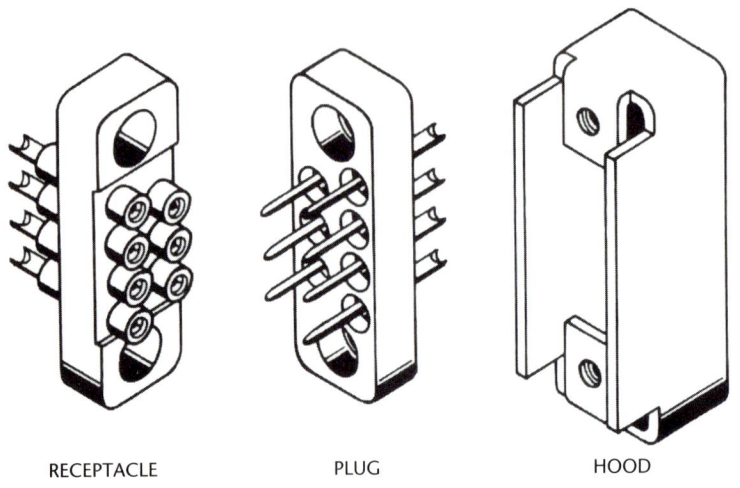

Figure 2-16. Miniature rectangular connectors

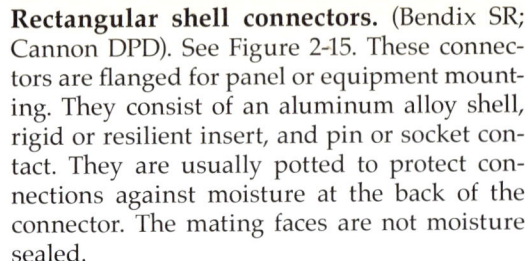

Figure 2-17. Removal of solid back shell

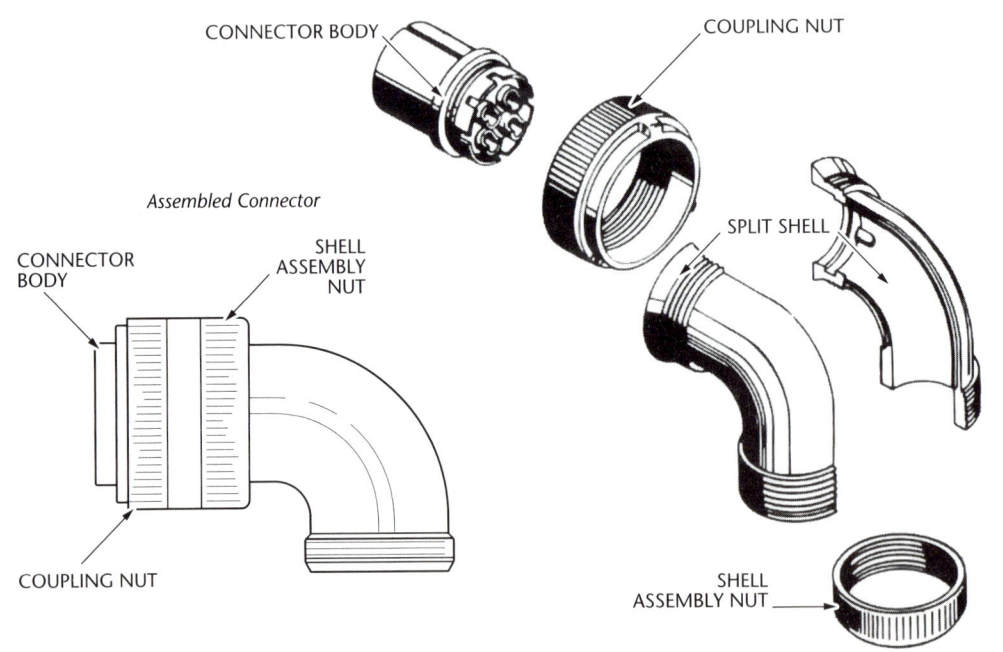

Figure 2-18. Removal of Cannon split back shell

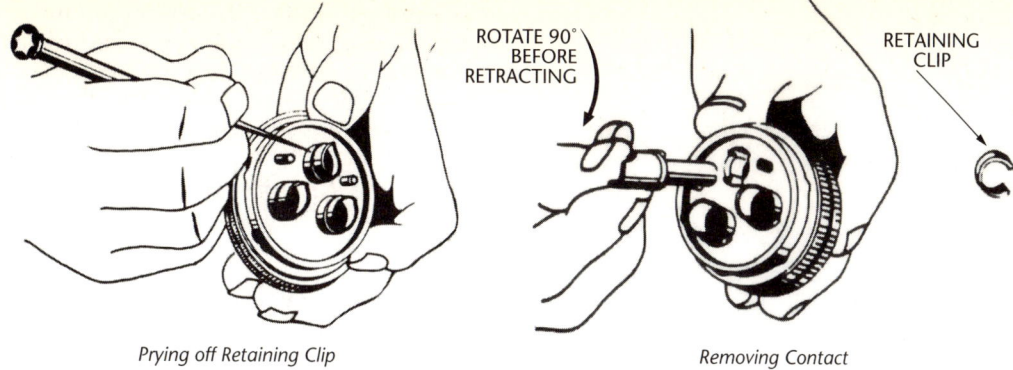

Prying off Retaining Clip   *Removing Contact*

Figure 2-19. Removal of Cannon clip held contacts

Environment-resisting fireproof connectors. (Cannon KE series). These connectors are similar to the MS-K series, with the addition of a moisture-resisting seal. They will mate with MS-K plugs or receptacles, but retain the moisture sealing feature only when mated with corresponding KE series plugs or receptacles.

Section 3
Disassembly of Connectors

Precautions and Procedures

Solder type contacts size 8 and smaller are usually not removed for assembly purposes. Large solder contacts (size 4 and larger) are removed from connectors with hard inserts to protect the insert against the greater amount of heat necessary to properly solder wires to the larger contacts. Large solder contacts may be removed from connectors with resilient inserts provided the connector is not a pressurized assembly with the contacts bonded into the insert. Crimp-type contacts are removed to enable the contact and wire assembly to be inserted into a crimping tool. Detailed instructions for disassembly of each MS connector variation are given in the following paragraphs. Because of the differences in connectors made by the manufacturers, the detailed disassembly instructions are given separately for each manufacturer where necessary.

Removal of back shells. Remove back shells, if present, from all connectors before attaching wires. Solid back shells of Classes A, C and K are removed by unscrewing from the front shell as shown in Figure 2-17.

Split back shells of Class B connectors are held together either by an assembly ring or by captive screws. See Figure 2-18 for details of disassembly.

Class E and Class R connectors require a special disassembly technique that is described in later paragraphs.

Removal of contacts. Solder-type contacts sizes 4 and 0 are removed before soldering provided they are not bonded in the connector (refer to previous page). Amphenol Class C pin contacts sizes 4 and 0 are threaded into the insert and can be removed safely. Size 8 contacts (if not bonded in the connector), may be removed if close spacing makes soldering difficult.

> **CAUTION:** *Avoid removing contacts from Class C, Class E and Class R connectors. Never remove inserts from Class C, E and R connectors.*

Solderless contacts such as supplied in Class K must be removed in order to crimp wire into contacts. Contacts are removed by the following procedures.

Cannon spring clip contacts (see Figure 2-19) are removed by prying off the clip with a small

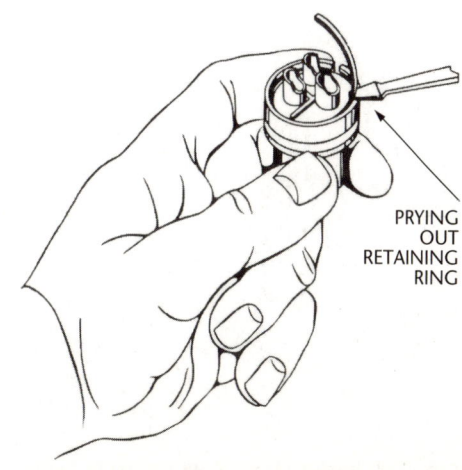

Figure 2-20. Removal of Amphenol insert assembly

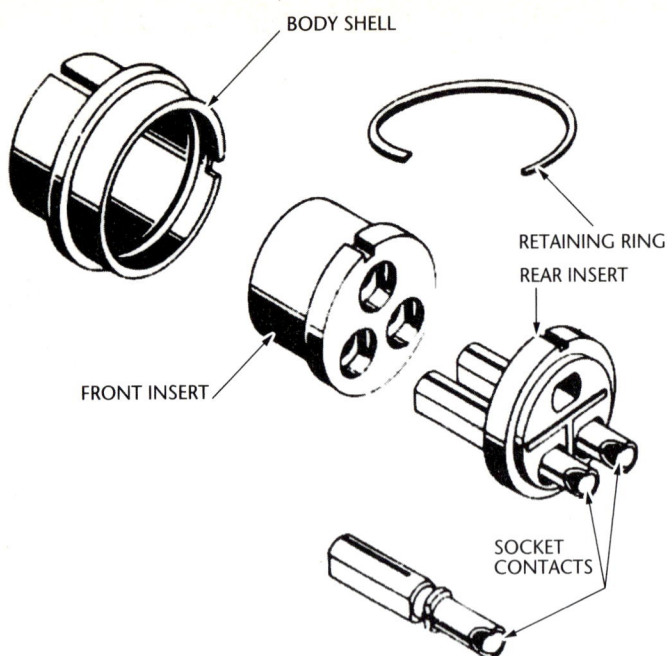

Figure 2-21. Amphenol contact and insert assembly - exploded view

screwdriver or scriber. Rotate contact 90° and lift out.

Amphenol two piece inserts have contacts which are held in place by the insert halves. See Figures 2-20 (previous page) and 2-21. Pry out retaining ring which holds insert in place with a small screwdriver. Remove inserts and separate them. This will free contacts.

Some Bendix and Cannon connectors have resilient inserts. Contacts are removed from these resilient inserts by pushing against the solder cup end with a round phenolic rod a little smaller in diameter than the solder cup. An arbor press, as shown in Figure 2-22 is helpful in removing large size contacts.

CAUTION: *Never use pliers to remove contacts. This will damage the contacts.*

Disassembly Instructions for AN-MS Class E Connectors

Class E (environment-resisting) connectors are made in two forms. Those made by Bendix have a separate cable clamp similar to MS3057B. Those made by Cannon use a cable clamp that is part of the back shell and similar to MS3057A (see Figure 2-23).

Disassembly of a Bendix Class E connector is accomplished as follows:

1. Unscrew cable clamp from back shell, using strap wrench or padded jaw connector pliers if necessary.
2. Remove tapered grommet compression sleeve and grommet from back shell.
3. Unscrew back shell from body assembly using strap wrench or padded jaw pliers if necessary.
4. Remove contacts size 8 and larger as described in the earlier section on removal of contacts.

NOTE: *Avoid removing contacts from Class E receptacles.*

Disassemble Class E connectors made by Cannon as follows:

1. Remove cable clamp saddle by removing two screws. Unscrew cable clamp if MS 3108E.
2. Unscrew back shell from body assembly, using strap wrench if necessary.
3. Remove AN3420 telescoping bushing(s).
4. Remove grommet follower (grommet retainer) and grommet.
5. Remove contacts size 8 and larger as described on the previous page.

CAUTION: *Do not remove inserts of moisture-proof connectors. Removal will destroy the moisture-proofing. Do not remove contacts smaller than size 8 except for replacement.*

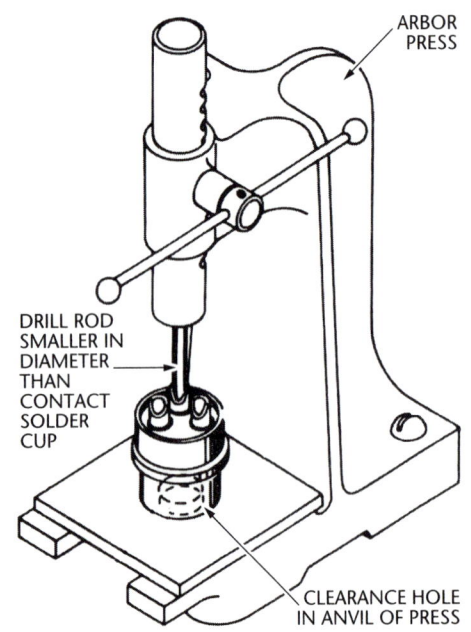

Figure 2-22. Removal of solder contacts from resilient insert

General Purpose Connectors | 2-17

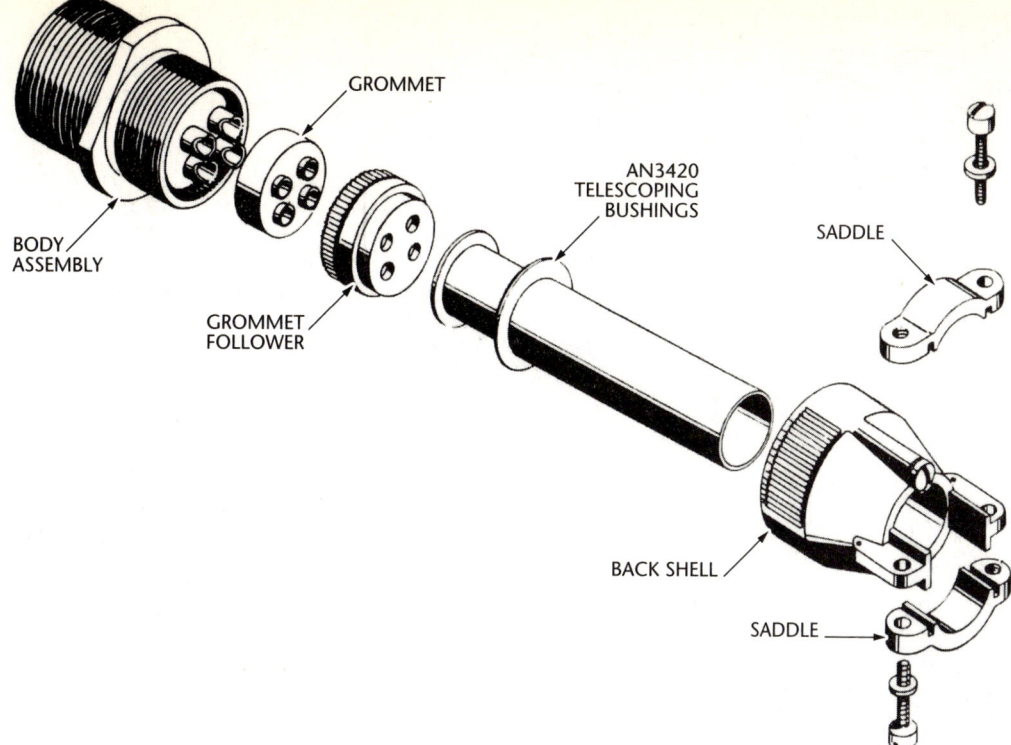

Figure 2-23. Class E connector - Cannon - exploded view

MS Connector Disassembly

MS Class R connector disassembly. These lightweight environment-resisting connectors differ in appearance from Class E connectors. See Figure 2-24. Class R connectors have no separate cable clamp. A grommet, grommet compression nut and ring are preassembled as a single unit, which is disengaged from the shell assembly before wiring contacts. The cable clamp has been replaced with a back nut and compression sleeve. The nut backs the grommet away from the contacts when removed. To disassemble, remove compression sleeve and grommet and unscrew back nut from body assembly. See Figure 2-25 (next page). The cable clamp has been retained on the MS3108R angle plug that is identical with MS3108E except for improved sealing. Otherwise disassembly instructions for Class R connectors are the same as for the Class E connectors. To effect a satisfactory seal, the grommet must seat against the rear face of the insert.

Disassembly of miniature MS connectors with solder type contacts. Miniature MS environment-resisting connectors, Class E, are disassembled as described in disassembly instructions for AN-MS connectors, class E, except that there is no back shell, and contacts are not removed. Class P potting connectors are disassembled by removing the plastic potting mold and retaining ring. Class H hermetically sealed connectors are not disassembled.

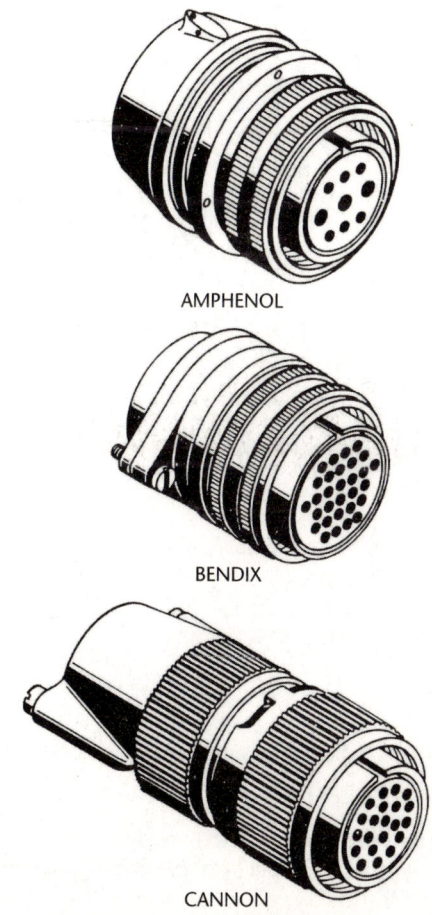

Figure 2-24. Class R connectors

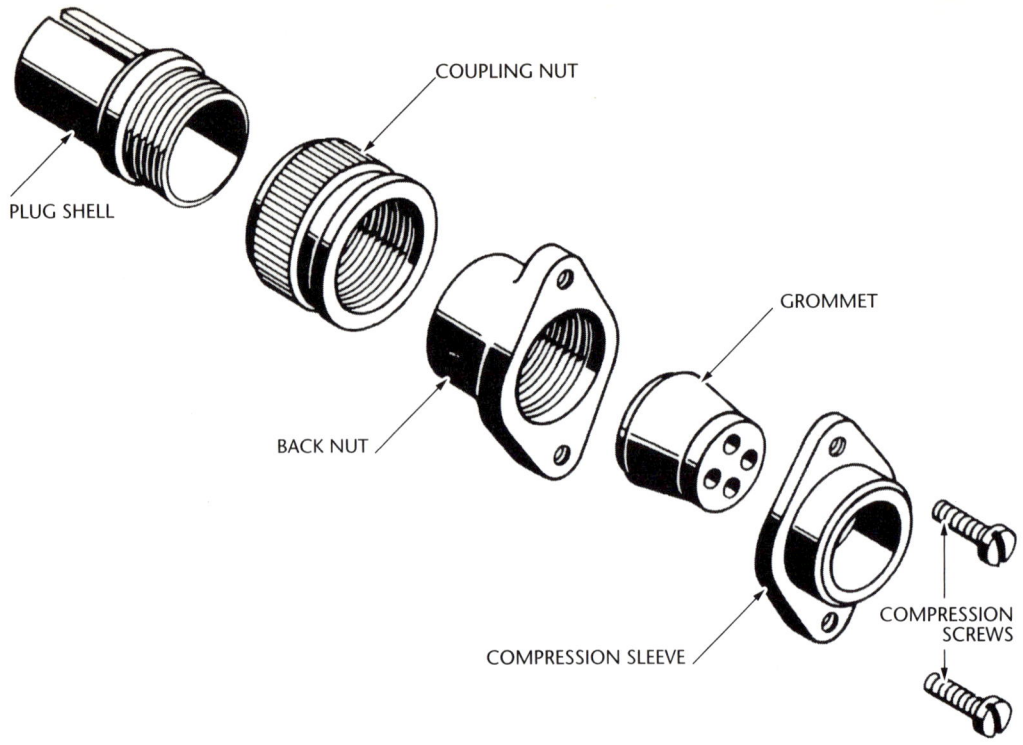

Figure 2-25. Class R connector - Bendix - exploded view

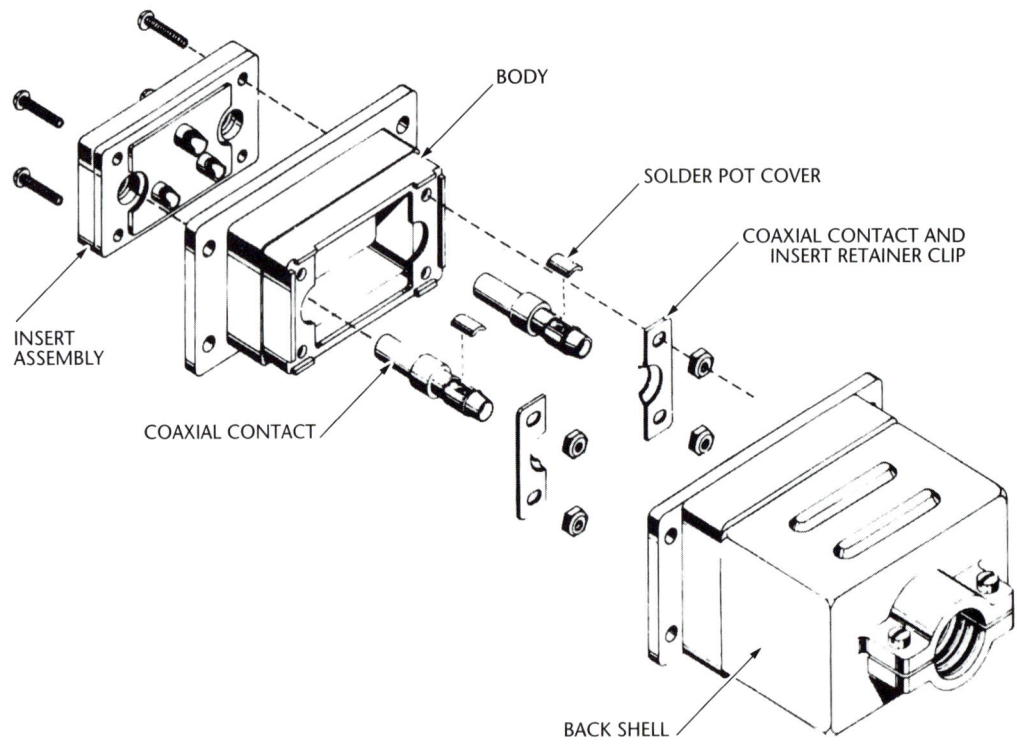

Figure 2-26. Rectangular shell connector - Cannon DPD (exploded view)

Instructions for MS connectors with removable crimp-type contacts. The contacts for this type of connector are usually packaged separately, so very little disassembly of the connector is necessary. For Class E or F connectors, remove the grommet compression nut and sleeve. The grommet is bonded to the insert and is not removable. On Class P connectors, remove the potting mold and retaining ring. If the contacts have been assembled into the connector, remove them as described in the later section on Removing Contact From Connector.

Disassembly instructions for Cannon DPD connectors. Cannon DPD connectors are used in installations where it is necessary to simultaneously disconnect coaxial cables and general purpose wires. See Figure 2-26. DPD connectors are disassembled as follows:

1. Remove four self-locking nuts.
2. Lift away coaxial contact and insert retainer clips.
3. Pull out coaxial contacts from rear of insert.
4. Remove insert from front of body.

Subminiature connector disassembly. (Cannon US series) Contacts are not removed from this type of connector. The connector is disassembled by removing the keeper, coupling nut, two-piece plastic sleeve and ferrule; these pieces are installed over the wire bundle in that order before soldering wires to the contacts. See Figure 2-27 for an exploded view of this connector.

Miniature rectangular connectors. No disassembly is necessary for these connectors.

Section 4
Assembly of Wires to Connectors

Wire Types

Five types of wire and cable are normally fastened to connectors. These are:

- Tin-coated copper wire. (MIL-W-5086)
- Silver-coated copper wire. (MIL-W-7139, MIL-W-8777, MIL-W-16878, MIL-W-22759)
- Nickel-plated copper wire. (MIL-W-22759, MIL-C-25038 and MIL-W-27300)
- Coaxial cable. (MIL-C-17)
- Thermocouple wire. (MIL-W-5845, MIL-W-5846)

The choice of wire is controlled by the installation requirements and is indicated on the engineering drawing and parts lists.

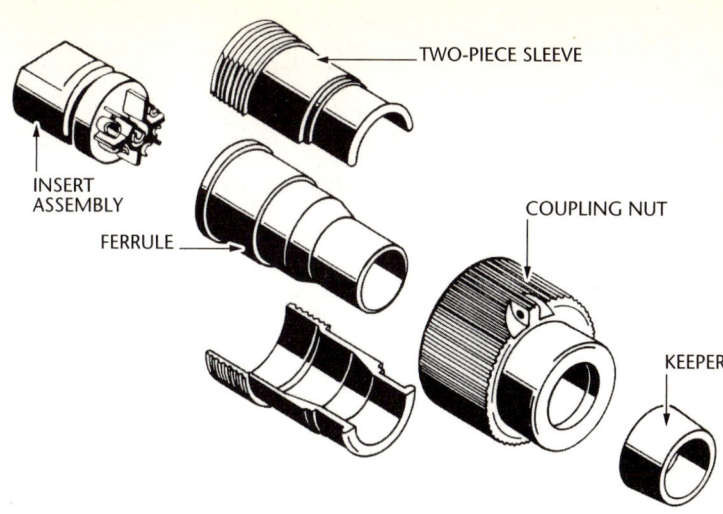

Figure 2-27. Subminiature connector (exploded view)

Solders

Solders and other fastening means are matched to the wire type and to the installation as follows:

- Soft solder-60/40 tin-lead (Federal Specification QQ-S-571, composition Sn 60) is used for tin-coated copper wire and for coaxial cable.
- Soft solder-lead-silver (Federal Specification QQ-S-571, composition Ag 2.5 or Ag 5.5) is used for silver-coated copper wire.
- Crimp connections are used for nickel-clad copper wire, and tin, silver or nickel coated copper wire.
- Thermocouple wires require special procedures that are detailed in Chapter 5.

Contacts. Contacts are supplied as two types:

- Solder cup - for all standard AN-MS connectors, except Class K, and for miniature MS connectors listed in Table 2-3.
- Crimp-type - for MS Class K connectors, and for MS miniature connectors with removable crimp-type contacts described in earlier in the section on Miniature MS Connectors: Crimp Contacts.

Solder cup contacts are silver or gold plated to provide low contact resistance. Silver plated contacts have pre-tinned solder cups. Gold plated contacts are not pre-tinned because the gold prevents oxidation and is therefore always easy to solder. Crimp type contacts are gold or rhodium plated, and are not pre-tinned.

General Purpose Connectors

Contact Size	Stripped Length (Inches)
20	1/8
16	1/4
12	5/16
8	5/8
4	5/8
0	3/4

Table 2-6. Stripping lengths for solder connections

Contact Size	Wire Size (AN)	Stripped Length (Inches)
20	20, 22 & 24	3/16
16	16, 18 & 20	1/4
12	12 & 14	5/16
8	8	9/16

Table 2-7. Stripping lengths for crimp connections

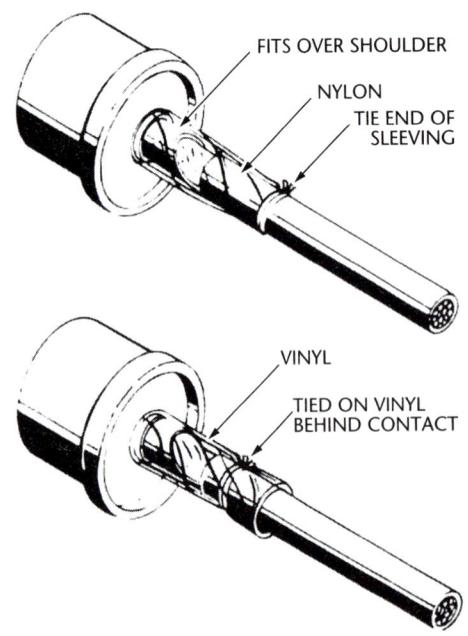

Figure 2-28. Insulating sleeving installed over solder cup

Preparation of wires before assembly. The preparation of wires before assembly is as follows: (See Chapter 1)

1. Cut wire to prescribed length.
2. Identify wire with proper coding.
3. Strip ends to the dimensions in Tables 2-6 or 2-7 as applicable.
4. Tin wires that are to be soldered to contacts. Do not tin wires that are to be crimped to contacts.

Insulating sleeves. Insulating sleeves are used over soldered connections to help protect the connection against vibration and to lengthen the arc-over path between contacts. Insulating sleeves are not used under the following conditions:

a. Insulating sleeves are not used when connectors are to be moisture-proofed by potting.

b. Insulating sleeves are not used in miniature MS connectors, nor in AN type Class E or Class R connectors, as the sealing grommets cover the soldered connection. Class E connectors made by Bendix need insulating sleeves, as the grommets do not cover the soldered connection. The insulating sleeve should not extend into the grommet.

NOTE: *No. 22 wire can be crimped into No. 16 contact provided the No. 22 wire is stripped to 1/2 inches and doubled back on itself.*

Selection of insulating sleeves. Select insulating sleeving from the materials listed in Table 2-8 to suit the temperature conditions in the area where the connector will be installed. Select the proper size from Table 2-9, so that the inside diameter of the sleeving will fit snugly over the solder cup.

Installation of insulating sleeves. Cut the sleeving into lengths, as given in Table 2-9, to cover the soldered connection completely from the insert to a little over the wire insulation. See Figure 2-28. Slip insulating sleeve of correct size, material and length over each prepared wire, far enough back from the stripped end to avoid heat from soldering operation (about one inch).

Soldering Procedure

Wires are soldered to contacts in electrical connectors by means of a soldering iron, resistance heating, or a torch. Safe connections are the result of clean parts carefully soldered together. See Chapter 7 for a description of soldering

Temperature Range	Material	MIL Specification
Up to 160°F	Vinyl, transparent Nylon, transparent	MIL-I-7444 or MIL-I-631
160°F-400°F	Silicone-impregnated fiberglass Silicone-rubber Fiberglass	MIL-I-3190 or MIL-I-18057
400°F-600°F	Extruded TFE TFE-impregnated fiberglass	—

Table 2-8. Insulating sleeving material

methods and procedures. When soldering wires to electrical connectors, observe the following precautions:

- Make sure that the wire and the contact are clean, and properly tinned.

- Use a soldering iron, or other heating method, of a heat capacity sufficient for the work to be soldered. Resistance soldering is recommended for sizes 8 thru 0. A soldering iron is recommended for sizes 12 thru 20; resistance soldering may also be used.

- Make sure that the iron has a smooth well-tinned tip. See Chapter 7 for detailed instructions on the care and maintenance of the soldering iron.

- Keep electric resistance pliers clean and free from flux and solder splatter. Use a brass wire hand brush to clean contacting surfaces.

- Select a soldering iron tip of a shape to provide good heat transfer. A large contact area touching the solder cup will help to produce a good connection quickly. See Figure 2-29 for suitable soldering iron tips.

- Use only rosin or rosin-alcohol as flux for soldering wires to connector contacts.

  CAUTION: *Do not use any corrosive flux for soldering in an electric connector.*

- Do not hold the hot iron against the solder cup longer than necessary; this will force solder up into the conductor and stiffen the wire. Stiff wires will break under vibration.

- Avoid having solder run on the outside of solder cup or dripping to insert face. Do not move the soldered connection until the solder has hardened.

- Solder has little mechanical strength. Do not depend on solder to keep a wire from pulling out of a contact. Use a cable clamp, grommet seal or potting to give mechanical strength.

AN Wire Size	Number	ID (inches)	Length (inches)
16-14	7	.148	3/4
12	5	.186	3/4
10	3	.234	3/4
8	1	.294	1
6	0	.330	1-1/4
4	7/16	.438	1-1/4
2	1/2	.500	1-1/4
0	5/8	.625	1-1/4

Table 2-9. Insulating sleeving sizes

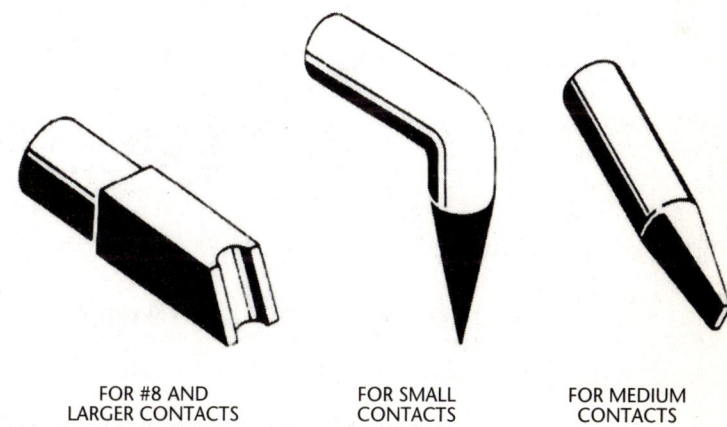

FOR #8 AND LARGER CONTACTS   FOR SMALL CONTACTS   FOR MEDIUM CONTACTS

Figure 2-29. Soldering iron tip shapes

Electrical Resistance Soldering

Resistance soldering will yield excellent results for both very large and very small contacts.

Large contacts are soldered to wires by the use of resistance soldering pliers. See Figure 2-30 (next page). The contact, removed from insert, is held in

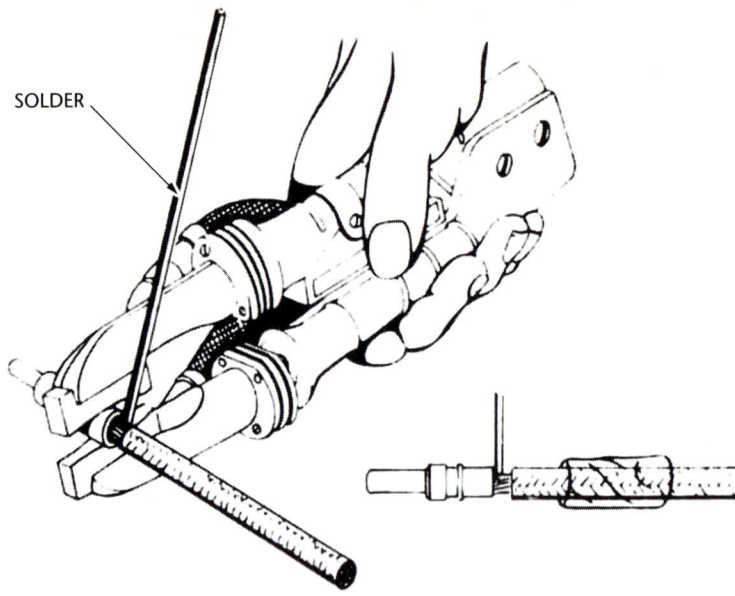

Figure 2-30. Resistance soldering pliers for large contacts

so that the heating current will pass through the wall of the cup. When the solder in the cup flows, insert the pre-tinned wire. Continue to apply heat to connection until solder flows to form smooth fillet, then stop current and allow joint to cool without movement.

Torch Soldering

A torch can be used to solder wire into a large contact that has been removed from its insert. (See Figure 2-32). The contact is held in a non-metallic block to avoid heat loss, and the torch is played over the solder cup area until the solder melts.

> **CAUTION:** *Do not overheat. Excessive heat will destroy the plating and soften the contact.*

When the solder in the cup has melted, insert the wire slowly into the cup and add more 60/40 rosin-core solder if necessary. Continue to heat the connection until the solder flows into a smooth fillet, then remove the flame. Allow the joint to cool without movement.

Soldering Iron Procedure

Soldering with an electrically heated iron is the most common procedure. For convenience either the iron or the connector is fastened to the bench as described on the next page in the paragraph that describes holding connectors for soldering. Soldering is accomplished in one of two ways depending upon whether or not the contact has been removed from its insert.

Large contacts that have been removed from inserts are held in a non-metallic block and soldered by first heating the solder cup with the specially shaped tip as shown in Figure 2-33. Then while heat is still applied, the pre-tinned wire is slowly inserted into the solder

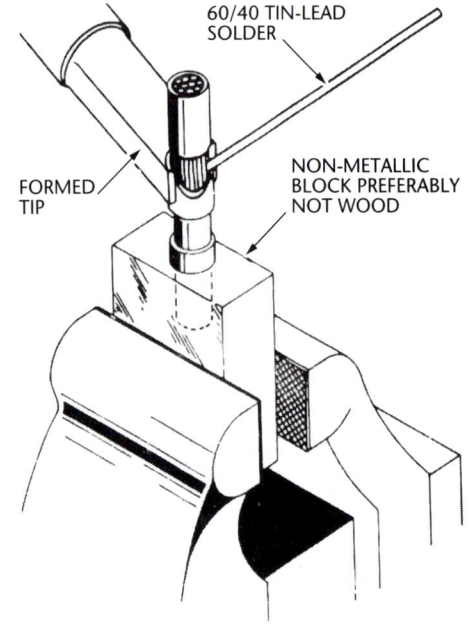

Figure 2-31. Resistance soldering pencil for small contacts

the jaws of the pliers and current is applied until the solder in the solder well has melted. Then the pre-tinned wire is inserted slowly into the solder cup while current is still being applied. After the wire is fully inserted, continue heating until the solder flows to form smooth fillet. Allow joint to cool and harden without movement.

Small contacts are heated for soldering by use of pencil type resistance soldering tool shown in Figure 2-31. The two electrodes of the tool are placed in contact with the side of the solder cup

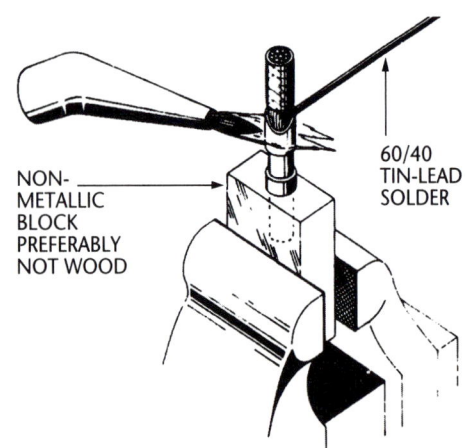

Figure 2-32. Torch soldering large contact

General Purpose Connectors | 2-23

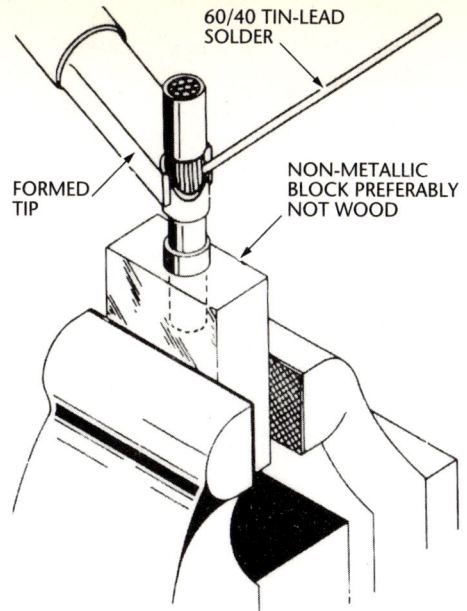

Figure 2-33. Soldering large size contacts

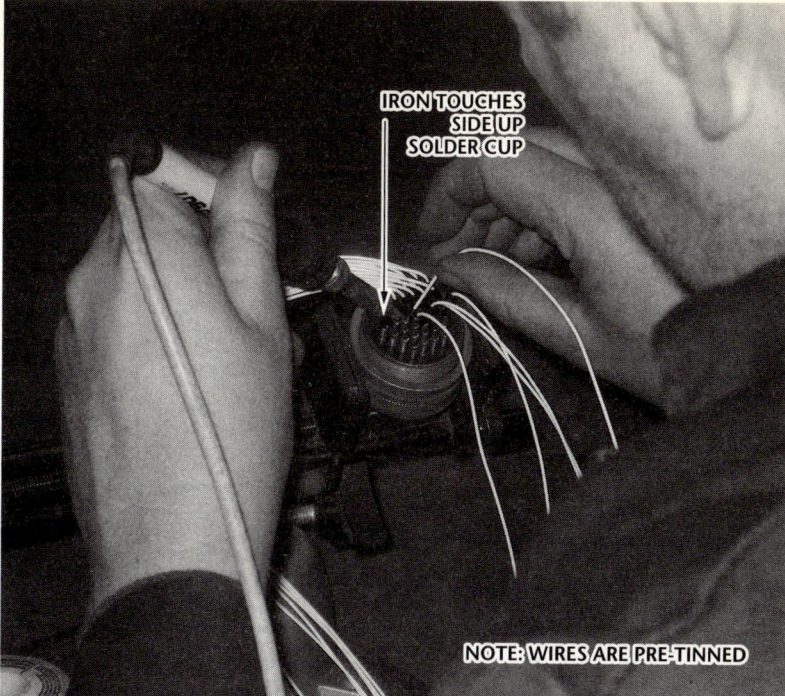

Figure 2-34. Soldering small size contacts

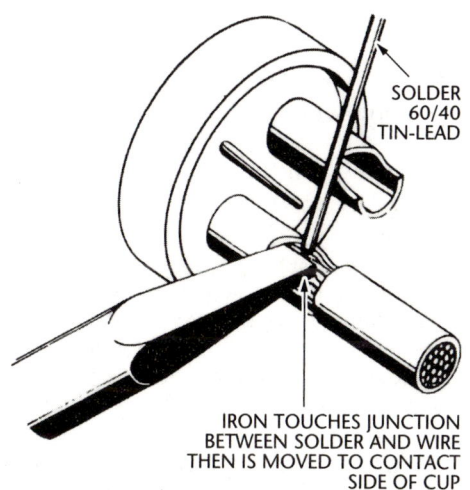

Figure 2-35. Soldering medium size contacts

CAUTION: *Do not allow solder to collect outside of the solder cup. This will reduce the arc-over distance between contacts and can result in connector failure.*

Holding connectors for soldering. To facilitate soldering wires to contacts that have not been removed from connectors, it is helpful to either work to a fixed soldering iron or to fasten the connector into a holding fixture. If the iron is fastened to the bench, secure it into a safety screen such as shown in Figure 2-36. The

cup until it bottoms. Extra 60/40 rosin-core solder is added to the solder cup if necessary. Hold the hot iron to the solder cup until the solder has flowed into a smooth fillet, then allow to cool.

Contacts that have not been removed from inserts are soldered as shown in Figures 2-34 and 2-35. The solder is flowed by placing the iron alongside the solder cup as the wire is being inserted into it. Medium size contacts such as #8 and #12 will solder more easily if the iron is held at the point where the wire touches the cutaway of the solder cup as shown in Figure 2-35. Adding a small quantity of 60/40 rosin-core solder at this point will aid in carrying the heat into the joint.

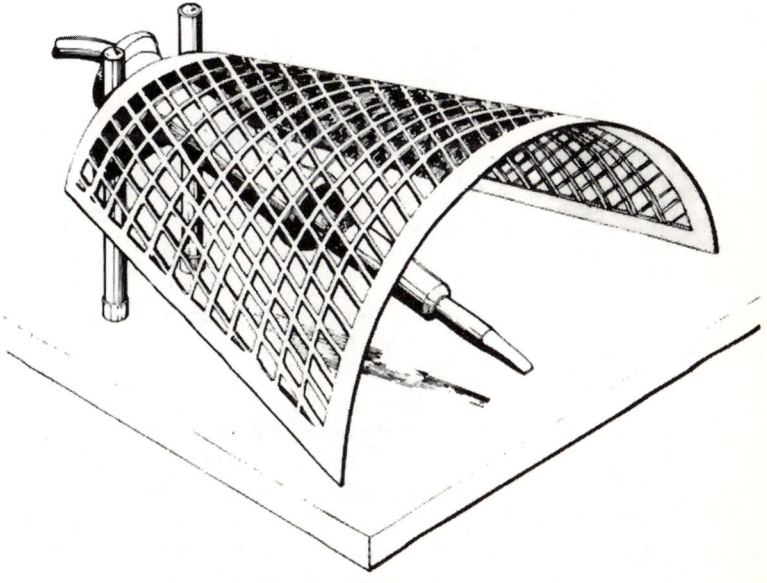

Figure 2-36. Soldering iron in safety screen

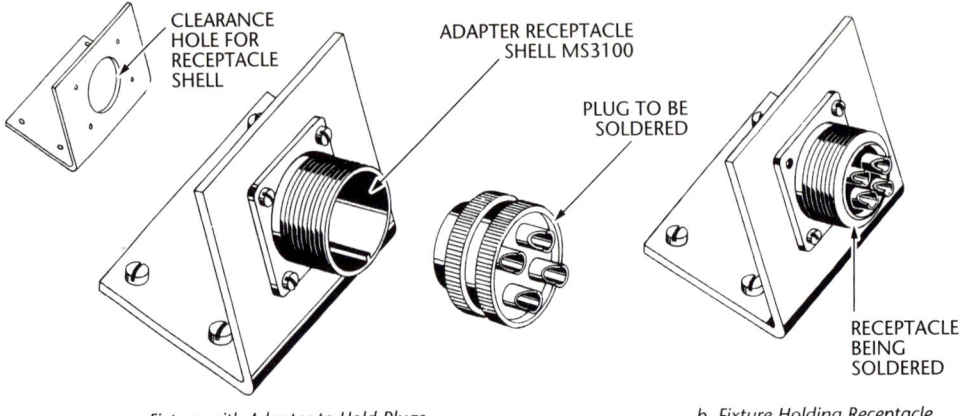

a. Fixture with Adapter to Hold Plugs   b. Fixture Holding Receptacle

Figure 2-37. Holding fixtures for connectors

screen is made from expanded or perforated steel, painted to retard corrosion. To solder connectors with a fixed iron, it is necessary to hand hold the connector. If the connector is to be fastened to the bench, a steel bracket bent

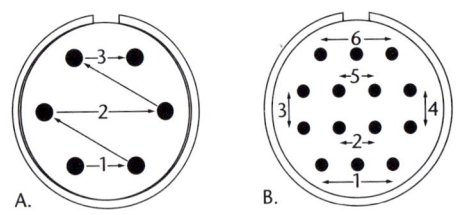

Figure 2-38. Connector soldering sequence

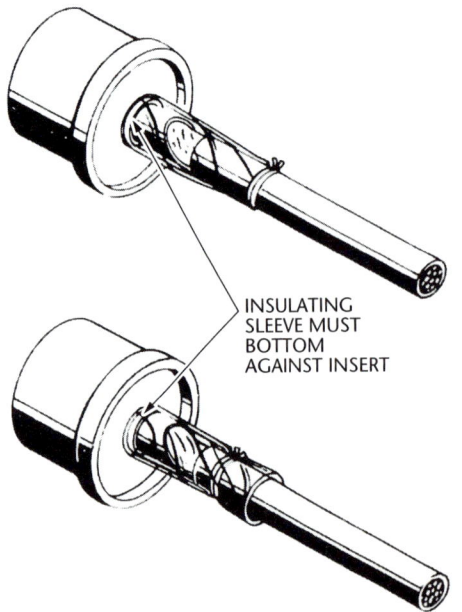

Figure 2-39. Insulating sleeve bottomed against insert

to a 60° – 75° angle as shown in Figure 2-37 is very useful. To hold a plug, use an empty shell from a mating receptacle. To hold a receptacle use two screws to mount the receptacle with the threaded portion inserted through the hole in the bracket. This will place the solder cups in position for easy soldering.

Soldering sequence. Follow a rigid sequence in soldering wires to a connector. This helps avoid errors in wiring and also prevents burning or scorching the insulation of wires already soldered. Two useful sequences are shown in Figure 2-38.

The soldering of the connector in Figure 2-38a is started at the right or left lower edge, depending on whether the mechanic is left or right handed, and follows the bottom row across. The row above is next and is done in the same direction as the bottom row. This will permit the insert to cool between soldering operations. The operation is repeated for each row in sequence until all contacts are soldered.

> **NOTE:** *If wires are being soldered to a connector with a large number of contacts, plan the work to allow a cooling-off period after each series of twenty contacts in order to prevent heat buildup.*

The sequence for the connector shown in Figure 2-38b also starts with the bottom row from the right or left. The next step is to solder to the center contacts working out to each edge. The final operation is to solder wires to the top row of contacts.

The two sequences above are a suggested procedures that work well in many aircraft plants. They are not mandatory but it is important for the mechanic to develop a fixed sequence and then not to deviate from that sequence.

Cleaning soldered connections.
After all connections have been made, examine the connector for excess solder, cold joints and flux residues. Take following corrective measures if any of the above are found:

1. Remove excess solder by using a soldering iron that was carefully wiped clean with a heavy clean cloth.
2. Disassemble cold joints. Shake out all old solder and remake the connections using new 60/40 rosin-core solder.
3. Remove flux residues with denatured ethyl alcohol or approved proprietary solutions applied with a bristle brush. Blow the connector dry with compressed air.

Insulating sleeve positioning.
All connectors except Class R, E, F, P, Miniature MS or those that will be moisture proofed by potting, have insulating sleeves installed over the individual wires prior to assembly to solder cups. After the connections are cleaned, push the insulating sleeves down over the contact until they bottom against the insert as shown in Figure 2-39. Tie the insulating sleeves in position with nylon braid to prevent sliding back on the wires. Tie nylon sleeves individually because of nylon's stiffness. Tie all other sleeves in groups or as a complete bundle. Make sure that tie will not interfere with the MS3057 cable clamp.

Preshaping wires.
Preshape large diameter wires (No. 14 and larger) before soldering to contacts. This will avoid strain on soldered connection when MS3057 cable clamp is installed. See Figure 2-40.

> **CAUTION:** *Preshaping is a necessity for connectors using resilient inserts. Side strain on the contacts will cause contact splaying and prevent proper mating of pin and socket contacts.*

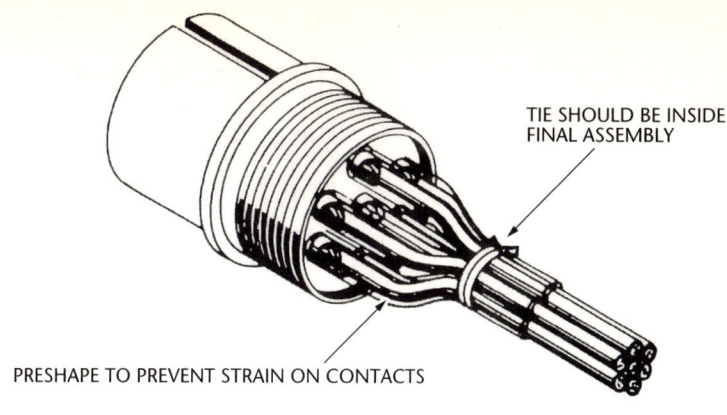

Figure 2-40. Tying sleeves and pre-shaping wires

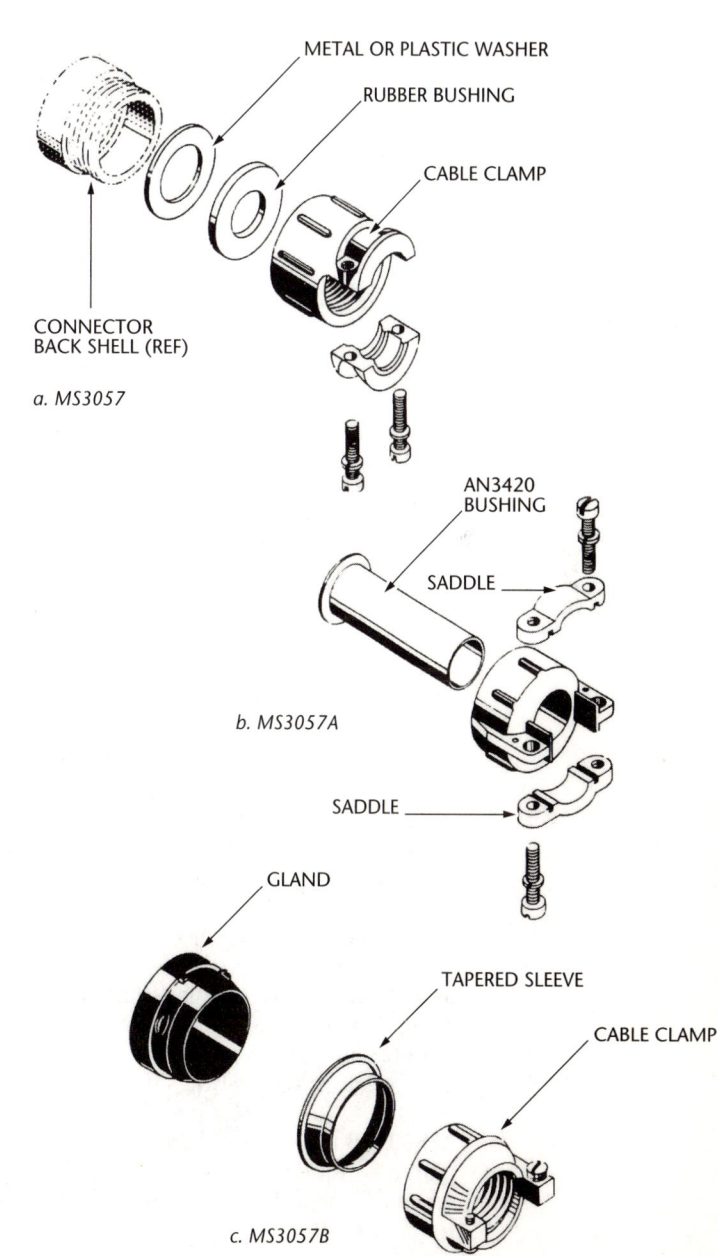

Figure 2-41. MS3057 Connector cable clamp types-exploded view

Section 5

Installation of MS3057 Series Connector Cable Clamps

MS3057 Cable Clamp Types

There are three types of MS3057 connector cable clamps. (See Figure 2-41).

MS3057 has a single saddle held by two screws. It contains a metal or plastic washer and a flat

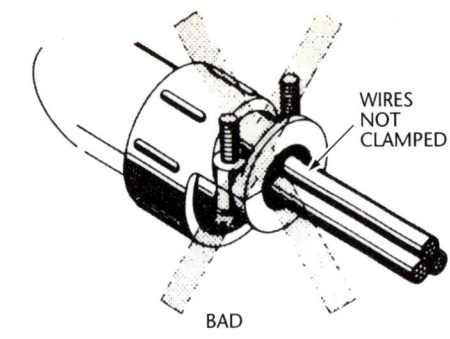

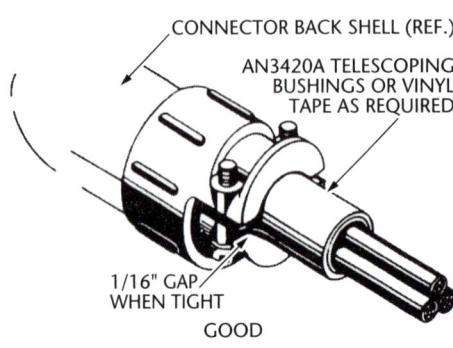

Figure 2-42. Installation of MS3057 clamp

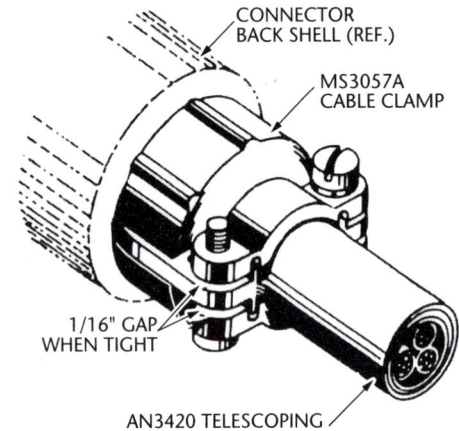

Figure 2-43. Installation of MS3057A clamp

rubber bushing. For shipping to prevent loss of metal or plastic washer, it is placed inside the clamp and held in place by the rubber bushing. Before using, reverse position, to that shown in Figure 2-41a, so that metal or plastic washer will contact back shell of connector.

MS3057A has two saddles separated by a centering bar. This cable clamp is supplied with AN3420 bushing to protect wire bundle under clamp. Add extra bushings if necessary.

MS3057B has a gland and tapered sleeve. The gland is squeezed around the wire bundle when the cable clamp is screwed to the connector back shell.

CAUTION: *Before installing MS3057 cable clamps onto connectors, screw mating part onto connector with coupling nut. This will help to prevent contacts splaying when cable clamp is tightened.*

Installation of MS3057 Cable Clamp

The MS3057 cable clamp is installed as follows: (See Figure 2-42).

1. Slide AN3420A telescoping bushings on wire bundle if bundle diameter is too small to be effectively gripped by the saddle.
2. Slide MS3057 cable clamp without saddle on wire bundle followed by rubber bushing and metal or plastic washer.
3. Assemble wires to connector and tighten back shell. (See later paragraphs for details for each connector class.)
4. Push MS3057 cable clamp towards back shell and hand tighten. Use strap wrench to tighten fully.
5. Push AN3420A bushings, if required, into cable clamp until past saddle.
6. Attach saddle with both screws and tighten until 1/16 inch space is left between saddle and body as shown in Figure 2-42. AN3420A bushing, if used, should bulge slightly when saddle is tight.

NOTE: *A wrap of vinyl, nylon or fiberglass tape can be used instead of AN3420A bushings. If tape wrap is used, secure with nylon braid behind saddle.*

Installation of MS3057A Cable Clamp

The MS3057A cable clamp is installed as follows: (See Figure 2-43)

1. Slide MS3057A cable clamp without saddles on wire bundle before wires are connected.
2. Slide AN3420 telescoping bushing on wire bundle before wires are connected.
3. Assemble wires to connector and tighten back shell (see later paragraphs for details for each connector class).
4. Push MS3057A cable clamp together with inserted AN3420 bushing towards back shell and hand tighten. Use strap wrench to tighten fully.
5. Attach both saddles with supplied screws and lockwashers.

CAUTION: *When replacing saddles, observe that screw heads are placed so that* *pushing on screws will tighten cable clamp to back shell.*

6. Tighten saddles until 1/16 inch remains between each saddle and the centering bar as shown. Use extra AN3420 telescoping bushings in original assembly if saddles are not tight with this 1/16 inch opening. See Table 2-10 for telescoping dimensions.

Installation of MS3057B Cable Clamp

The MS3057B cable clamp is installed as follows: (See Figure 2-44)

1. Slide AN3420A telescoping bushings, if required, on wire bundle before wires are connected. See Table 2-10 for telescoping dimensions.

2. Slide MS3057B cable clamp assembly on wire bundle before wires are connected.

   CAUTION: *Proper lubrication of gland and tapered sleeve is important for this assembly. In handling parts, keep them clean and free of dirt. Wash away dirt with clean denatured ethyl alcohol and relubricate with petrolatum (Federal Specification VV-P-236) as shown in Figure 2-45 (next page). Do not apply petrolatum to inside of gland or to serrated face of gland.*

3. Assemble wires to connector and tighten back shell.

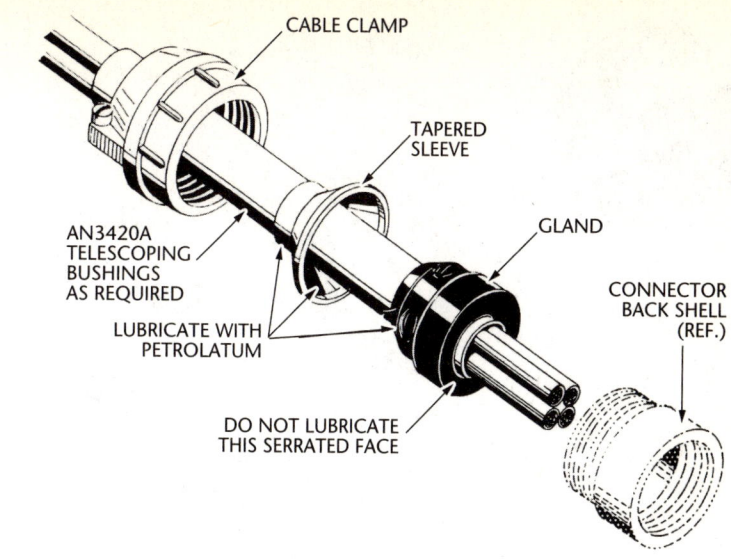

Figure 2-44. Installation of MS3057B cable clamp

4. Push AN3420A bushings, if used, through gland of MS3057B so that end of bushing is flush with serrated face of gland.

5. Slide entire assembly toward back shell and hand tighten. Use strap wrench or padded jaw connector pliers to fully tighten until MS3057B bottoms onto back shell. While tightening, hold wire bundle and telescoping sleeves until gland seats on back shell.

   NOTE: *Shield braid (and jacket if present) must end outside of MS3057B cable clamp to retain moisture resistant qualities of clamp. Ground braid to either grounding screw on cable clamp.*

Connector Size	*Group I Bushing No.	*Group II Bushing No.	*Group III Bushing No.	Inside Diameter in inches	
				Free	Closed
10	AN3420-3A	AN3420-3A	AN3420-3A	.125	.000
12	-4A	-4A	-4A	.219	010
14	-6A	-6A	-6A	.312	.114
16	-8A	-8A	—	.438	.222
18	—	—	-10A	.438	.200
20	-12A	-12A	—	.541	.270
24	-16A	-16A	—	.750	.433
30	—	-18A	—	.938	.504
32	-20A	—	—	.938	.620
36	—	-24A	—	1.125	.682
40	-28A	—	—	1.250	.816

*Bushings listed in each group will telescope respectively when free.

Table 2-10. Telescoping bushings

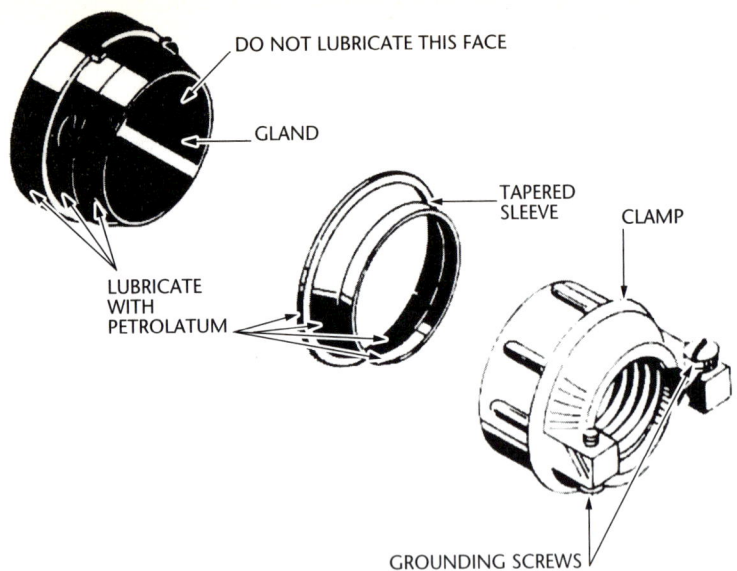

Figure 2-45. Lubrication of MS3057B clamp

Section 6

Disassembly and Reassembly of Connectors

Amphenol and Cannon MS Class A Connectors

Amphenol and Cannon connectors are installed as follows: (See Figure 2-46)

1. Remove back shell by unscrewing from body assembly. If the back shell is too tight to be loosened by hand, attach connector to mating connector shell held in fixture as illustrated in Figure 2-37; and use strap wrench to loosen back shell. Do not remove coupling nut from body.

   CAUTION: *Never use pliers to disassemble or reassemble connectors. Use strap wrench.*

2. If all contacts are size 12 or smaller, no further disassembly is required. For removal of larger contacts see the previous section on removing contacts (page 2-15).

3. Install the following items on wire bundle in listed order.

   a. Cable clamp without saddle
   b. Rubber bushing
   c. Metal or plastic washer
   d. Back shell
   e. Insert retaining ring, if removed in step 2.

4. Slide insulating sleeves over each wire in bundle. Sleeves should be placed one inch back from wire ends to avoid burning during soldering operation. See page 2-20 for data about insulating sleeves.

   NOTE: *Steps 5 through 8 apply only if large contacts were removed in step 2.*

5. Solder wires to large contacts removed in step 2.

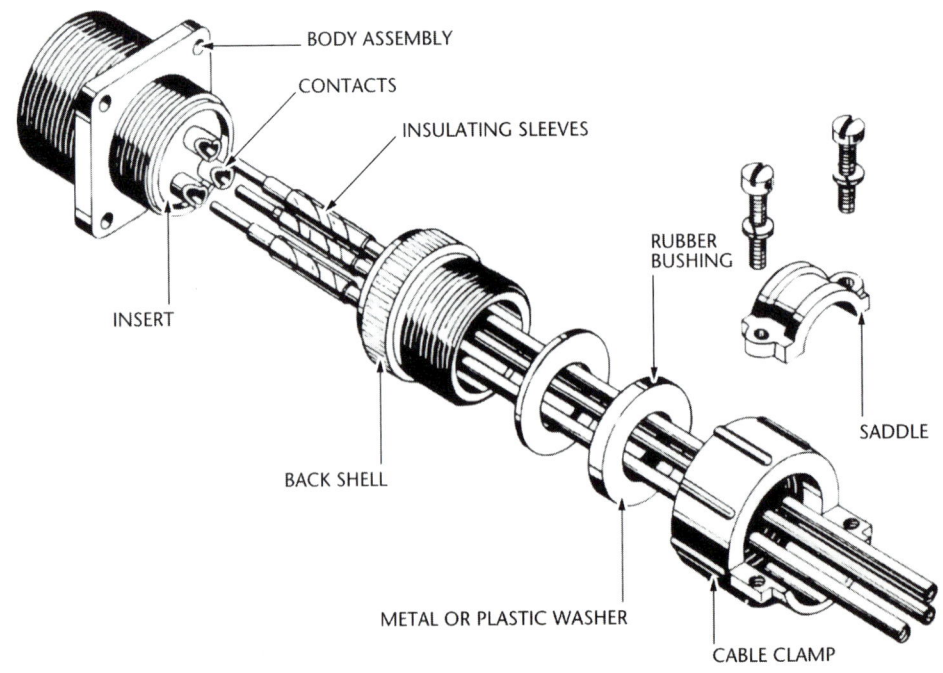

Figure 2-46. Installation of Amphenol Class A connector

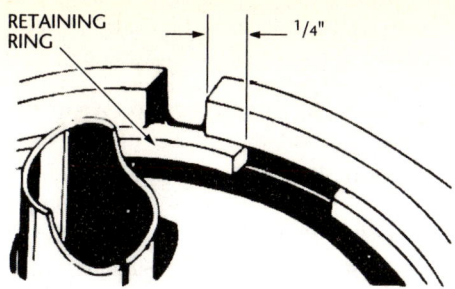

Figure 2-47. Location of end of retaining ring

c. Metal or plastic washer

d. Back shell

e. Insert retaining ring, if removed in step 2.

4. Slide insulating sleeves over each wire in bundle. Sleeves should be placed one inch back from wire ends to avoid burning during soldering operation.

NOTE: *Steps 5 through 8 apply only if large contacts were removed in step 2.*

5. Solder wires to large contacts removed in step 2.

6. Reinstall large contacts by threading through rear insert and inserting contacts into front insert.

7. Reassemble inserts and contacts into body assembly. Be careful to align keyway with key. Do not force the assembly because a damaged keyway will ruin a connector.

8. Replace retaining ring so that end of ring is about 1/4 inch from removal slot as shown in Figure 2-47. This will simplify future disassembly.

9. Solder wires to remaining contacts.

10. Clean connections and slide insulating sleeves over contacts until they bottom against insert.

11. Tie sleeves to wires using nylon braid as shown in Figure 2-40.

12. Reassemble split shell. Be careful not to pinch wires.

NOTE: *Angle back shell can be assembled at 45° angles. See engineering drawing for proper setting for each installation.*

6. Reinstall large contacts by threading through rear insert and inserting contacts into front insert.

7. Reassemble inserts and contacts into body assembly. Be careful to align keyway with key. Do not force the assembly because a damaged keyway will ruin a connector.

8. Replace retaining ring so that end of ring is about 1/4 inch from removal slot as shown in Figure 2-47. This will simplify future disassembly.

9. Solder wires to remaining contacts.

10. Clean soldered connections and slide insulating sleeves over contacts until they bottom against insert.

11. Tie sleeves to wires using nylon braid as shown in Figure 2-37.

12. Slide back shell down over wire bundle and hand tighten to body assembly. Use strap wrench to tighten back shell 1/8 turn beyond hand tight.

13. Install cable clamp as described on pages 2-26 through 2-27.

Amphenol MS Class B Connectors

Amphenol Class B connectors are installed as follows: (See Figure 2-48)

1. Remove back shell by loosening captive assembly screws. Do not remove coupling nut from body.

2. If all contacts are size 12 or smaller, no further disassembly is required. For removal of larger contacts see the previous section on removing contacts (page 2-15).

3. Install the following items on wire bundle in listed order:

    a. Cable clamp without saddle

    b. Rubber bushing

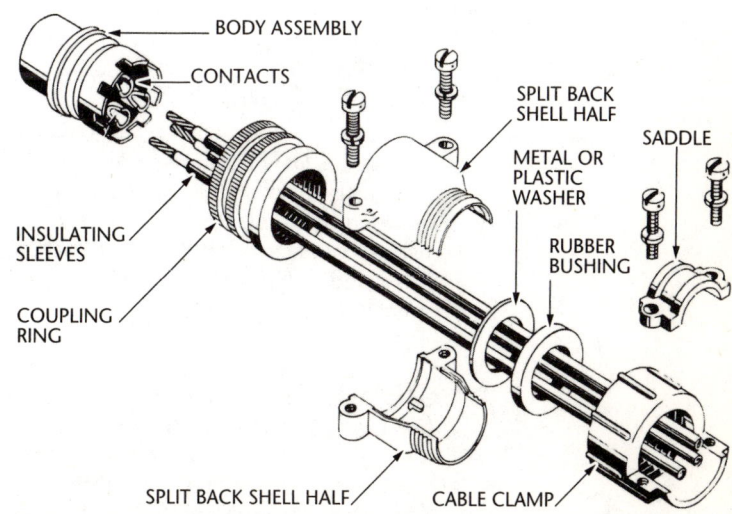

Figure 2-48. Installation of Amphenol Class B connector

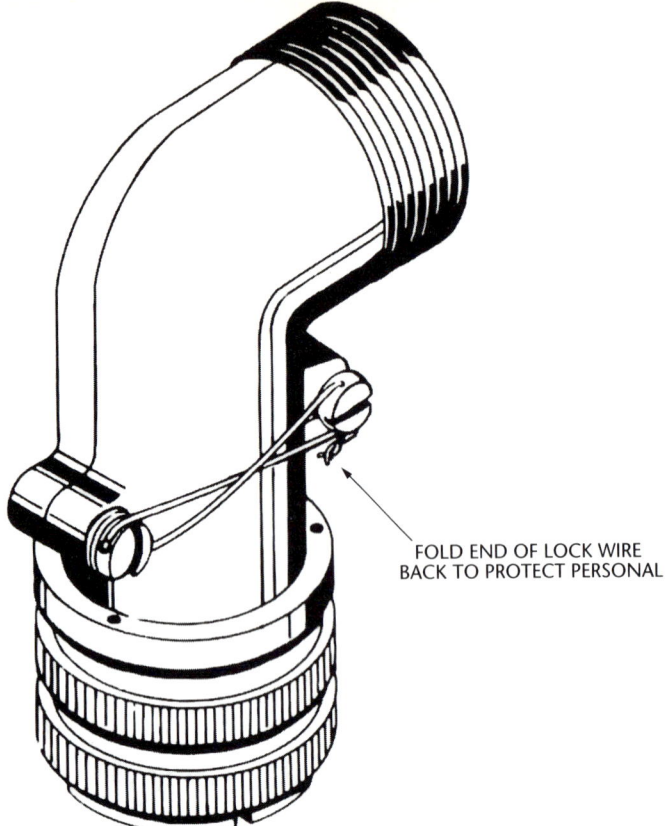

Figure 2-49. Safety wiring Class B connector

13. Install cable clamp.

14. Safety wire the split back shell holding screws, if required by engineering, by passing the wire through screw heads (See Figure 2-49), crossing it and completing with a twist.

Amphenol MS Class C Connectors

Amphenol Class C connectors are installed as follows (see Figure 2-50):

1. Remove back shell by unscrewing from body. If the back shell is too tight to be loosened by hand, attach connector to mating connector shell held in fixture as illustrated in Figure 2-37; and use strap wrench to loosen back shell. Do not remove coupling nut from plug body.

   CAUTION: *Never use unpadded pliers to disassemble or reassemble connectors.*

2. Large contacts (8, 4 and 0) are threaded into insert. Unscrew the contact for soldering. Do not twist or attempt to remove small contacts.

CAUTION: *Do not remove Class C inserts. These connectors are pressurized and insert removal will break the pressure seals.*

3. Install the following items on wire bundle in listed order:

    a. Cable clamp without saddle

    b. Rubber bushing

    c. Metal or plastic washer

    d. Back shell

4. Slide insulating sleeves over each wire in bundle. Sleeves should be placed one inch back from wire ends to avoid burning during soldering operation.

5. Reinstall large contacts. Tighten contact until it is seated firmly against lead pressure washer. Use curved long nose pliers if necessary.

6. Solder wires to remaining contacts.

7. Clean connections and slide insulating sleeves over contacts until they butt against the insert.

8. Tie sleeves to wires using nylon braid as shown in Figure 2-40.

9. Slide back shell down over wire bundle and hand tighten to body assembly. Use strap wrench to tighten back shell 1/8 turn beyond hand tight.

10. Install cable clamp.

Amphenol Class E and Class R Connectors

Amphenol Class E and R connectors are installed as follows:

1. Disengage the grommet compression nut assembly from the shell.

2. Thread pre-tinned wires through the proper holes in the grommet.

3. Solder wires to contacts.

4. Work the grommet compression nut assembly back up the wire bundle and engage nut with shell assembly, making sure that the grommet is drawn up flush with the insert.

5. Fill all unused grommet holes with grommet sealing plugs.

Amphenol fireproof connectors. Amphenol fireproof connectors have crimp type contacts to withstand high temperature operating requirements. Installation is as follows:

General Purpose Connectors | 2-31

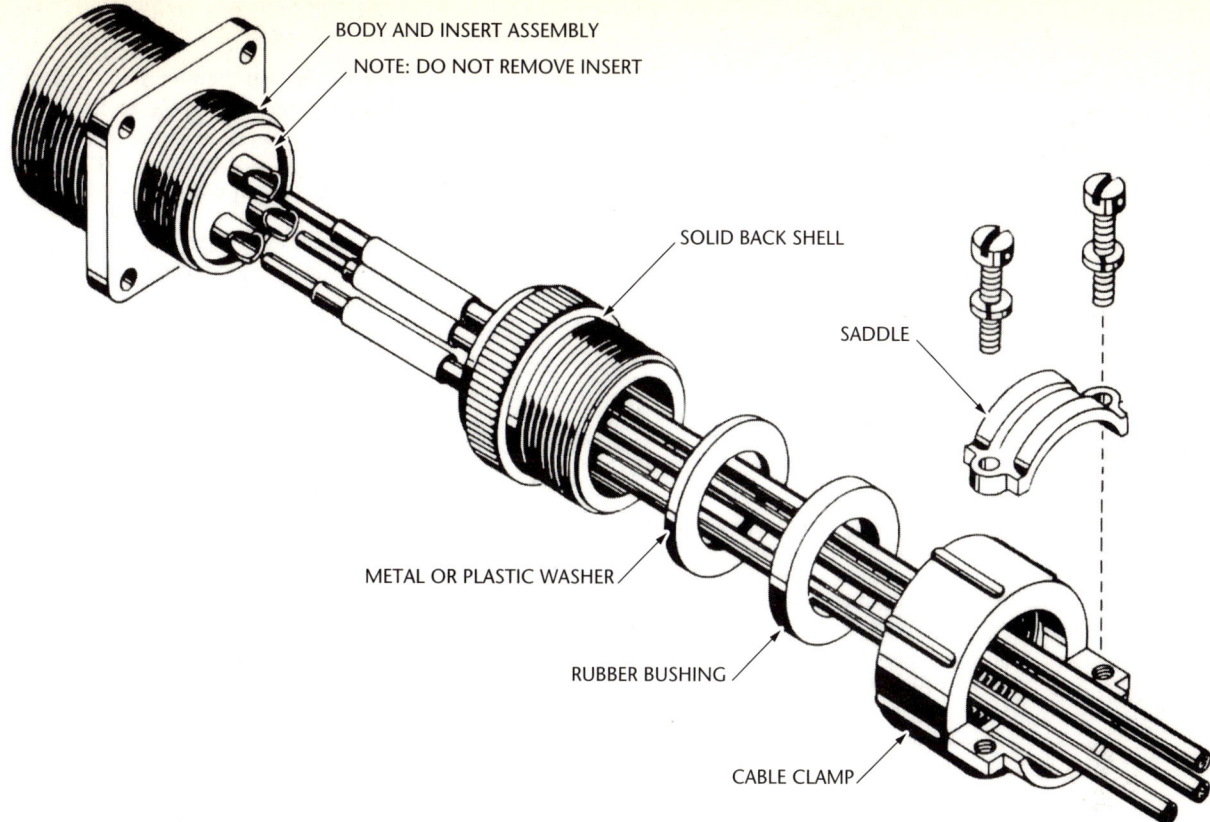

Figure 2-50. Installation of Amphenol Class C connectors

1. Use spanner wrench to remove spanner nut. See Figure 2-51.

2. Tap body assembly lightly in palm of hand to remove and separate contact and insert assembly.

3. Install the following items on wire bundle in listed order:

   a. Conduit with coupling nut or other required fittings

   NOTE: *Fitting threads are not identical with standard MS connectors of same size. Use fitting sizes listed in Table 2-11.*

   b. Spanner nut

   c. Ceramic rear insert

   d. Silicone rubber gasket

4. Crimp wires into contacts using any of the methods described in Chapter 4.

5. Reassemble by (1) sliding contacts into ceramic front insert; (2) pushing silicone rubber gasket down over contacts; and (3) sliding ceramic rear insert over contacts.

6. Examine contact and insert assembly to see that parts butt and then slide assembly into body. Tighten spanner nut into place until flush with rear of body shell.

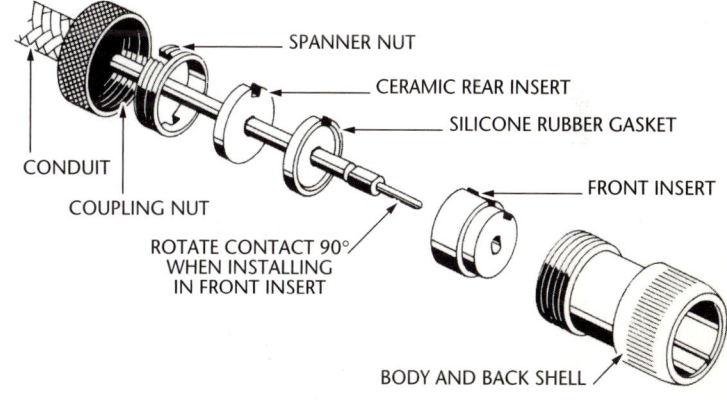

Figure 2-51. Installation of Amphenol fireproof connector

Connector Size	Fitting	
	Size	Thread
18	12	1-3/16 - 18
22	16	1-7/16 - 18
32	24	2 - 18
36	28	2-1/4 - 16

Table 2-11. Amphenol fireproof connector fittings

2-32 | General Purpose Connectors

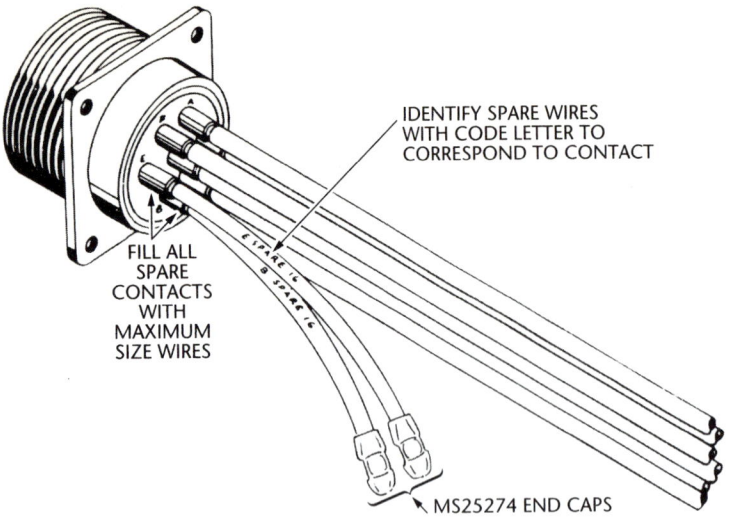

Figure 2-52. Spare wires for potting connector

7. Tighten conduit coupling nut or other required fitting over body assembly. Use strap wrench to tighten 1/8 turn beyond hand tight.

Potting connectors. Potting connectors are supplied with a plastic potting mold. Installation is as follows:

1. Slide the plastic mold over the wire bundle.
2. Solder wires to contacts.

   CAUTION: *Do not install insulating sleeves over individual wires. Potting compound will not cure properly in contact with vinyl sleeving.*

3. Install spare wires on all unused pins. Use largest gage wire that would normally be attached to each contact. Spare wires are approximately 9 inches long. (See Figure 2-52).

4. Clean the complete connector assembly by scraping off rosin and then brush vigorously in new unused Stoddard's Solvent followed by second rinse in clean Stoddard's Solvent. See Figure 2-53.

5. Rinse area to be potted with methylene chloride applied from hand operated laboratory wash bottle or similar device.

   CAUTION: *Do not breathe methylene chloride fumes. Use only in well ventilated area.*

   NOTE: *Complete potting within two hours after cleaning.*

6. Slide plastic mold into position.

   CAUTION: *Mate connectors before potting either part to avoid splaying contacts during the potting operation.*

7. Insert potting compound prepared in accordance with directions given in Chapter 8. Fill back of connector by inserting nozzle down between wires until it almost touches back of insert (See Figure 2-54). Fill slowly while moving nozzle back from insert and watch compound to be sure no air bubbles are trapped. Fill to top of mold. Tamp down the compound, if necessary, with a wooden or metal 1/8 inch dowel. Tap connector assembly on a resilient surface or vibrate mechanically to help flow the compound into all spaces and to release trapped air.

8. Insulate the ends of all spare wires. See Figure 2-52. The preferred method of insulating a spare wire is to crimp it into an MS25274 wire end cap with tool MS25037-1A. See next page for installing procedure. Nonstandard end caps are also available for either stripped or unstripped wire.

9. Immediately after filling each connector, tie the wires together loosely about 6 inches back from connector. Make sure that wires are centrally located in the connector so that each wire is completely surrounded by potting compound. Suspend the assembly by placing the tie over a nail as shown in Figure 2-54 and allow to air cure for at least 1 1/2 hours at 75°F without any movement.

   WARNING: *The accelerator contains a toxic lead compound. Avoid excessive skin contact. Clean hands thoroughly after using. Use gloves.*

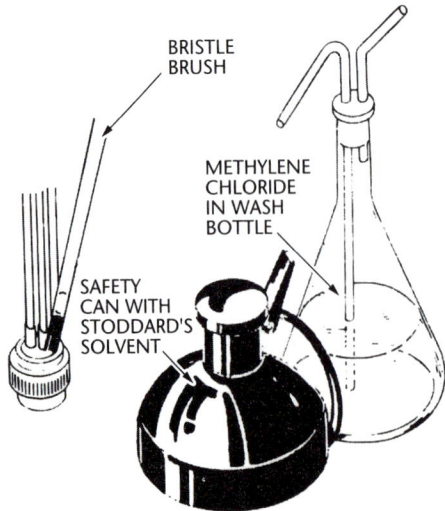

Figure 2-53. Cleaning connector prior to potting

General Purpose Connectors | 2-33

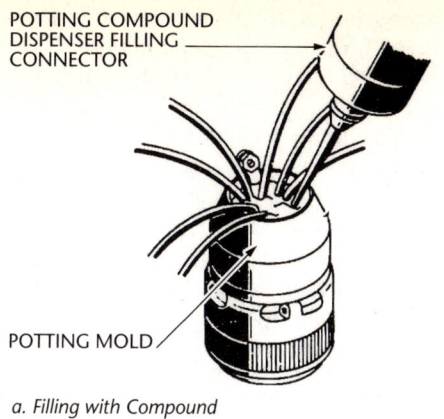

a. Filling with Compound

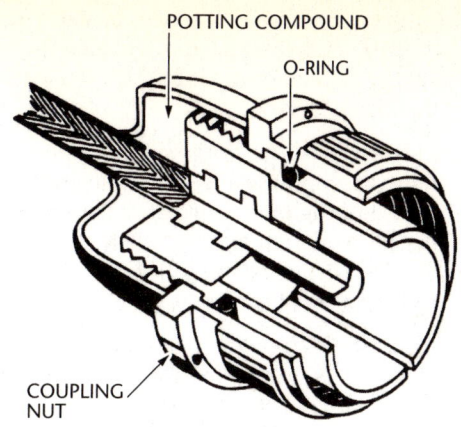

Figure 2-55. Installation of O-ring on potted connector

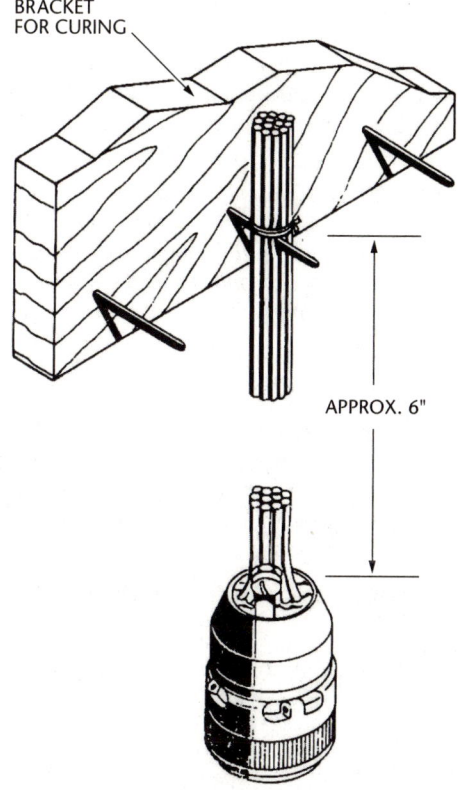

b. Curing

Figure 2-54. Filling and curing potting connector

Plug Size	O-Ring Thickness	O-Ring ID	MS29513 Dash Nos.
8S	.070 inches	.312 inches	-10
10S & 10SL	.070 inches	.364	-12
12 & 12S	.070	.489	-14
14 & 14S	.070	.489	-14
16 & 16S	.070	.614	-16
18	.070	.739	-18
20	.070	.864	-20
22	.070	.989	-22
24	.070	1.114	-24
28	.070	1.364	-28
32	.070	1.364	-28
36	.103	1.737	-132
40	.103	1.987	-136
44	.103	2.237	-140
48	.103	2.487	-144

Table 2-12. O-ring sizes for AN type connectors

10. Carefully place assembly still suspended from nail into drying oven for 3 to 4 hours at 100°F, or air cure at 75°F for 24 hours.

    WARNING: *The accelerator contains a toxic lead compound. Avoid excessive skin contact. Clean hands thoroughly after using. Use gloves.*

    NOTE: *Full cure with maximum electrical characteristics is not achieved until 24 hours after potting. Do not perform any electrical insulation resistance tests during this period.*

11. Apply a light film of lubrication oil to all exterior metal surfaces after potting compound is completely cured.

12. If the plug does not have an O-ring or gasket seal on the barrel as shown in Figure 2-55 then install an MS29513 O-ring selected from Table 2-12. Roll O-ring tightly against shoulder of plug inside coupling ring. Plug barrel and O-ring must be clean and dry before assembly.

    CAUTION: *Do not use two rings. The added thickness of a second ring will prevent proper mating of contacts.*

Procedure for crimping wire end caps onto spare wires. The procedure for crimping wire end caps with standard tool MS25037-1A is as follows:

1. Select an end cap of the correct size for the wire to be insulated after referring to Table 2-13, and crimp it to the wire with the MS25037 tool.

2. Make sure the locator is properly positioned behind the lower nest. Position the wire end cap in the correct die nest (color of cap insulation matches color coding on tool handle) with the closed end of the cap resting against the locator.

3. Insert the stripped wire so that the end of the stripped wire is seated against the closed end of the cap, and the insulation against the metal sleeve of the cap.

4. Close tool handles to crimp end cap to wire, until ratchet releases and the tool opens. Remove the crimped assembly.

Bendix-Scintilla Connectors

Bendix-Scintilla MS Class A connectors. Bendix Class A connectors are installed as follows (see Figure 2-56):

MS Number	Color	Wire Size
MS25274-1	Yellow	26-24
MS25274-2	Red	22-18
MS25274-3	Blue	16-14
MS25274-4	Yellow	12-10

Table 2-13. Wire end caps and crimping tools

Contact Size	Rod Diameter (Inches)
0	.450
4	.312
8	.187

Table 2-14. Contact removal tool diameter

1. Remove back shell by unscrewing from body. If the back shell is too tight to be loosened by hand, attach the connector to mating connector shell attached to fixture as illustrated in Figure 2-37; and use a strap wrench or padded jaw pliers to loosen back shell. Do not remove coupling nut from plug body.

CAUTION: *Never use pliers with unpadded jaws to disassemble or reassemble connectors.*

2. If all contacts are size 12 or smaller, no further disassembly is required. Larger contacts such as size 8, 4 or 0 may be removed by applying pressure on solder well end by means of steel or bakelite rod slightly smaller in diameter than the solder well. See Figure 2-22 for arbor press fixture suitable for this operation. Table 2-14 lists diameter of rods for removing contacts.

CAUTION: *Hold the connector so that pressure is applied in a straight line with contacts. Pushing at an angle may damage the contacts. Bent contacts must be replaced. Do not attempt to straighten damaged contacts.*

3. Install MS3057B cable clamp.

4. Slide insulating sleeves over each wire in bundle. Sleeves should be placed one inch back from wire ends to avoid burning during soldering operation. See Table 2-9 for sleeving sizes.

5. Solder wires to larger contacts removed in step 2.

6. Reinstall large contacts by pushing them through rear of insert until seated. Use ethyl

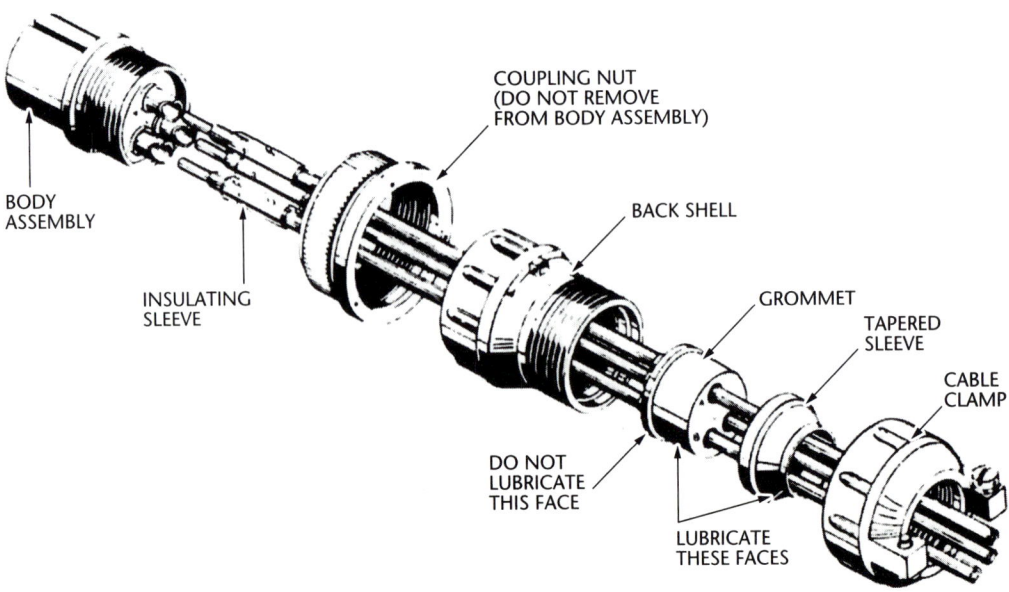

Figure 2-56. Installation of Bendix Class A and C connector

alcohol as a lubricant, if necessary. Install each contact, when cool, before proceeding to solder next contact. This will help avoid errors. Use a bakelite screwdriver to aid in seating contacts. See Figure 2-57.

CAUTION: *Use care not to fold thin lip of rubber into hole with contact.*

7. Solder wire to remaining smaller contacts.
8. Slide insulating sleeves over cooled connections until they bottom against insert.
9. Slide back shell down over wire bundle and hand tighten to body. Use strap wrench or padded jaw pliers to tighten back shell until it bottoms.
10. Slide telescoping sleeves, if required, through gland until flush with inside edge. Hold telescoping sleeves in back of cable clamping nut while engaging threads. Do not release telescoping sleeves until gland seats on back shell.
11. Tighten cable clamp with strap wrench or padded jaw pliers until it bottoms. Mate connector with mating shell in fixture while tightening cable clamp.

CAUTION: *Keep all parts free of dirt and foreign material. Clean dirty parts with ethyl alcohol and relubricate all threads with Mil-Spec MIL-G-3278 grease. Relubricate the indicated parts of gland and tapered sleeve as shown in Figure 2-45 with petrolatum.*

Bendix-Scintilla MS Class C connectors. Bendix Class C connectors are pressurized and care must be given to avoid breaking the pressure seal. Installation procedure for Bendix Class C connectors is the same as for Bendix Class A connectors. (See Figure 2-56).

CAUTION: *Never remove insert or contacts from Class C connector as this would break the pressure seal incorporated in the unit at the time of factory assembly.*

Bendix-Scintilla MS Class E connectors. Bendix Class E connectors are installed as follows:

1. Remove cable clamp by unscrewing from back shell. Slide cable clamp over wire bundle.

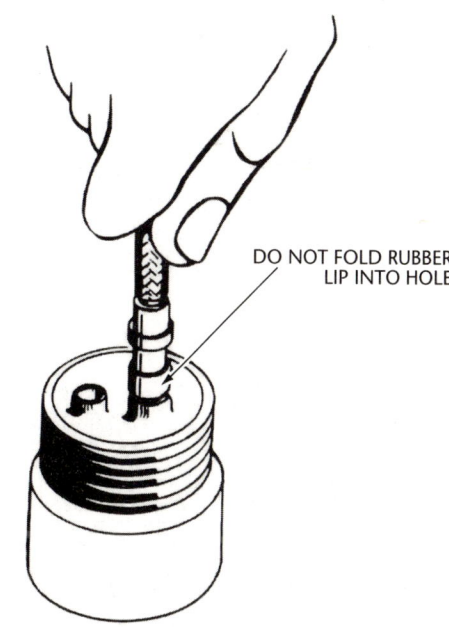

Figure 2-57. Reinstalling contact in Bendix resilient insert

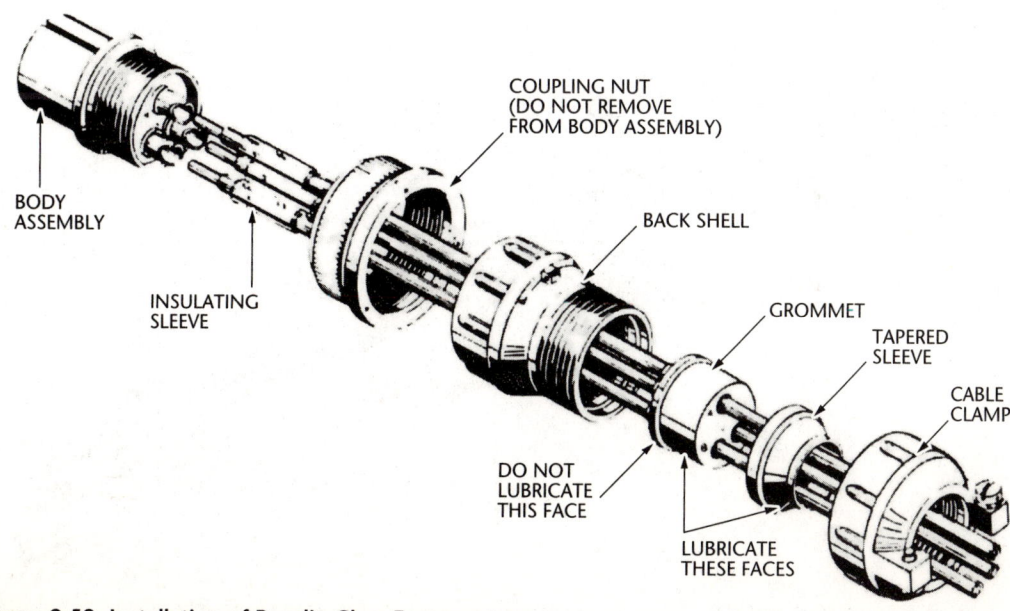

Figure 2-58. Installation of Bendix Class E connector

2-36 | General Purpose Connectors

2. Check tapered sleeve and grommet for thin film petrolatum lubricant on indicated surfaces. (See Figure 2-58).

3. Slide tapered sleeve over wire bundle.

4. Insert pre-tinned wires through proper holes in grommet.

NOTE: *Grommet is coded to match insert coding.*

Use alcohol as a lubricant if necessary. After wires are threaded through grommet, use air blast to dry alcohol.

5. Unscrew back shell from body assembly and slide over wire bundle. If coupling nut is removed, push nut back on wire bundle with threaded end forward.

6. Install insulating sleeves on wires.

7. Solder wires to contacts.

CAUTION: *Do not remove contacts from receptacles for soldering.*

8. Slide insulating sleeves over cooled connections until they bottom against insert.

Wire Size	MS Part No.
22-16	MS25251-16
14-12	MS25251-12
10-8	MS25251-8

Table 2-15. Sizes of grommet sealing plugs

9. Slide back shell down over wire bundle and hand tighten to body shell.

10. Examine insulating sleeves. They should not project over shoulder in back shell.

11. Use strap wrench or padded jaw pliers to tighten back shell until it bottoms.

12. Carefully push grommet down over wire until it is seated in shoulder of back shell.

CAUTION: *Do not allow wires to fold inside back shell.*

13. Fill all unused holes with MS25251 grommet sealing plugs. Sizes are listed in Table 2-15.

14. Slide tapered sleeve and cable clamp over grommet and hand tighten.

15. Tighten cable clamp with strap wrench or padded jaw pliers until it bottoms.

CAUTION: *When tightening cable clamp insure that tapered sleeve and grommet do not rotate with cable clamp. This will cause tension and possible breakage of wire at connection.*

Bendix-Scintilla MS Class R connectors.
Bendix Class R connectors are installed as follows (see Figure 2-59):

1. Remove compression sleeve from back nut by unscrewing the two compression screws, and slide sleeve over wire bundle.

2. Insert pre-tinned wires through the

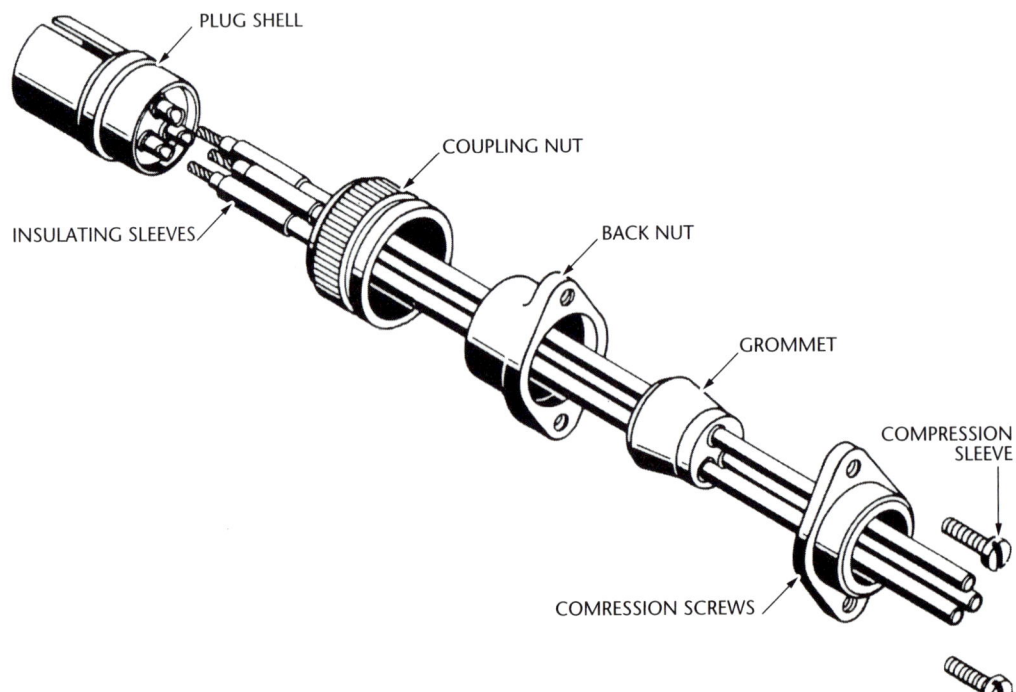

Figure 2-59. Installation of Bendix Class R connector

proper holes in grommet and slide along the wire bundle.

NOTE: *Use identifying letters on the rear face of the grommet as a guide in threading wires. Align letters with corresponding letters on rear face of the insert.*

3. Unscrew the back nut from the body assembly and slide it over the wire bundle. If coupling nut is removed, make sure it is pushed back on the wire bundle with the threaded end forward.

4. Solder wires to contacts.

5. Mate connector with its mating piece and slide back nut down over wire bundle and hand tighten to shell. Use a strap wrench or padded jaw connector pliers to tighten back nut until it bottoms.

6. Push grommet carefully down over wires until it is seated in the shoulder of the back nut.

7. Fill all unused grommet holes with MS25251 grommet sealing plugs; sizes are listed in Table 2-15.

8. Slide the grommet compression sleeve down over the wire bundle and tighten the two compression screws until they bottom.

Bendix-Scintilla fireproof connectors. Bendix fireproof connectors have crimp-on type contacts to withstand high temperature operating requirements. Installation is as follows (see Figures 2-60 and 2-61):

1. Remove back shell by unscrewing from body.

2. Remove threaded insert retaining ring with fingers. If necessary, use spanner wrench (Bendix 11-4045).

3. Remove insert assembly by pushing on front insert toward the rear of the connector.

4. Remove front insert from contacts. Rotate contacts 90° and remove from rear insert and sealing insert.

NOTE: *Use care to prevent tearing or cutting sealing insert*

5. Strip wires.

6. Attach wires to contacts by silver soldering crimping with appropriate tool.

7. Install the following on the wire bundle in the order given:

   a. Elbow and adapter or other fittings as required.

   b. Threaded insert retaining ring.

8. To install contacts with wires attached, rotate contacts 90° and insert contacts through proper holes of rear insert.

NOTE: *Smallest outside diameter of rear insert must face wires.*

9. Push sealing insert over contacts until it butts against front face of rear insert.

10. Align keyway of front insert with keyway of rear insert and slide front insert (largest diameter first) over contacts down to sealing insert.

11. Align keyway of assembled inserts with keyway in body. Carefully push assembly into back of body. If sealing insert catches on key, depress insert with a dull instrument while it passes under the key.

12. Start spanner nut into connector body by

Connector Size	Spanner Wrench (Bendix)	Torque
10S	11-4045-10	10 lb inch
12S & 12	-12	12
14S & 14	-14	14
16S & 16	-16	16
18	-18	18
20	-20	20
22	-22	22
24	-24	24
28	-28	28
32	-32	32

Table 2-16. Torque value for fireproof connectors

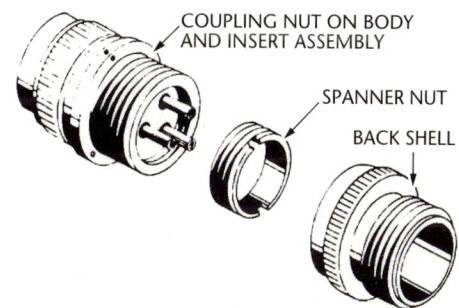

Figure 2-60. Bendix fireproof connector partial disassembly

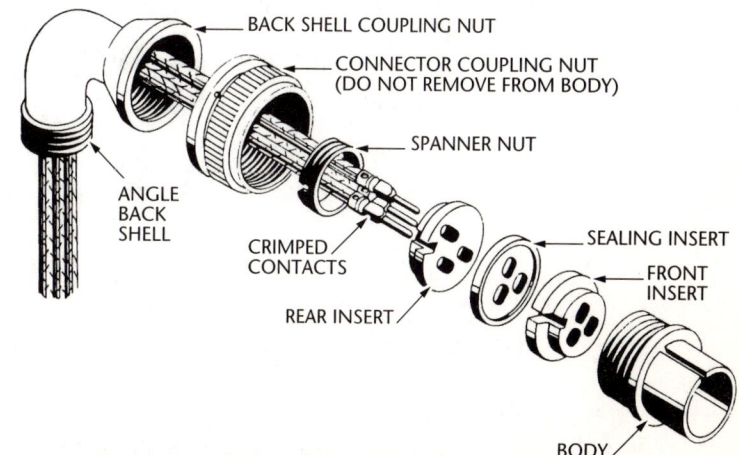

Figure 2-61. Installation of Bendix fireproof connector

2-38 | General Purpose Connectors

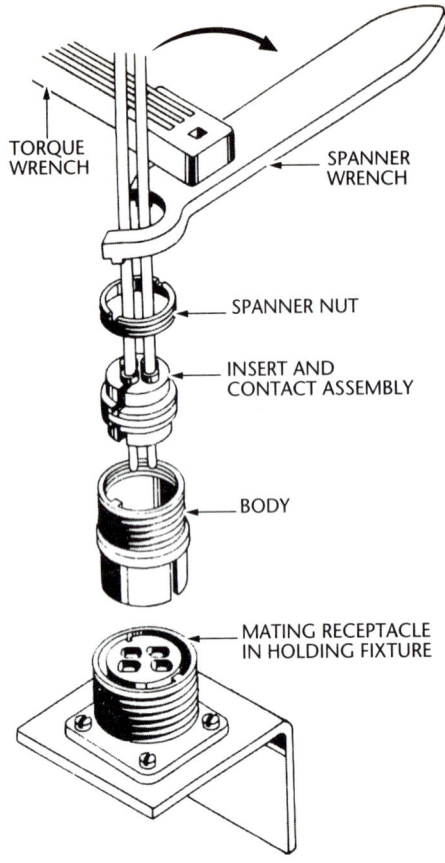

Figure 2-62. Torque wrench used on Bendix fireproof connector

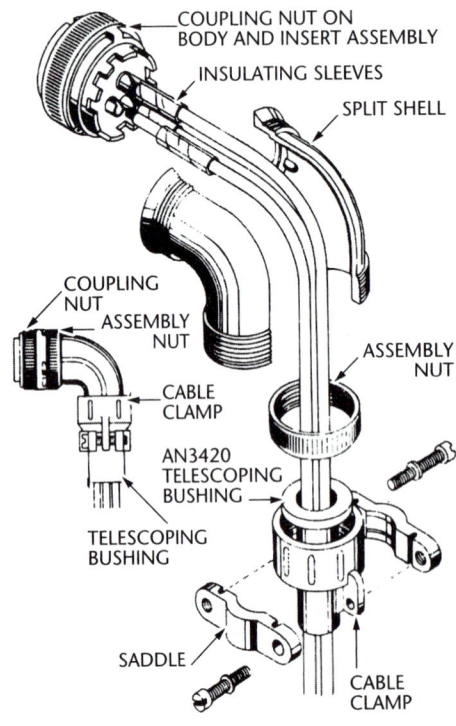

Figure 2-63. Installation of Cannon Class B connector

hand and tighten with spanner wrench (see Table 2-16), using standard torque wrench handle (1/4 inch drive), as shown in Figure 2-62. Tighten to torque values in Table 2-16.

13. Tighten back shell, or elbow if used, to body. Use strap wrench or padded jaw pliers to tighten until back shell is bottomed.

Cannon Connectors

Cannon MS Class B connectors. Cannon Class B connectors are supplied with a split shell held together by a shell assembly nut as shown in Figure 2-18. Cannon Class B connectors are installed as follows:

1. Remove split shell by loosening shell assembly nut. Do not remove coupling nut from plug body.

2. If all contacts are size 12 or smaller, no further disassembly is required. For removal of larger contacts see page 2-15 through 2-16.

3. Install the following items on wire bundle in listed order as shown in Figure 2-63.

   a. Cable clamp with loosened saddle.

   b. AN3420 telescoping bushing (at least one, extra telescoping sections are added if necessary).

4. Slide insulating sleeves over each wire in bundle. Sleeves should be placed one inch back from wire ends to avoid burning during soldering operation.

5. Solder wires to contacts.

6. Clean connections and slide insulating sleeves over contacts until they bottom against insert.

7. Tie sleeves to wires using nylon braid as shown in Figure 2-40.

8. Reassemble split shell. Be careful not to pinch wires. Rotate and lock angle back shell in proper position.

9. Install cable clamp.

Cannon MS Class C connectors. Cannon MS Class C connectors are supplied only as receptacles MS3100C and MS3102C.

CAUTION: *Do not remove Class C inserts or contacts. Removal will break the pressure seal.*

Install Cannon MS3100C as follows (Figure 2-64):

1. Remove back shell by unscrewing from body. Slide MS3057A cable clamp, AN3420 telescoping bushing and back shell on the wire bundle in that order.

2. Slide insulating sleeves over each wire in bundle. Sleeves should be placed one inch back from wire ends to avoid burning

during soldering operation.

3. Solder wires to contacts.

4. Clean connections and slide insulating sleeves over contacts until they butt against the insert.

5. Tie sleeves to wires using nylon braid as shown in Figure 2-40.

6. Slide back shell down over wire bundle and tighten body. Use strap wrench to tighten back shell 1/8 turn beyond hand tight.

7. Install cable clamp. Install Cannon MS3102C in the same manner as Cannon MS3100C except omit steps 1 and 6.

Cannon MS Class E connectors. Install Cannon Class E connectors as follows (see Figure 2-65):

1. Unscrew back shell. Do not pull AN3420 bushings out of back shell. Loosen saddles but do not remove them from back shell.

2. Slide back shell and bushing assembly over wire bundle.

3. Thread pre-tinned wires through grommet follower and through grommet. Note relationship of grommet follower and grommet in Figure 2-65.

   CAUTION: *Insert proper wires in holes of grommet and grommet follower. These are coded to match markings on back of insert.*

Use alcohol as a lubricant if necessary. After wires are threaded through grommet, use air blast to dry alcohol.

   NOTE: *Do not install insulating sleeves on wires.*

4. Solder wires to contacts. If contacts are larger than number 12 size, they are removed as described on pages 2-15 through 2-16.

5. Reinstall large contacts by pushing them through rear of insert until seated. (See Figure 2-57.) Use alcohol as a lubricant, if necessary. Install each contact, when cool, before proceeding to solder next contact. This will help avoid errors. Use a bakelite screw-driver to aid in seating contacts.

6. Slide grommet down over wire bundle.

   CAUTION: *Pull lightly on each wire as grommet is seated to prevent wires folding between insert and grommet.*

7. Slide grommet follower down over wire bundle. Use same caution as in step 6. See Figure 2-66.

8. Slide back shell and bushing assembly

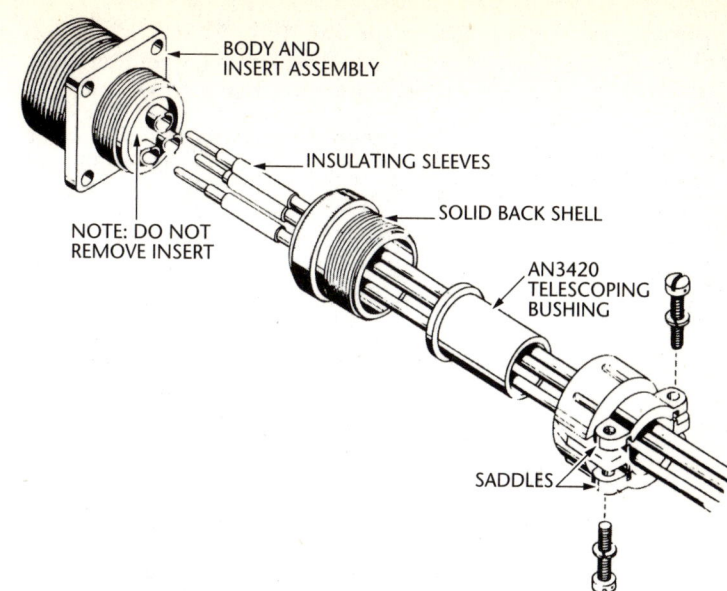

Figure 2-64. Installation of Cannon Class C connector

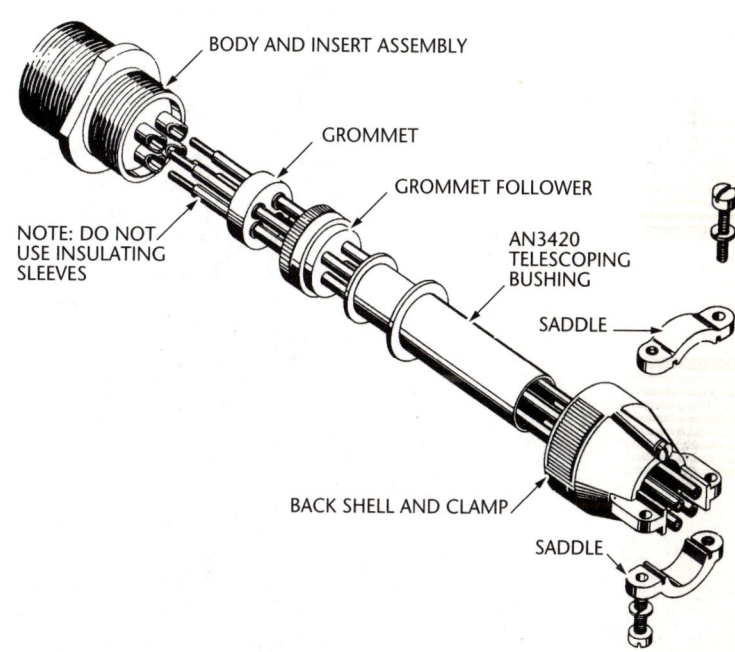

Figure 2-65. Installation of Cannon Class E connector

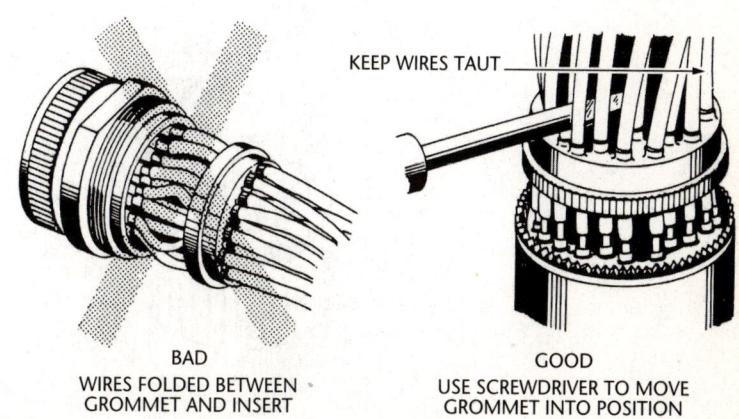

Figure 2-66. Cannon Class E connector - grommet installation

over wire bundle and hand tighten to body. Use strap wrench to tighten back shell until it bottoms.

9. Tighten saddle screws until 1/16" clearance remains between each saddle and mounting tab of back shell.

Cannon MS Class R connectors. The installation of Cannon Class R connectors is the same as that for Cannon Class E connectors.

Cannon fireproof connectors. Cannon fireproof connectors have crimp type contacts to withstand high temperature operating requirements. See Figure 2-67. Installation is as follows:

1. Release back shell by removing four holding screws. Do not remove coupling nut from plug body.

2. Use retaining ring pliers to remove internal retaining ring.

3. Tap connector body in palm of hand to remove inserts. Separate inserts to release contacts.

4. Carefully pull retaining clips from contacts. Do not lose clips.

5. Install the following items on wire bundle in listed order:

    a. Back shell, with conduit if required.

    b. Retaining ring.

6. Silver solder, or crimp wires into contacts using any of the methods described in Chapter 4.

7. Slide contacts through rear insert. Be careful to get each contact into its proper location by observing the identification letters on the insert.

8. Reinstall the retaining clips to lock contacts into inserts.

9. Push front insert over contacts. Observe identification letters to assure proper location of contacts.

10. Examine contact and insert assembly to see that insert halves butt, and then slide assembly into body.

11. Reinstall retaining ring.

12. Tighten back shell by replacing four holding screws.

13. Tighten conduit coupling nut or other required fitting. Use strap wrench to tighten 1/8 turn beyond hand tight.

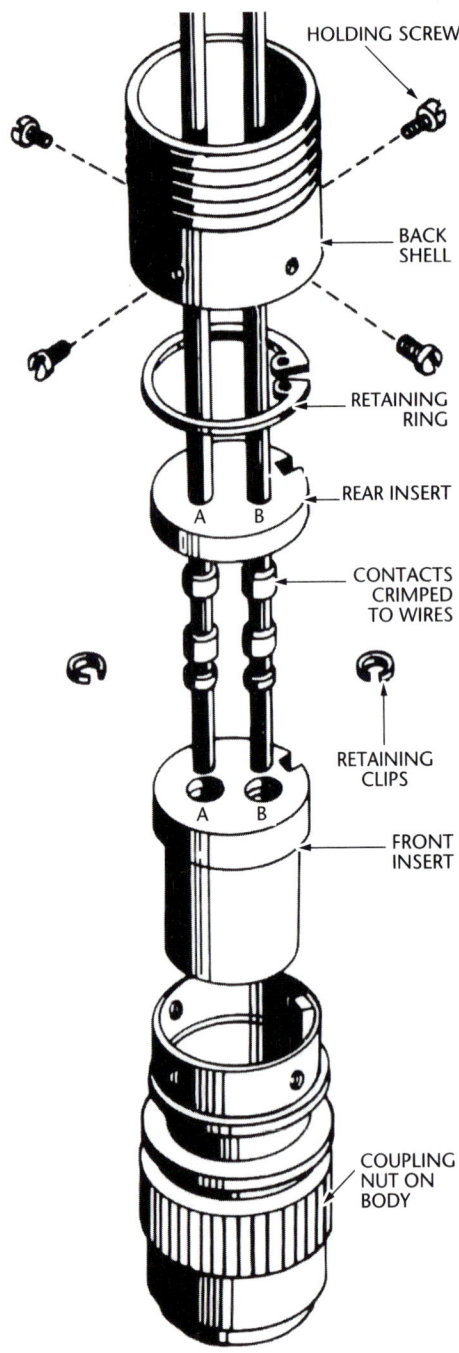

Figure 2-67. Installation of Cannon fireproof connector

MS Miniature Connectors

Miniature MS connectors with solder type contacts. Miniature MS connectors Class E and F are installed as follows:

1. Remove retaining nut by unscrewing it from the body shell, and slide the nut over the wire bundle.

2. Slide tapered sleeve, (grommet follower) over wire bundle.

3. Thread wires through the proper grommet holes.

4. Strip and tin wires following the procedures outlined in Chapter 1. Strip insulation 1/4 inch for sizes 12 and 16 contacts; strip insulation 1/8 inch for size 20 contacts.

5. Insert tinned wire into the solder cup of the contact. Apply heat with a resistance soldering unto as shown in Figure 2-31; a soldering unit with a foot switch is recommended. Heat until solder is liquified; apply more solder if necessary. When solder is liquid, press the foot switch to turn off the heat, but do not remove the resistance unit probes until the solder has solidified. Do not move the joint until the solder has cooled.

6. Slide the grommet forward over the solder cups of the contact against the face of the insert.

7. Slide the tapered sleeve (grommet follower) over the end of the grommet.

8. Bring the retaining nut forward and tighten it to the body shell.

CAUTION: *Do not remove inserts from these connectors.*

Install Class P miniature MS connectors with solder type contacts as follows:

1. Remove potting mold and retaining ring from body shell, and slide over wire bundle.

2. Prepare wires and solder to contacts as described on the previous page, steps 4 and 5.

3. Slide retaining ring and potting mold down wire bundle, and screw to body shell.

NOTE: *Make sure that retaining ring is fully tightened. A loosely coupled ring affects the bayonet action of the coupling ring.*

4. Pot connector as described on pages 2-32 through 2-34.

Class H miniature MS connectors are not disassembled for soldering. Prepare wires and solder to contacts as described on page 2-41, steps 4 and 5.

Section 7
Contact Tools

Miniature MS Connectors with Removable Crimp-type Contacts

These connectors are received from the manufacturer with the contacts separately packaged, so that the wires can be crimped to the contacts with a crimping tool before they are assembled into the connector insert.

Hand Crimping Tools for Connector Contacts

The tool for crimping electric connector contacts is MS3191, a hand operated tool, cycle controlled by means of a ratchet which will not release until the crimping cycle is finished. See Figure 2-68 (next page). The tool has a separate positioner for each contact size, color coded as follows:

Contact Size No. 20: Red

Contact Size No. 16: Blue

Contact Size No. 12: Yellow

Use this tool for assembling wires to crimp type removable contacts of all MS series miniature connectors.

CAUTION: *Do not disassemble this tool. Do not tighten or loosen nuts on back of tool.*

MS3191-1 crimping tool. As shown in Figure 2-68a (next page), this tool has one contact positioner installed in the tool, and the other two stored in a cavity in the tool handle. The three contact sizes, and the wire sizes each contact will accommodate are marked and color coded on a data plate on the face of the tool.

MS3191-3 crimping tool. In this tool the color coded positioners are fixed in a rotating turret contained in the tool head. (See Figure 2-68b, next page). Wire sizes are marked on a selector plate on the face of the tool. Correct indentor closure for wire size being used is set by moving thumb button pointer to desired wire size.

NOTE: *Either MS3191-1 or MS3191-3 may be used to crimp contacts shown on drawing MS3191.*

Tool Inspection

The standard MS3191 crimping tool is checked for proper adjustment of the crimping jaws by means of the appropriate MS3196-3 gage. Do this before each series of crimping operations.

NOTE: *If tool fails to gage correctly, or if the ratchet fails, return the tool for repair.*

Inspecting MS3191-1 crimping tool. This tool is gaged separately for each positioner. The procedure is as follows:

1. Make sure the positioner is locked in place.

2. Select the proper gage for the positioner being used.

3. With the tool fully closed, insert the GO

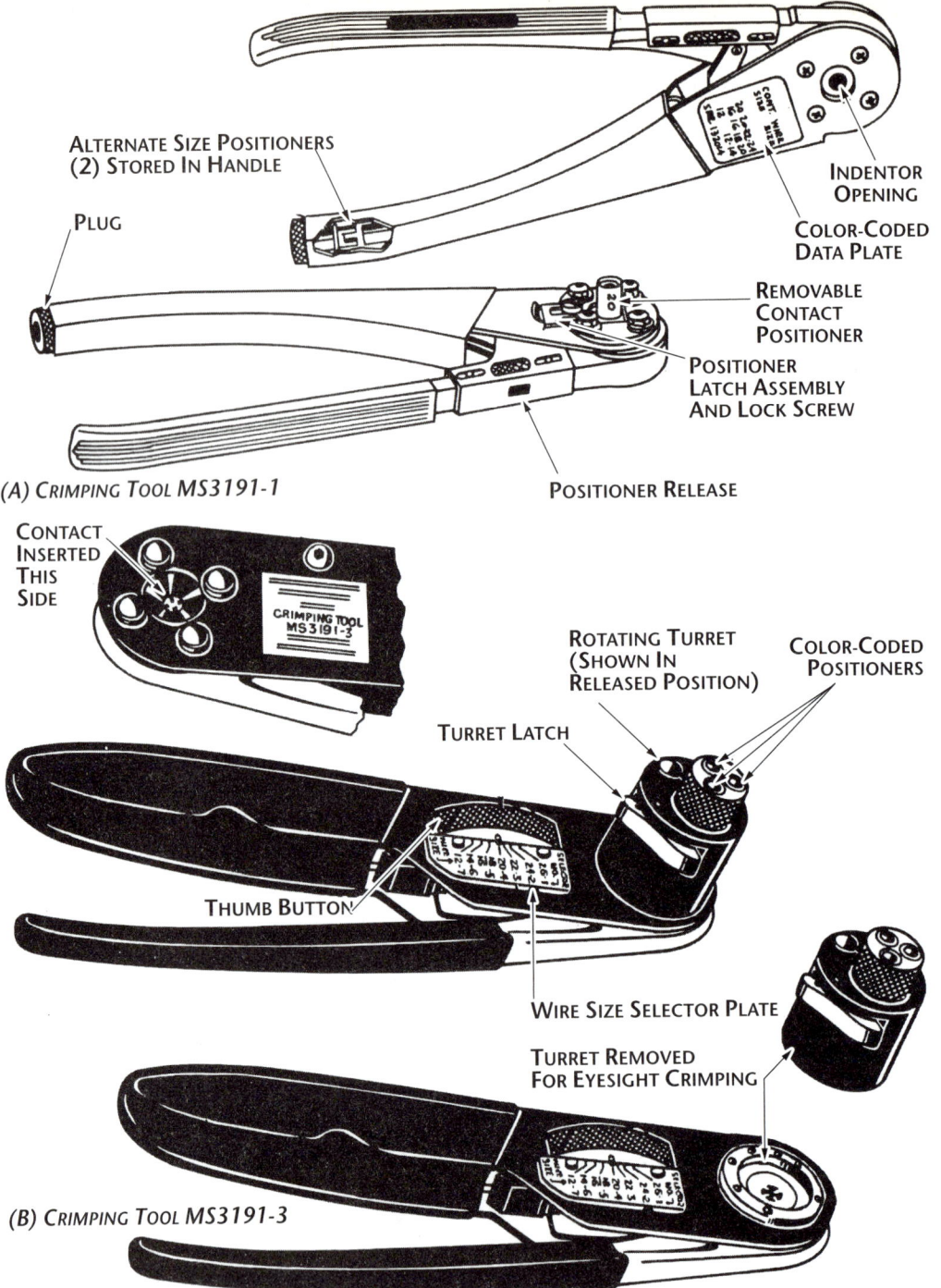

Figure 2-68. MS3191 standard crimping tools

gage through the positioner and into the crimping dies. The GO gage should enter the positioner and dies freely so that the gage handle seats firmly on the top of the positioner. See Figure 2-69.

4. The NO-GO gage should not be able to enter into the crimping dies, and the gage handle does not seat on the positioner.

CAUTION: *Do not crimp down on the gage pin as this will prevent the tool from cycling to ratchet release position.*

Inspecting MS3191-3 crimping tool. Only the No. 20 wire size indent or setting is gaged for this tool. The procedure is as follows:

1. Slide the thumb button until the pointer is in line with the No. 20 wire size on the selector plate. (Refer to Figure 2-68b).

2. Close the handles to fully closed position.

3. The GO pin of MS3196-20 gage should pass freely through the indentor tips.

General Purpose Connectors | 2-43

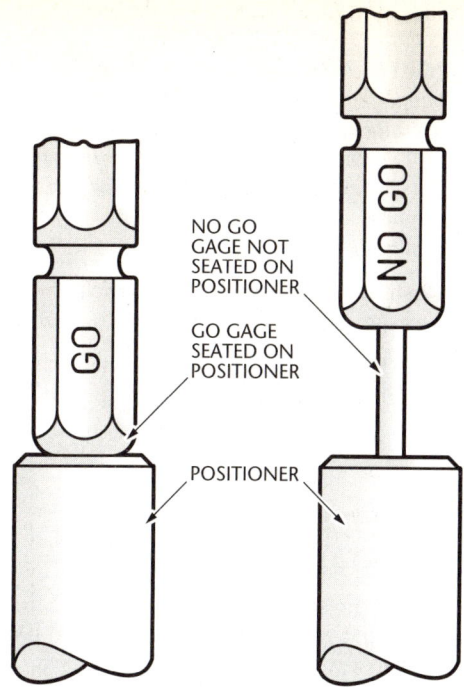

Figure 2-69. Gaging MS3191-1 crimping tool

4. The NO-GO gage pin should not pass between indentor tips.

CAUTION: *Do not crimp down on the gage pin as this will damage the indenters.*

Insertion Tools for Crimp-type Contacts

See Figure 2-70a. The MS tool for inserting contacts into connector inserts is MS24256A. There is a separate tool for each contact size as listed in Table 2-17. Inserting tools for contact sizes 16 and 12 have a hollow lip that fits snugly over the contact crimping barrel. The tool for size 20 contact fits over the insulation support. Contact sizes 16 and 12 do not have insulation support. An indicating band on the working end of the tool determines correct depth of tool insertion. Use this tool to insert contacts in all MS miniature connectors with removable contacts.

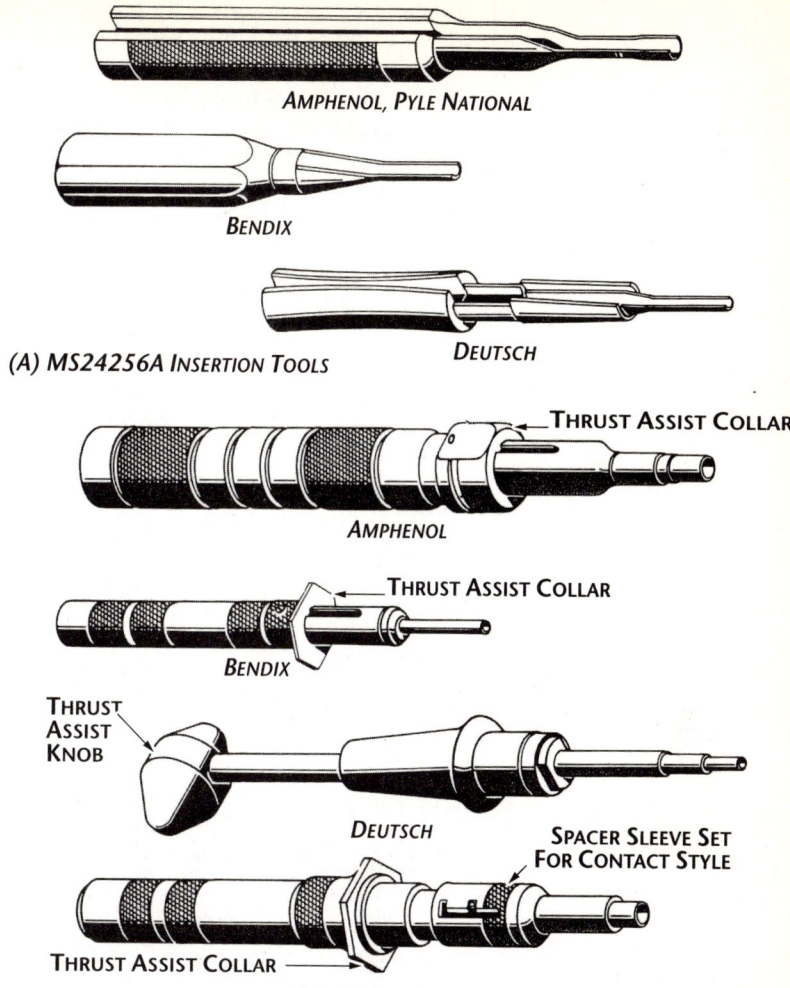

Figure 2-70. Insertion and extraction tools for crimp-type contacts

Extraction Tools for Crimp-type Contacts

See Figure 2-70b. The MS tool for extracting contacts from connector inserts is MS24256R. The sizes for each contact are listed in Table 2-17. This tool has a hollow cylindrical probe that fits snugly over the pin or socket end of the contact, and releases the insert retention clip

Connector and Contact Size		Color Code	Crimping Tool	Insertion Tool	Extraction Tool
MIL-C-26482	20	Red	MS3191-1 or 3	MS24256A-20	MS24256R-20
	16	Blue	MS3191-1 or 3	-16	-16
	12	Yellow	MS3191-1 or 3	-12	-12
MIL-C-26500	20	Red	MS3191-1 or 3	MS24256A-20	MS24256R-20
	16	Blue	MS3191-1 or 3	-16	-16
	12	Yellow	MS3191-1 or 3	-12	-12

Table 2-17. Connectors with crimp-type contacts and assembly tools

when pushed over the contact. Two indicating bands determine correct depth; the band nearest the working end of the tool is for pin contacts, the other for socket contacts. The extraction tool has a thrust assist collar (or slide) that is pushed forward to eject the contact from the insert retention clip by means of an internal plunger. Use this tool to remove contacts from all MS miniature connectors with removable contacts.

Crimping Procedure

Crimping procedure for MS3191-1 tool. The procedure for assembling wires to contacts with the MS3191-1 hand crimping tool is as follows (see Figure 2-71):

1. Open the crimping tool by exerting pressure sure on the handles until the ratchet releases.

2. Loosen the latch locking screw, and pull the latch to the open position, (refer to Figure 2-68a).

3. Pull the positioner release all the way down against spring pressure and insert the correct positioner for the contact being crimped. Insert the positioner so that the flat on its flange mates with flat on the handles, and flange is flush with the handle.

NOTE: *Positioners are stored in the tool handle, and are stamped with contact number, and color coded to match the color code on the data plate on the face of the tool. Pull the spring loaded plug to remove (or replace) the positioners. Store positioners not being used in handle.*

4. Push the positioner latch to the closed position, and tighten the latch locking screw.

5. Strip the wire using any of the methods described in Chapter 1. Stripping lengths are:

   Contact size 20 - .157 to .186

   Contact size 16 - .250 to .284

   Contact size 12 - .250 to .284

6. Insert the stripped wire into the contact until end shows through inspection hole and insert both through the indent or opening into the positioner.

NOTE: *When crimping size 20 contacts, make sure that insulation extends into insulation support of contact.*

7. Squeeze the tool handles with one firm stroke until the positive stop is reached. The ratchet will then release and the tool will open. Remove the crimped contact and wire.

Crimping procedure for MS3191-3 tool. The procedure for assembling wires to contacts with the MS3191-3 hand crimping tool is as follows:

1. Release turret by depressing turret latch with thumb, and rotate turret until correct positioner for contact being crimped lines up with index line on head assembly. (Refer to Figure 2-68b).

2. Push turret down into latched position.

3. Slide thumb button until pointer is in line with wire size being used, as marked on selector plate.

4. Strip wire as described in step 5 above.

5. Insert the stripped wire into the contact until end shows through assembly hole. Insert wire and contact through the indent or opening into the positioner.

NOTE: *When crimping size 20 contacts, make sure that insulation extends into insulation support of contact.*

6. Squeeze the tool bandies as far as they will go. The ratchet will then release and the tool will open.

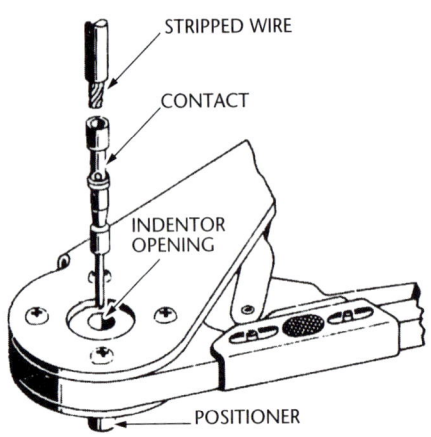
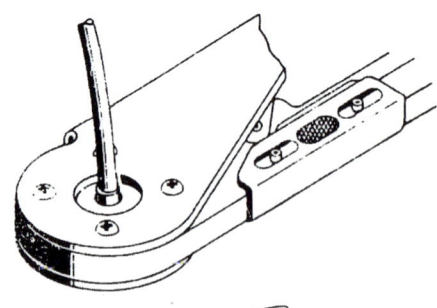
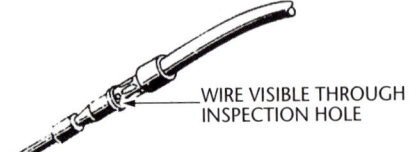

Figure 2-71. Assembling wires to crimp-type contacts

7. Remove crimped contact and inspect.

   NOTE: *Provision is made for safety-wire locking both head and thumb button for production runs of same size contact and wire.*

Eyesight crimping. Short contacts may be crimped in the MS3191-4 tool (MS3191-3 with the turret removed.) The procedure is as follows:

1. Release the turret by depressing the turret latch.

2. Loosen the two retainer screws with the 1/8 Allen wrench supplied with the tool. Remove the head from the tool frame.

3. Slide the thumb button until the pointer lines up with the wire size to be crimped.

4. Insert the contact into the tool, and slowly close the handles, at the same time positioning the contact so the indentors will crimp midway on the crimp barrel.

5. Hold the contact lightly with the indent or tips, and insert the stripped wire into the contact.

6. Make sure the wire is bottomed in the contact, and close the handles all the way.

7. Remove the crimped contact, and inspect.

8. Check to make sure that the wire is visible in the inspection hole.

Assembling Wired Contacts into Connector

Insert the crimped contact into the connector as follows (see Figure 2-72):

1. Slide rear accessories back onto wire bundle.

2. Select the correct insertion tool from Table 2-17. Insert the crimped end of the contact into the hollow end of the insertion tool, and lay the wire into the slot in the tool handle.

3. Guide the contact into the correctly numbered grommet hole in the rear face of the insert, and feed the contact carefully into the hole.

4. Push the tool straight in at right angles to the grommet surface, until the contact is fully seated. At the indicator band on the tool enters the grommet hole the contact retention clip will snap into place on the contact with a slight audible click.

5. Withdraw the tool, keeping it perpendicular to the grommet face.

6. Fill all unused holes with sealing plugs of appropriate size.

CAUTION: *Do not attempt to reseat a contact once the insertion tool has been removed. Remove contact and start over again with contact barrel properly located in tool. Failure to follow this precaution will cause insertion tool to shear barrel while inside grommet. Sharp edge of sheared material will cut through grommet web and cause short circuit.*

Alternate contact assembly procedure. If desired, the following procedure maybe used to insert wired contacts into the connector:

1. Push the wired contact carefully into the correct grommet hole. Do not push all the way in.

2. Slide the insertion tool over the contact barrel. When installing a size 20 contact, the tool internal shoulder will butt against the rear of the contact insulation support. When installing size 16 or 12 contacts the tool tip will butt against the contact shoulder.

3. Complete the procedure by following steps 4, 5 and 6 on the previous page. Always observe the related caution note.

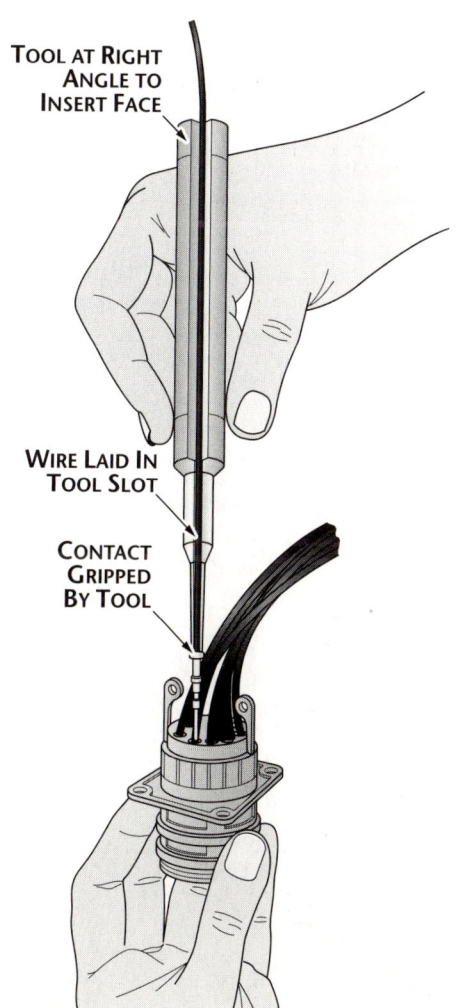

Figure 2-72. Assembling wired contacts into connector

Removing contacts from connector. Remove contacts from the connector as follows (See Figure 2-73):

1. Select the correct extraction tool for the contact to be removed from Table 2-17.

2. Slide rear accessories back on the wire bundle.

3. Working from the front or mating end of the connector, slip the hollow end of the extraction tool over the contact, with the tool parallel to the contact, and squarely perpendicular to the insert face.

NOTE: *The Pyle-National extraction tool has a spacer sleeve with positions for either male or female contacts. Set to correct position before installing tool on contact. Refer to Figure 2-70b.*

4. Push the tool toward the rear of the connector with a firm steady push until the tool comes to a positive stop and bottoms in the insert hole. A slight rotation of the tool may aid the tool insertion.

5. Push the thrust assist collar or slide forward as far as it will go.

6. Withdraw the tool from the contact, keeping the tool perpendicular to the insert face.

7. Remove the contact from the back of the connector.

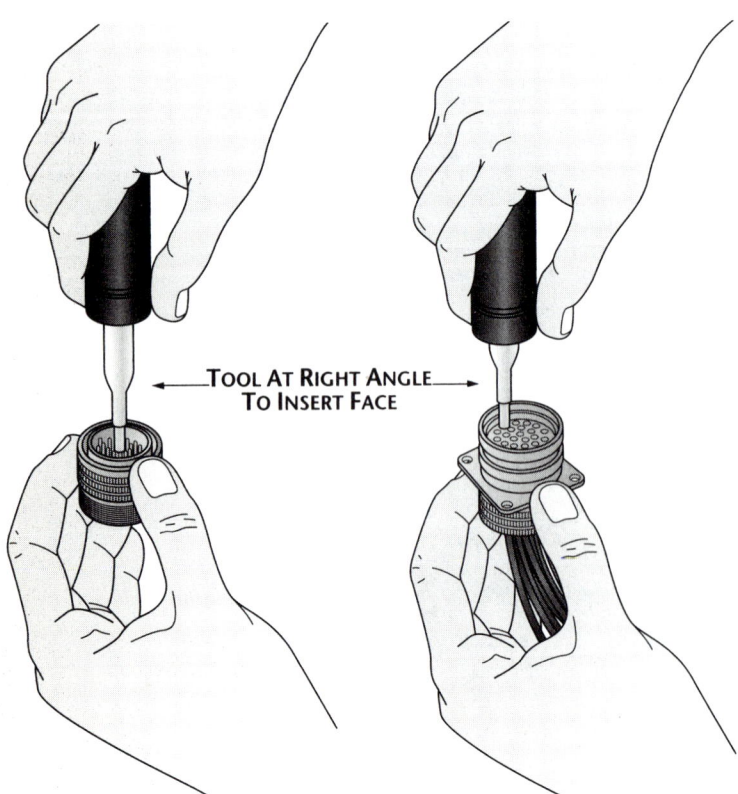

Figure 2-73. Removing crimp-type contacts from connector

CAUTION: *Make sure the extraction tool is always exactly aligned with the contact to avoid damage to the contact or to the insert.*

Non-standard connectors with removable crimp-type contacts. In addition to the MS connectors described on pages 2-4 and 2-6, there are four series of non-standard connectors also having removable crimp-type contacts. These are assembled in the same way as the corresponding MS standard connectors, except that tools used are made by the manufacturers of the connectors. These non-standard connectors are as follows:

- **Amphenol #69 series.** Similar to the large standard MS connectors, and mating with them. Applicable tools are listed in Table 2-18.

- **Bendix #10-214000 series.** Similar to the large standard MS connectors, and mating with them. Crimping tools are listed in Table 2-19.

CAUTION: *Size 16 pin and socket contacts are available in long and short lengths. Care should be exercised to prevent mismating one to the other.*

- **Cannon EXA series.** Similar to the AN type Class R connectors and mating with them. Tools listed in Table 2-17 are used for assembling these connectors.

- **Bendix CE series.** A miniature series of connectors, available in shell styles similar to those of MS miniature connectors made by Bendix. Applicable tools are listed in Table 2-20.

Installation of Bendix "CE" series connectors is as follows:

1. Strip the wire insulation to 1/8 inch for size 20 contacts, and to 1/4 inch for size 16 contacts.

2. Install the correct positioner from Table 2-20 into the crimping tool and crimp the contact to the wire.

3. To insert contacts into the connector select the correct insertion tool by referring to Table 2-20. Slide rear accessories on the wire bundle. Grasp the crimped end of the contacts with the pliers so that the shoulder in the tip of the tool butts against the end of the wire well.

4. Align the contact with the hole in the rear face of the insert and push forward in line with the hole until the contact is felt to snap in position. A slight increase in resistance may be noticed just before the contact reaches its seated position.

NOTE: *The contact may be positioned by hand in the corresponding hole in the grom-*

met before final seating with the insertion tool, to insure proper alignment of the contact during insertion.

5. Remove the insertion tool tips from the grommet by releasing the holding pressure on the handles and pulling straight to the rear.

6. Fill all unused insert holes with grommet sealing plugs of appropriate size.

7. To remove contacts, push contacts from the front face of the connector back through the grommet using the correct tool from Table 2-20.

Cannon DPD connectors. Cannon DPD connectors have rectangular shells. Their main use is for rack and panel plug-in assemblies. These connectors have many applications where it is desirable to combine coaxial cable disconnects with general wiring disconnects in one unit. See Figure 2-74. Installation is as follows:

1. Thread bushings (AN3420) if required, into end bell from inside so that flange of bushing rests flush against inside of rear wall of end bell.

2. Slide end bell assembly on wire bundle.

NOTE: *Steps 3 through 8 are only for connectors which have coaxial contacts.*

Contact Size	Crimping Tool	Insertion Tool	Extraction Tool
16	MS3191	294-39	294-40
12	—	—	—

Table 2-18. Installation tools for crimp-type contacts - Amphenol #69 series

Contact Size	Crimping Tool	Locator
16 M	11-6941-1	11-6932-1
16 M long	11-6941-1	11-6932-2
12	11-6941-1	11-6932-3

Table 2-19. Crimping tools for crimp-type contacts Bendix #10-214000 contacts series

Contact Size	Crimping Tool	Positioner	Insertion Tool	Removal Tool*	
				Pins	Sockets
20	MS3191-1	11-7771-5	11-6782	11-6783	11-6784
16	MS3191-1	11-7771-6	11-6781	11-3697	11-3698

*With Handle 11-6911 or 11-3699

Table 2-20. Installation tools for crimp type contacts - Bendix CE series

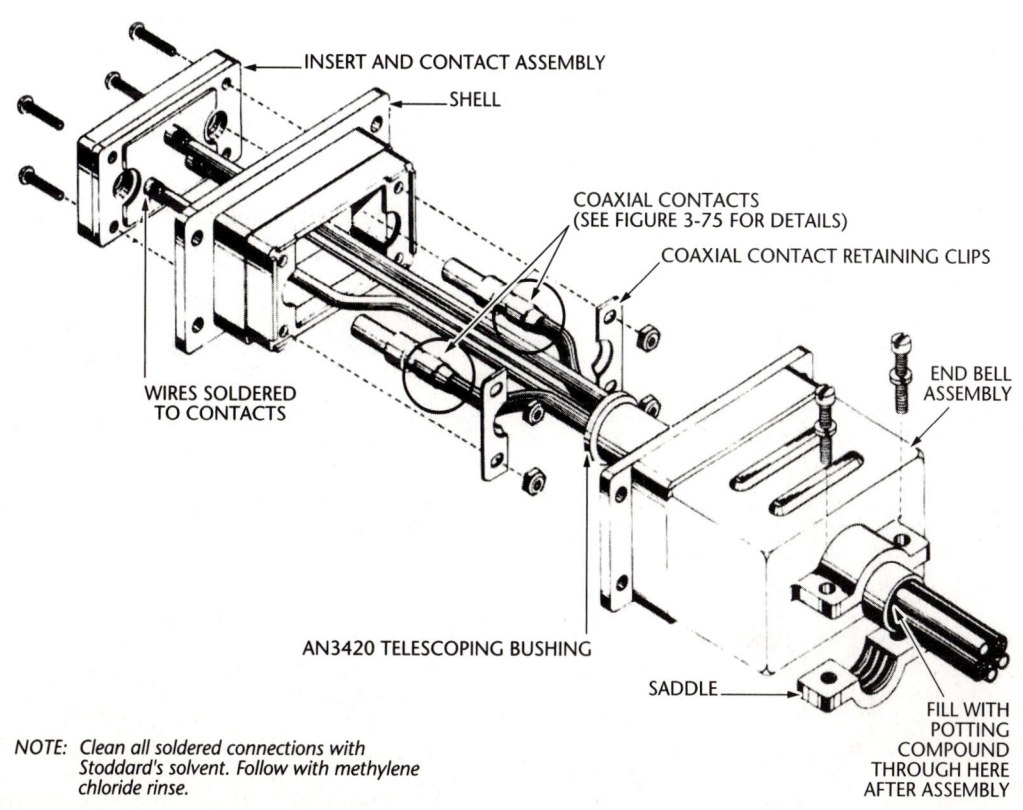

NOTE: Clean all soldered connections with Stoddard's solvent. Follow with methylene chloride rinse.

Figure 2-74. Installation of Cannon OPD connector

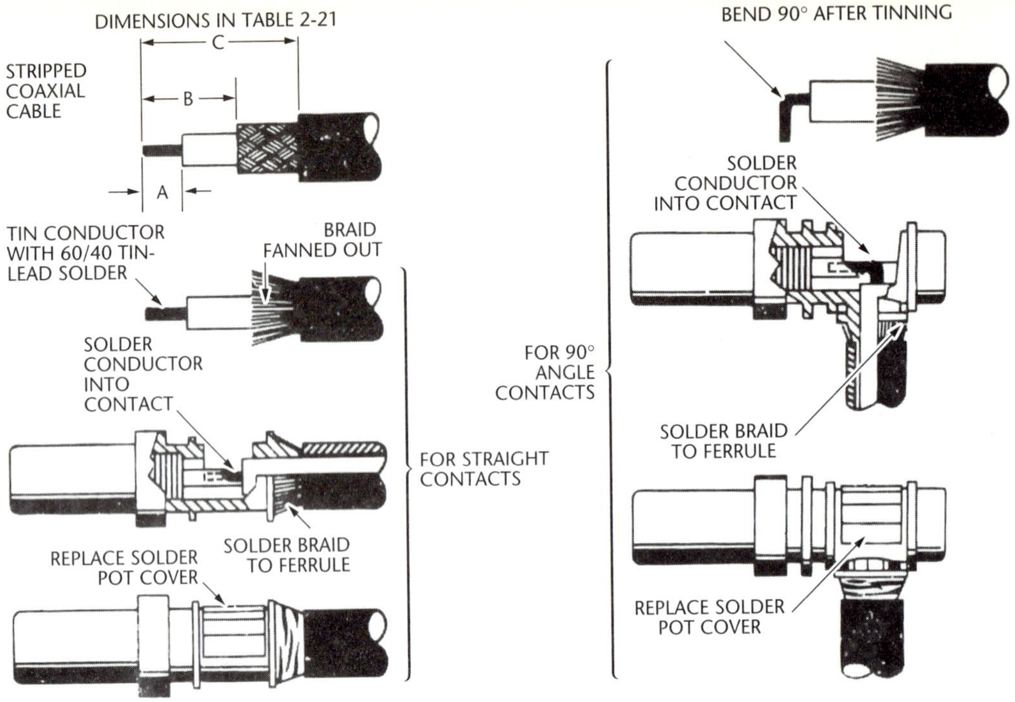

Figure 2-75. Soldering coaxial cable to contacts for DPD connector

3. Remove stop nuts on rear of shell to free coaxial contact retaining clips. Coaxial contacts are removed from rear of insert. See Figure 2-75.

4. Strip and prepare coaxial cable following procedures given on page 3-3. See Table 2-21 for stripping dimensions.

5. Remove solder pot cover from contact by prying out with knife.

6. Insert cable and solder center conductor to contact. The dielectric should butt against contact solder pot. Remove flux with Stoddard's solvent.

7. Replace solder pot cover and solder shield to ferrule.

8. Remove flux with solvent. Do not allow solvent to flow under solder pot cover.

9. Solder general purpose wires to remaining contacts in insert.

10. Clean rear of insert by brushing with Stoddard's solvent. If rear of connector is to be potted, follow cleaning with a methylene chloride wash as described in Figure 2-53.

CAUTION: *Methylene chloride vapors are toxic.*

11. Reinstall coaxial contacts and replace retaining clips. Replace stop nuts to hold insert and coaxial contacts in shell.

12. Slide end bell into position and mount to flange of shell.

NOTE: *Steps 13 and 14 apply only to connectors that are to be potted.*

13. Push nozzle of potting gun down through telescoping bushing (AN3420) and fill connector back with potting compound. Fill slowly to be sure all air is expelled.

14. Allow potting compound to cure for 24 hours at room temperature (70°F to 75°F).

15. Tighten saddle to compress telescoping bushing (AN3420).

Subminiature connectors. Cannon connectors, UCS, UDS and UES are simple subminiatures. Their disassembly was described on page 2-19. Solder wires carefully to the contacts using a pencil type resistance soldering iron as shown in Figure 2-31. Clean off excess flux, and reassemble connector by reversing the disassembly procedure. Potting compound is forced through the back shell to provide a moisture

Contact Type	Cable	Strip Length (Inches)		
		A	B	C
Straight	RG-58/U	1/8	15/32	5/8
	RG-59/U	1/8	15/32	5/8
	RG-62/U	1/8	15/32	5/8
90° Angle	RG-58/U	3/16	15/32	19/32
	RG-59/U	3/16	15/32	19/32
	RG-62/U	3/16	15/32	19/32

Table 2-21. Stripping dimensions for coaxial cable

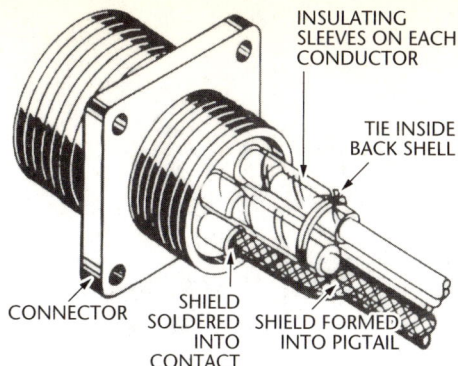

Figure 2-76. Terminating shielded wire at MS connector

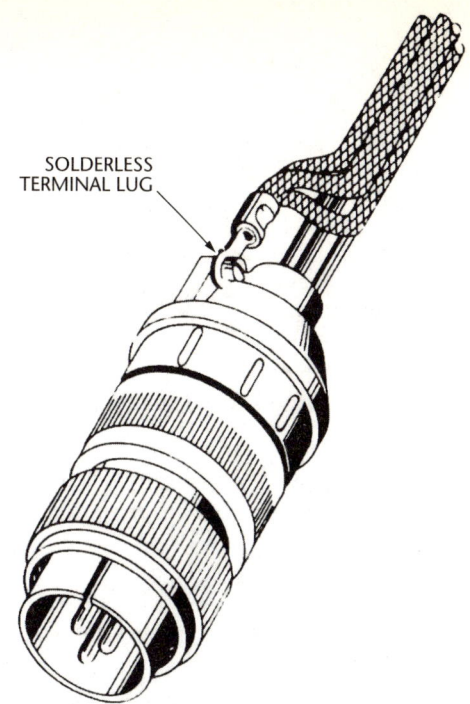

Figure 2-77. Grounding shields outside connector

and vibration proof seal. Directions for potting are described on pages 2-32 through 2-34.

Miniature rectangular connectors. These connectors consist of a one piece body into which the contacts are molded. There is no back shell, and no disassembly is required. A hood with cable clamp is used on either plug or receptacle to provide protection and support for the wires. Install the hood over the wire bundle, and solder wires to contacts using a resistance soldering unit. After the solder connection has cooled attach the hood to the connector with screws of the proper size. These connectors are also available with taper pin contacts.

Shield and Multiple Connections

Connecting single shielded wire to MS and potted connectors. Terminate single shielded wires as described Chapter 1, page 1-19. For connection to Class A, B, C or K connectors, shield must end inside back shell as shown in Figure 2-76. For connection to Class E, Class R or potted connectors shield must end outside seal. Crimp pigtail into terminal lug and ground to screw as shown in Figure 2-77, or use permanent splice to join pigtail to short length of AN wire which is then terminated inside connector to contact in regular manner, as shown in Figure 2-78.

Connecting several shielded wires to MS and potted connectors. Potted connectors that contain shielded wires and all other connectors which have many shields must terminate shields outside the connector. The procedure follows:

1. Form pigtail from shield outside connector area. This procedure is described in Chapter 1 on page 1-22.

    a. For potted connector, pigtail should start 1" from end of wire.

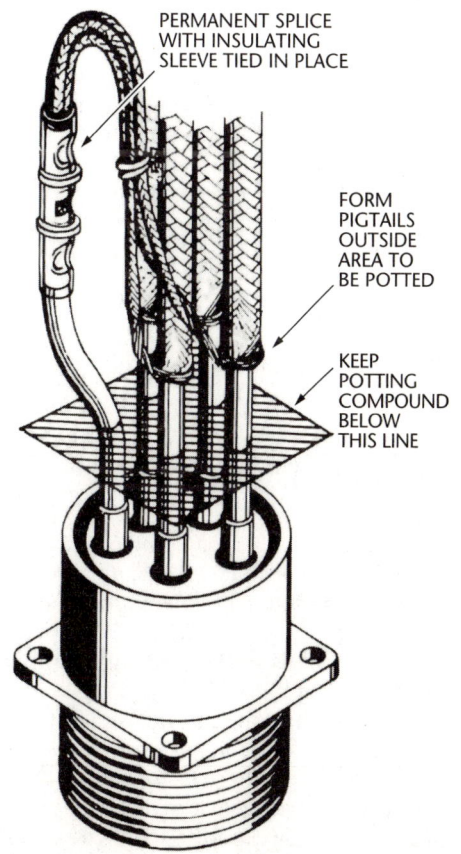

NOTE: Potting mold omitted for clarity.

Figure 2-78. Terminating shielded wires at potted connector

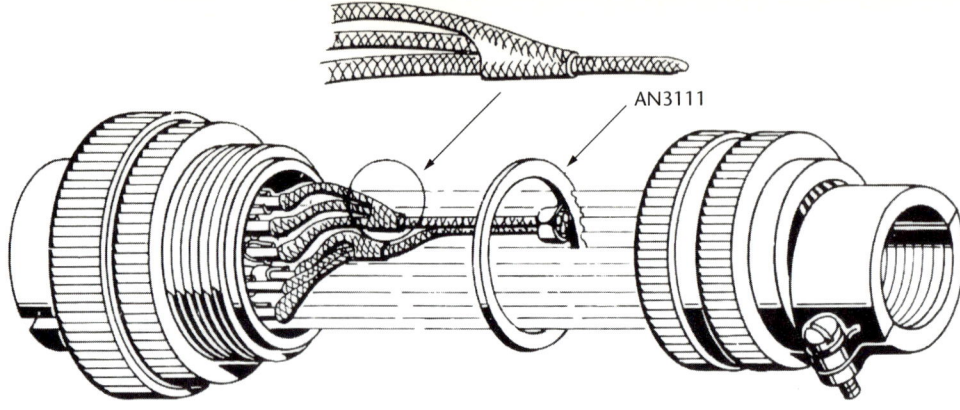

Figure 2-79. Installing AN3111 bonding ring

    b. For other connectors, pigtail should start far enough back to remain outside cable clamp.

2. Solder each wire to its contacts.
3. Crimp pigtails together into one end of permanent splice.
4. Crimp single wire, (doubled if necessary- see Chapter 4) into other end of permanent splice.
5. Slide insulating sleeve over splice and tie in place as shown in Figure 2-78.
6. Solder single wire to proper contact in connector as shown in Figure 2-78.
7. Complete connector assembly in normal manner.

An alternate method for grounding utilizes a screw on the cable clamp to ground the shields as follows:

1. Follow steps 1 and 2, above.
2. Crimp pigtails together into solderless terminal lug.
3. Attach terminal lug under screw, as shown in Figure 2-77.

NOTE: *If all shields will not fit into one solderless terminal lug, use several terminal lugs and distribute them under both screws.*

Grounding shields with bonding ring. When specified on the applicable engineering drawing, an AN3111 bonding ring maybe used to ground eight or more shields at an AN type connector.

1. Remove washers from the connector cable clamp, and slide clamp back on wire bundle.
2. Install the AN3111 bonding ring on the wire bundle between the connector back shell and the cable clamp, with the bonding ring lug toward the solder contacts.
3. Make a hole in one shield and expand it to hold up to three other shields as shown in Figure 2-79. Tighten the expanded shield around the others and sweat solder together. Repeat as necessary for number of shields to be grounded.
4. Pull one of the shields through to form a jumper. Install a length of insulating sleeving over the jumper shield.
5. Clamp the ears of the bonding ring lug around the jumper(s), and solder. Pull the insulating sleeve over the soldered connection.

Connecting two wires to one contact. Connect two wires to one contact by using one of the following methods:

1. If both wires can be fitted into contact solder cup, proceed as with single wire. Slide insulating sleeve over both wires together and insert them into solder cup. Make sure all strands are inside cup

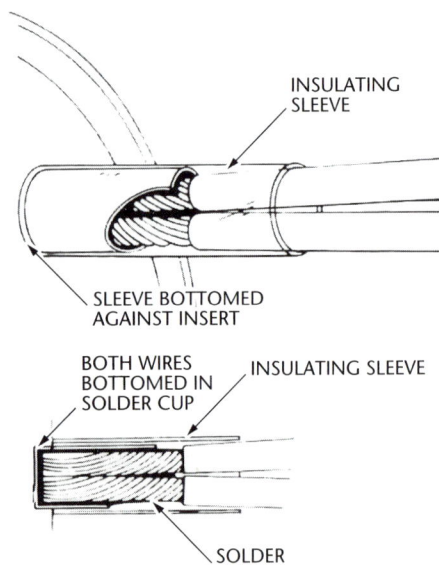

Figure 2-80. Terminating two wires at one contact

before soldering. When solder has cooled push insulating sleeve down until it butts against insert. See Figure 2-80.

CAUTION: *Avoid connecting two wires to one contact in Class E and R connectors; this will cause loss of moisture proofing.*

2. If both wires cannot fit into contact solder cup, use permanent splice to join both wires to a third wire which can fit into solder or crimp cup. See page 4-19 for splicing procedure and Figure 2-81 for illustration of this connection.

CAUTION: *The use of a single wire to terminate two wires at a connector must be approved by engineering.*

Reducing wire size at MS connector. Reduction of wire size to enable a larger diameter wire to be soldered to a smaller diameter contact solder cup is sometimes required. A safe method of making the reduction is as follows:

1. Select a permanent splice that will accommodate the larger wire. Crimp this splice to the stripped wire as described in Chapter 4 on page 4-18.
2. Select a six inch length of wire which will fit the cup of the contact. Strip one end sufficiently long to be able to double the stripped portion back on itself as shown in Figure 2-82.
3. Crimp this doubled wire into the free end of the permanent splice.

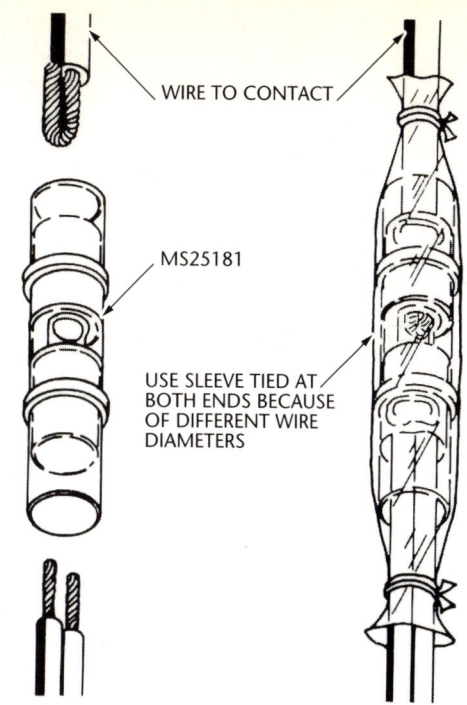

Figure 2-81. Permanent splice for terminating two wires at one contact

CAUTION: *Reduction of wire size needs engineering approval. Current carrying capacity of smaller wire or contact must not be exceeded.*

A second method for reducing wire size at an MS connector is by the use of an adapter as shown in Figure 2-82. Select an adapter to suit the reduction requirements from Table 2-22.

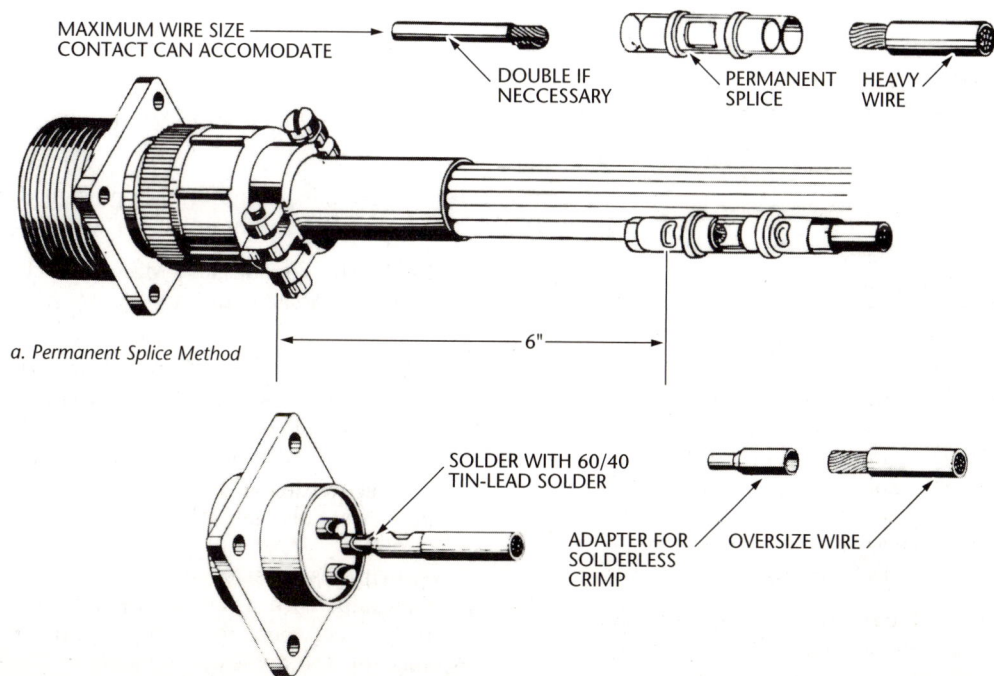

Figure 2-82. Reducing wire size at MS connector

Part Number	Wire Size	Contact Size	Crimping Tool
Thomas & Betts			
75-14586-1	#8 or #10	12	WT 130
C 503	#10 or #12	16	WT 130
75-14586-2	#12 or #41	16	WT 130
675-50588	#16 or #18	20	WT 111M
Bendix			
10-74696-6	#22	16	
-4	#16	12	
-14	#18	12	
-15	#20	12	
-1	#10	8	
-5	#12	8	
-12	#14	8	
-13	#16	8	
Burndy			
YE 8C12	#8 or #10	12	M8ND, N8CT-4 Die
YE 1216	#12 or #14	16	Y14 MRP
YE 1620	#16 or #18	20	Y16 TMR
YE 1220	#12 or #14	20	Y14 MRP

Table 2-22. Installation tools for wire-to-contact adapters

The procedure is as follows:

1. Strip wire to a length that the adapter wire well will accommodate.
2. Crimp the wire into the adapter, using the tool listed in Table 2-22 for that adapter.
3. Solder the adapter stem to the contact.

Protection of electrical connectors. See Figure 2-83. Protect all unmated MS connectors with protective covers. MS series protective covers and application to connectors is as follows:

- **MS3180.** For miniature MS plugs with bayonet coupling
- **MS3181.** For miniature MS receptacles with bayonet coupling
- **MS3182.** For miniature MS plugs with push-pull coupling
- **MS3183.** For miniature MS receptacles with push-pull coupling
- **MS25042.** For standard MS plugs (external thread coupling)
- **MS25043.** For standard MS receptacles (internal thread coupling)

Use a protective cover with MS dash number corresponding to the shell size of the connector to be protected. Protective covers are available with or without an attaching chain. Plastic dust caps to fit MS plugs (MS25177) and (MS25178) receptacles are also available.

Continuity test. Test all wires and wire groups as fabricated, with terminations attached, for continuity between the termination points specified on the applicable schematic. During the continuity test procedure observe the following precautions:

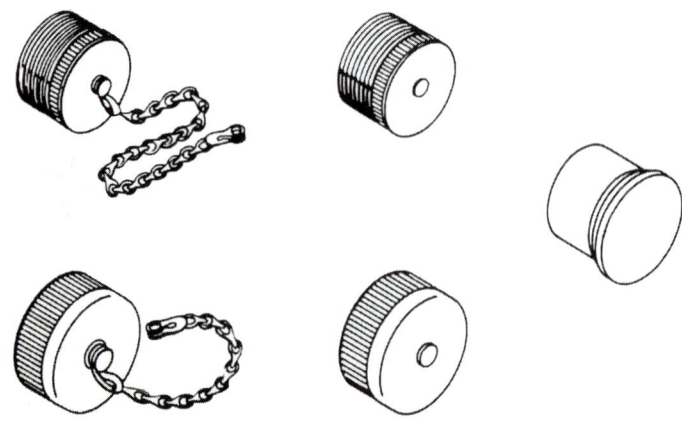

a. Metal Caps b. Plastic Cap

Figure 2-83. Protective connector caps

- Do not use lead pencils to count pins in connectors; points can break off and lodge in the connector, leading to arcing, shorting, and system malfunction.

- Do not use oversize prods in connector sockets during testing; this may result in splayed or damaged sockets.

- Do not puncture wire insulation with a probe, or attach clamps to wire insulation while continuity testing or troubleshooting.

Continuity test procedure. Use the ohmmeter section of an approved multimeter, such as the shown in Figure 2-84 to determine circuit continuity. Continuity for short runs, where conductor resistance is not a factor, is defined as "zero" resistance. The procedure for determining continuity using the multimeter is as follows:

1. Set the function control to OHMS and the range control to the 0-1,000 ohms range. Zero the instrument as directed in the operating manual for the instrument used.

2. Apply the test leads to the terminations of the wire run.

3. Note reading on the ohms scale. A reading of .25 ohms, ±.25 ohms is considered verification of circuit continuity.

NOTE: *The test lead extremities contacting the terminations under test must provide adequate constant contact, and must not damage the termination.*

Test leads. For ground points and terminal lugs, use test leads with alligator clips. For connector pins and sockets, use a special lead ending in a sleeve-insulated pin or socket of the same size as that being tested.

CAUTION: *Do not insert an oversize test probe into a connector socket, as this will result in a splayed or damaged contact. Do not hang a test lead from a pin contact as this will result in a bent pin.*

Figure 2-84. Multimeter for continuity test

Chapter 3

RF CONNECTORS *and* cabling

Coaxial cable assemblies are used to carry RF (radio frequency) power from one point to another with a known rate of loss. An assembly consists of RF connectors attached to coaxial cable. The coaxial cable described in this chapter is of the flexible, solid dielectric type, relatively small to medium size. The characteristic impedance of most of the cable is 50 ohms, but several of the cables listed have a characteristic impedance of 48, 53, 75 or 93 ohms.

This chapter describes and illustrates the MS series RF connectors and coaxial cable most commonly used in aircraft and the recommended methods for assembling coaxial cable to the connectors. Instructions for assembling miniature RF connectors to coaxial cable, and for RF connectors used in fuel quantity indicating systems are also covered. General procedures for installation of coaxial cable assemblies into aircraft are given in Chapter 10.

Learning Objectives:

- *RF Connectors*
- *BNC Connectors*
- *C Series Connectors*
- *HN Series Connectors*
- *N Series Connectors*
- *Pulse Series Connectors*
- *TNC Series Connectors*
- *RF Connectors*

Section 1

RF Connectors

Types of RF Connectors

RF connectors. RF connectors are available as plugs, jacks and panel jacks, and receptacles. Plugs and jacks are attached to the ends of coaxial cables; panel jacks and receptacles are mounted to panels or chassis. Some panel jacks are fastened by means of four screws through holes in a plate integral with the jack body (see Figure 3-1, next page). Other panel jacks are fastened by means of a single nut threaded over the jack body (see Figure 3-4). Panel jacks and recep-

Left. The assembly procedure for connectors differs with the type of cable used.

tacles may be either front or rear-mounted. Plugs always have male contacts; jacks and panel jacks always have female contacts. This section covers the following series of RF connectors:

- **BNC series.** See Figure 3-1. A small, lightweight, bayonet type, quick-connect/-disconnect connector, is used with small coaxial cables, where peak voltage is not more than 500 volts.

- **HN series.** See Figure 3-2. A high voltage (up to 5,000 volts), threaded coupling connector used with medium size coaxial cables.

- **N series.** See Figure 3-3. A general purpose, threaded coupling connector used with medium size coaxial cables.

- **C series.** See Figure 3-4. A bayonet type, quick-connect/-disconnect connector is used with medium size coaxial cables. It is electrically similar to the N series.

- **Pulse series.** See Figure 3-5. A high-voltage connector for pulse or DC applications. Designed for use with rubber-dielectric pulse cables, but may be used with equivalent-size cables of other construction where high voltage is not required. With ceramic inserts, peak voltage is 15,000 V at sea level. With rubber inserts, peak voltage is 5,000 V at 50,000 feet, but higher voltages may be used at lower altitudes.

- **TNC series.** See Figure 3-6. A small lightweight connector similar to the BNC series, but having a threaded coupling, used where a positive coupling under vibration and a low noise level is desirable.

- **SC series.** A connector used with medium size coaxial cables is similar to the C series, but has a threaded coupling.

Assembly Precautions and Procedures

Coaxial cable. Coaxial cable consists of an inner (center) conductor separated from the outer conductor, usually called a shield, by an insulating dielectric. The cable is protected against moisture and abrasion by a tough outer jacket (sometimes called a sheath). See Figure 3-7 for typical coaxial cables. The inner conductor is usually copper, either solid or stranded, and may be bare, tin plated or silver plated. The outer conductor (shield) is usually a copper braid, bare, tin plated or silver plated, woven over the dielectric. Some coaxial cables have a double outer conductor (double shield) to

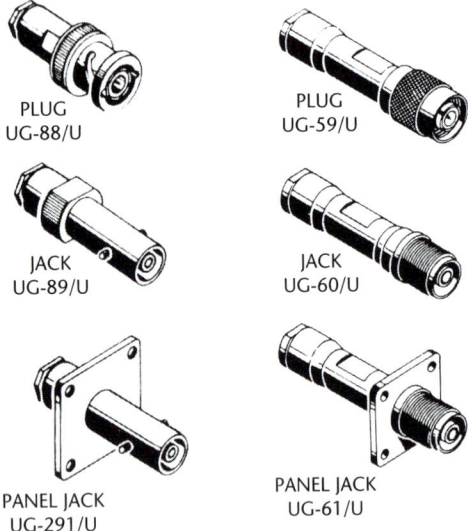

Figure 3-1. Typical BNC connectors

Figure 3-2. Typical HN connectors

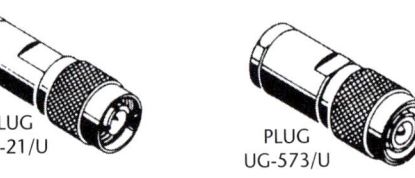

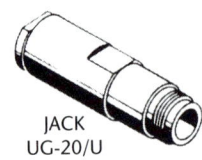

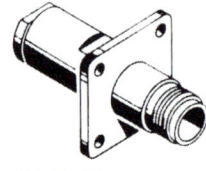

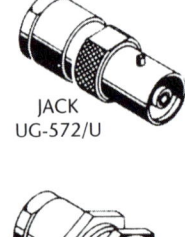

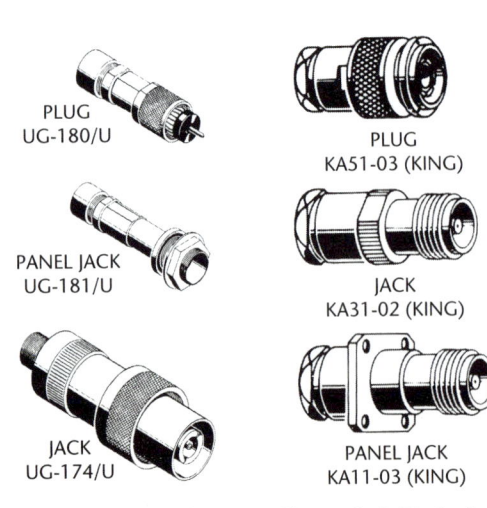

Figure 3-3. Typical N connectors

Figure 3-4. Typical C connectors

Figure 3-5. Typical pulse connectors

Figure 3-6. Typical TNC connectors

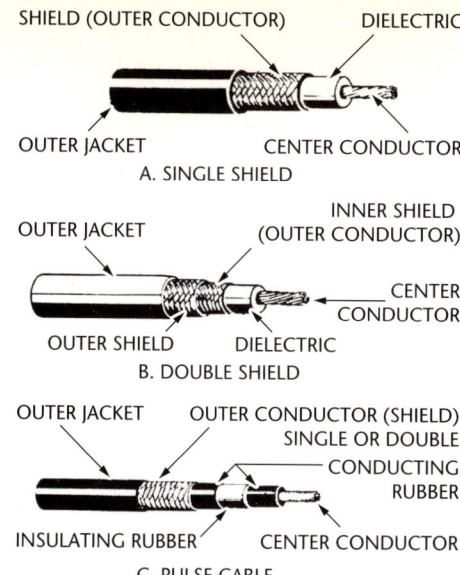

Figure 3-7. Typical coaxial cables

provide extra shielding. The dielectric has two functions: (1) it provides low loss insulation between the inner conductor and the outer conductor, (2) it maintains the relative position of the inner conductor inside the outer conductor and therefore keeps the capacitance between the two at a constant value.

General precautions. A good connection depends on holding coaxial cable and connectors to the design dimensions. Any change in these dimensions will cause added losses to the RF power being carried, and may cause radiation interference. It is important that the assembly directions given for each connection be followed carefully to avoid trouble. The following precautions are common to all assemblies of the coaxial cable and RF connectors.

When working with coaxial cable, never step on the cable, set anything heavy on it, or bend it sharply. This will flatten the cable and will change its electrical characteristics. Handle coaxial cable carefully at all times. Anything which damages it, or which might lead to its being damaged later, reduces the efficiency of the system.

Do not use pliers to assemble or disassemble RF connectors.

Contacts for RF connectors are usually packed unassembled. Do not misplace them.

When attaching connectors to coaxial cable having a double shield, make sure both shields are soldered together at connector.

Use care in starting the nut into a plug or jack body, in order to prevent cross threading.

Keep soldering iron clean, smooth, and well tinned at all times. See Chapter 7 for care of soldering iron.

General procedures. During the preparation of coaxial cable assemblies, observe the following general procedures:

1. Cut coaxial cable to length with long handled cable cutters or pruning shears, making sure cut is clean and square.

2. Identify cable by using the methods described in Chapter 1.

3. Strip outer jacket from cable by first making a cut carefully around circumference with a sharp knife. See Figure 3-8, step 1. Then make a lengthwise slit, step 2, and

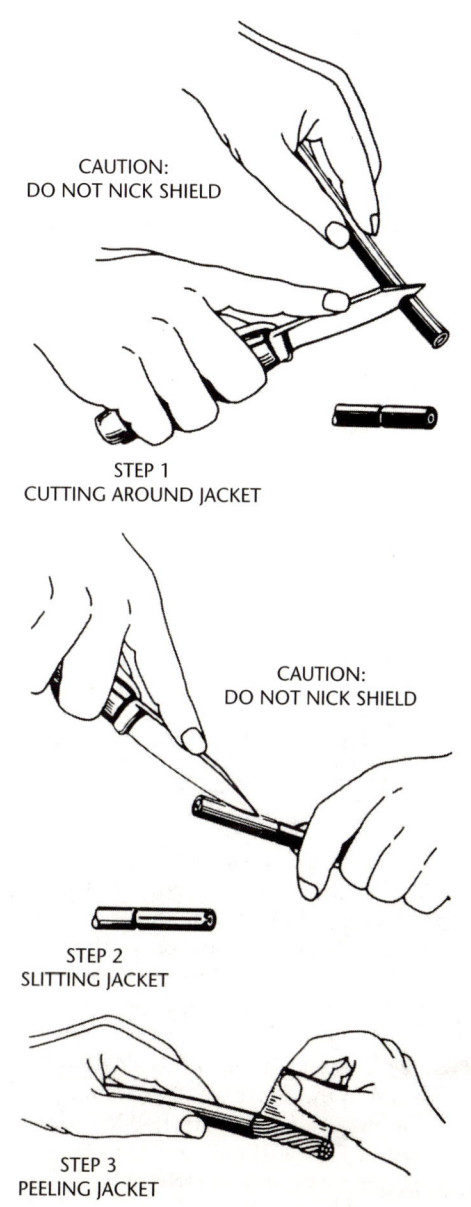

Figure 3-8. Stripping outer jacket from coaxial cable

3-4 | RF Connectors and Cabling

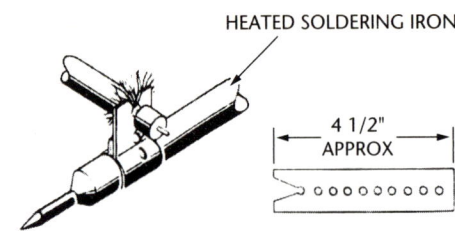

Figure 3-9. Improvised dielectric stripper

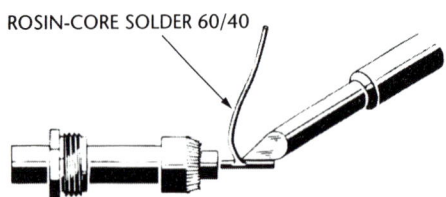

Figure 3-10. Tinning center conductor

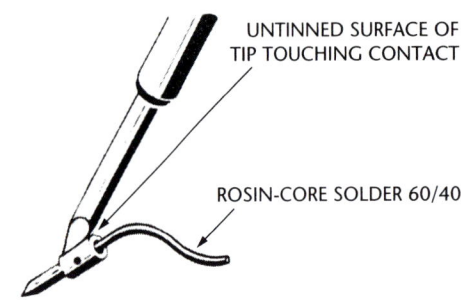

Figure 3-11. Tinning inside of contact

peel off jacket, step 3. Take care not to nick, cut or damage shield.

4. Comb out the braid by using a pointed wooden dowel or a scriber.

   CAUTION: *Do not damage dielectric or break shield strands.*

5. To remove dielectric, cut with sharp knife around circumference, not quite through to center conductor, taking care not to nick or cut strands, or otherwise damage conductor. Pull off dielectric. Another method of stripping the dielectric is to use a soldering iron to which a strip of sheet copper has been fastened as shown in Figure 3-9. The dielectric to be stripped is laid in the "V" and rotated. This will melt a clean break that permits the dielectric to be easily removed by hand.

6. Tin center conductor with soldering iron as shown in Figure 3-10.

7. Tin inside of contacts (male or female) with soldering iron as shown in Figure 3-11. (Use untinned face of tip to prevent depositing solder on outside of contact.)

8. Solder contacts (male or female) to center conductor with clean, well tinned, soldering iron using 60/40 tin-lead, rosin-core solder. See Figure 3-12.

   CAUTION: *Contact must butt flush against dielectric before and after soldering.*

9. When assembling the connector always start the clamping nut into the body by hand, and then hold the body assembly in a vise using lead or neoprene jaw protectors. See Figure 3-13. Hold body only on the flats. Do not use excessive pressure, since the body can be easily distorted. Tighten nut with end wrench.

Soldering Coaxial Cable to RF Connectors

Preparation of work. The work to be soldered must be clean and free from oxides. Remove

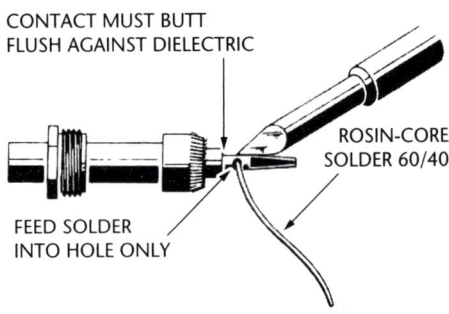

Figure 3-12. Soldering contact to coaxial cable

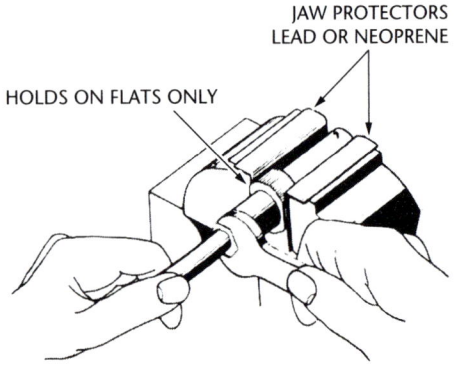

Figure 3-13. Tightening nut into plug or jack body

Figure 3-14. Correct shape for soldering iron tip

grease by cleaning with Stoddard's solvent or other approved cleaner. Oxides, if not too heavy are removed by the action of the rosin flux during the soldering operation. Heavily oxidized wire cannot be cleaned by the rosin flux and should be discarded.

Selection of soldering iron. For good soldering, it is important to select a soldering iron of the proper size and heat capacity. For soldering coaxial cable to RF connectors use an iron with a heating element rated at 65 to 100 watts, and a tip of about 1/4 inch diameter. The soldering tip should be shaped as shown in Figure 3-14. Maintain this shape by dressing the tip with a mill smooth file. Make sure the soldering iron is clean, smooth and well tinned. See Chapter 7 for detailed instructions on care and maintenance of the soldering iron.

> **NOTE:** *For soldering coaxial cable to RF connectors, tin only one face of the tip so that areas adjacent to that being soldered will not be coated with solder by accident.*

Soldering. See Chapter 7 for general soldering procedures and precautions for soldering cable in RF connectors. Use only 60/40 tin lead solder with a core of rosin flux. Heat the parts to be joined, and apply the solder at the junction of the soldering iron tip and the work as shown in Figure 3-12. Do not apply heat longer than is necessary to melt the solder. Excessive heat will swell the dielectric and make it difficult to insert into the body shell. Do not allow solder to flow over the outside of the contact. After the joint has cooled, remove excess flux by wiping with a clean cloth, using denatured alcohol as a solvent if necessary. Remove excess solder from the contact by scraping with a knife. Be careful not to cut into contact or dielectric.

Section 2
BNC Series Connectors

BNC Connector Types

There are two versions of BNC connectors, differing in the method of attaching coaxial cable to the connector body. See Figure 3-1 for typical examples of BNC connectors. Table 3-1 lists the more common connectors in the BNC series and shows the coaxial cables associated with each.

Improved version (see Figure 3-15). Consists of a plug or jack body assembled to coaxial cable with nut, grooved gasket and sleeve clamp. An insulation bushing is added where assembly is to RG-62/U or 71/U cables. Plug UG-88E/L and Jack UG-89C/U are typical of this version. The sleeve clamp has a sharp rear face, which cuts into the grooved gasket and thus makes a tight seal.

Plug	Jack	Panel Jack	For Use With Cable Types
Improved Version:			
UG-88E/U	UG89C/U	UG-291C/U	RG-55/U, 58/U and 223/U
UG-260D/U	UG-261C/U	UG-262C/U	RG-59/U, 62/U and 71/U
Captivated Contact Version (Amphenol):			
31-301	31-302	31-300	RG-55/U, 58/U, 141/U and 142/U
31-304	31-305	31-303	RG-59/U, 62/U, 71/U and 140/U

Table 3-1. BNC series connectors with associated cables

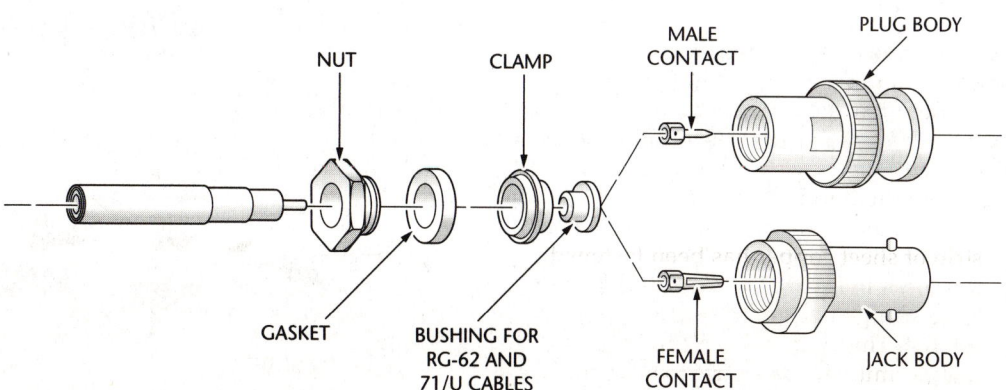

Figure 3-15. Improved BNC connectors - exploded view

Captivated contact version (see Figure 3-16). Similar to the improved version, but with the addition of bushing, front and rear insulators. When assembly is to RG-62 and 71/U cables, a bushing insulator is added between the bushing and rear insulator. This version does not have RG numbers.

NOTE: *Bushing and rear insulator used with RG-55, 58, 141 and 142/U coaxial cables differ from those used with RG-59, 62, 71 and 140/U cables.*

Assembly Precautions and Procedures

Attaching improved BNC connectors to coaxial cable. When attaching improved BNC connectors to coaxial cable (see Figure 3-17), follow this procedure:

NOTE: *While attaching connector, observe all general precautions and procedures listed on page 3-2.*

1. Remove 5/16 inch of outer jacket, exposing shield. See Figure 3-8.

CAUTION: *Do not nick shield.*

2. Comb out shield. Use care to prevent breaking shield strands.

3. Strip dielectric to 3/16 inch from edge of jacket, exposing center conductor.

CAUTION: *Do not nick center conductor.*

4. Disassemble nut, grooved gasket and sleeve clamp from plug or jack body. See Figure 3-15.

5. Taper shield toward center conductor and wrap a piece of thin pressure tape, wide enough to cover the combed out shield (one layer is sufficient) around the shielding, forming a cone with the narrow end toward the conductor.

6. Slide nut and gasket (V-groove away from nut) in that order over tapered shield onto jacket. Slide sleeve clamp over tapered shield until inside shoulder of clamp butts flush against cut end of jacket.

7. Remove tape from shield, comb shield back smoothly over sleeve clamp and trim to 3/32 inch with scissors.

8. Trim dielectric to 1/8 inch from shield, and cut off center conductor to 1/8 inch from edge of dielectric.

9. Tin center conductor as shown in Figure 3-10. Tin inside of contact (male or female) as shown in Figure 3-11.

10. Slip contact over center conductor so that contact butts flush against dielectric. For RG-62/U and 71/U add bushing. Solder, using a clean, well tinned, soldering iron-contact must still be flush against dielectric after solder has cooled; if it is not remake the joint. See Figure 3-12.

CAUTION: *Make sure that correct contact is used; a male contact always goes into a plug body, and a female contact always goes into a jack body.*

11. Push cable assembly into connector body as far as it will go. Make sure gasket is properly seated, with sharp edge of sleeve clamp entering gasket groove. Slide nut into connector body and fasten in vise. Start nut by hand and tighten with end wrench until enough pressure is applied to make a good seal by splitting the gasket. See Figure 3-13.

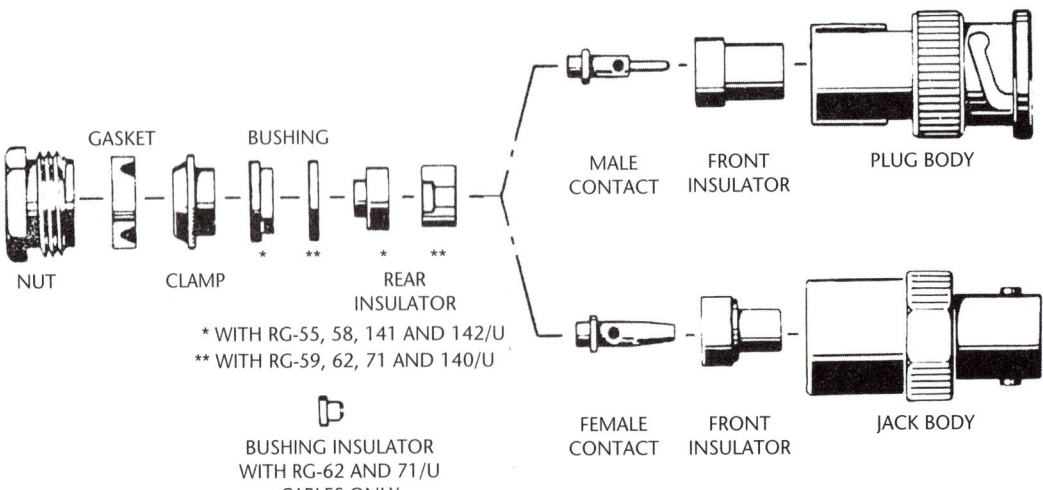

Figure 3-16. BNC connectors with captivated contacts-exploded view

RF Connectors and Cabling | 3-7

4. Disassemble nut, grooved gasket and sleeve clamp from plug or jack body. See Figure 3-16.

5. Taper shield toward center conductor, and wrap with tape as described on page 3-6. Slide nut and grooved gasket (V-groove away from nut) in that order over tapered shield onto jacket. Slide sleeve clamp over braid and push back against cable jacket.

6. Remove tape, comb shield back smoothly over sleeve clamp and trim to proper length; form evenly over clamp.

7. Tin center conductor as shown in Figure 3-10.

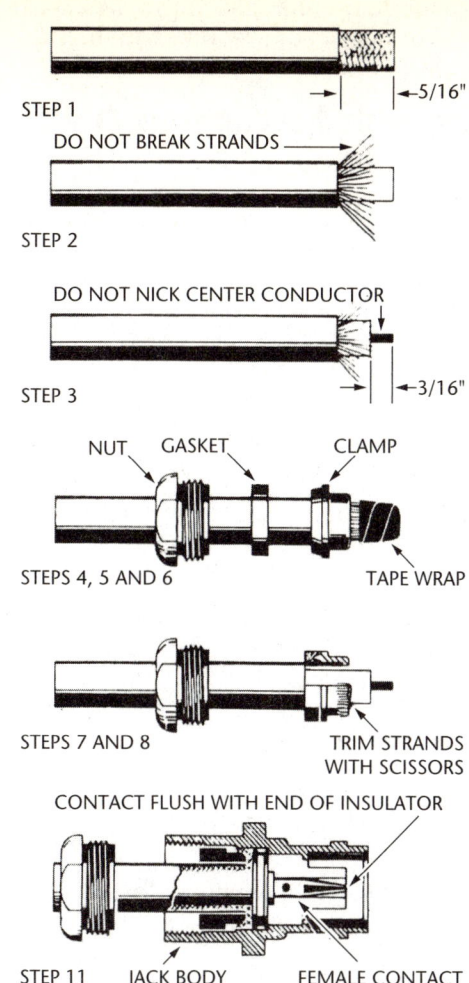

Figure 3-17. Attaching improved BNC connectors to coaxial cable

Attaching BNC connectors with captivated contacts to coaxial cable. When attaching BNC connectors with captivated contacts to coaxial cable (see Figure 3-18), follow this procedure:

> **NOTE:** *While attaching connector, observe all general precautions and procedures listed on page 3-2.*

1. Remove 3/8 inch of outer jacket exposing shield, for all except plugs 31-301 and 31-304; strip jacket for these plugs 27/64 inch. See Figure 3-8.

   CAUTION: *Do not nick shield.*

2. Comb out shield. Use care to prevent breaking shield strands.

3. Cut off cable dielectric to 3/16 inch for cables RG-55/U, 58/U, 59/U, 140/U, 141/U and 142/U. Cut to 5/32 inch for cables RG-62/U and RG-71/U.

   CAUTION: *Do not nick center conductor.*

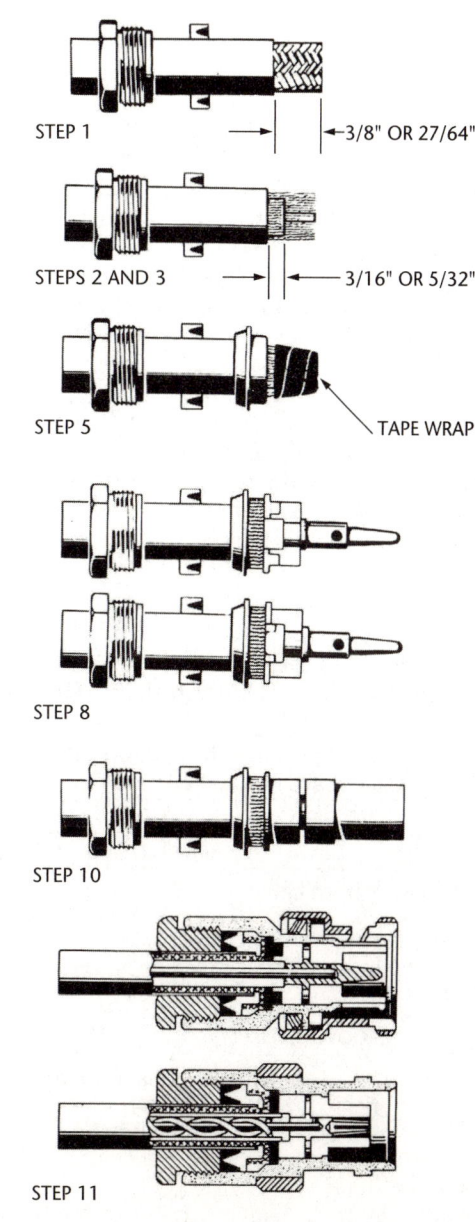

Figure 3-18. Attaching BNC connectors with captivated contacts to coaxial cable

8. Slide on bushing, rear insulator and contact. These parts must butt as shown. When attaching to cables RG-62/U and RG-71/U add insulator bushing.

9. Solder contact to center conductor. See Figure 3-12. Remove excess flux and solder from outside of contact.

10. Slide front insulator over contact and butt against contact shoulder. Do not reverse direction of insulator.

11. Insert cable assembly into connector body. Make sure that the sharp edge of the clamp seats properly in the gasket. Tighten the nut, holding the body stationary. See Figure 3-13.

Section 3
C Series Connectors

C connectors. (See Figure 3-4.) There is only one version of a C connector. It consists of a body assembled to coaxial cable by means of a nut, gasket and clamp. Table 3-2 lists the more common connectors in the C series and shows the coaxial cable associated with each. Plug UG-573B/U and Jack UG-572A/U are typical C Connectors. See Figure 3-19.

SC connectors. SC connectors are identical with the C connectors except for the coupling means. The procedure for attaching SC connectors to coaxial cable is the same as for C connectors. This procedure is described in the following paragraphs, and in Figure 3-20.

Assembly Precautions and Procedures

Attaching C connectors to coaxial cable. When attaching C connectors to coaxial cable (see Figure 3-20) follow this procedure:

NOTE: *While attaching connector, observe all general precautions and procedures listed on page 3-2.*

1. Disassemble nut, gasket and sleeve clamp from plug or jack body. See Figure 3-19.

2. Slide nut and gasket, in that order, onto jacket. Make sure grooved face of gasket is away from nut.

3. Strip outer jacket to "A" dimension given in Table 3-3, exposing the shield. See Figure 3-8.

CAUTION: *Do not nick shield.*

4. Comb out shield and cut dielectric to "B" dimension given in Table 3-3.

CAUTION: *Do not nick center conductor.*

5. Taper shield toward center conductor and wrap a piece of thin pressure tape, wide enough to cover the combed-out shield (one layer is sufficient) around the shielding, forming a cone with the narrow and toward the conductor.

6. Slide sleeve clamp over tapered shield until inside shoulder of clamp butts against cut end of jacket.

7. Remove tape and fold shield strands back over sleeve clamp taper without overlaps. Trim shield with scissors so that strands end at end of clamp taper.

8. Check that dielectric is exposed exactly to "C" dimension in Table 3-3.

9. Tin center conductor as shown in Figure 3-10.

10. Tin inside of contact (male or female) as shown in Figure 3-11.

11. Slip contact over center conductor so that contact butts flush against dielectric. Solder, using a clean, well tinned soldering iron; contact must still be flush against dielectric after solder has cooled; if it is not, remake the joint. See Figure 3-12.

CAUTION: *Be sure that correct contact is used; a male contact always goes into a plug body, and a female contact always goes into a jack body.*

Plug	Jack	Panel Jack	For Use With Cable Types
UG-573B/U UG-701B/U	UG-572A/U	UG-570A/U UG-571A/U	RG-8/U, 9/U, 213/U and 214/U
UG-626B/U	UG-633A/U	UG-629A/U UG-630A/U	RG-5/U, 6/U, and 212/U
UG-707A/U	—	—	—
—	—	—	RG-14/U and 217/U

Table 3-2. C series connectors with associated cables

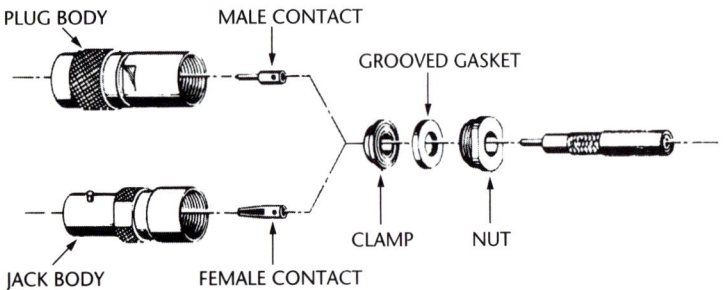

Figure 3-19. C connectors-exploded view

RF Connectors and Cabling | 3-9

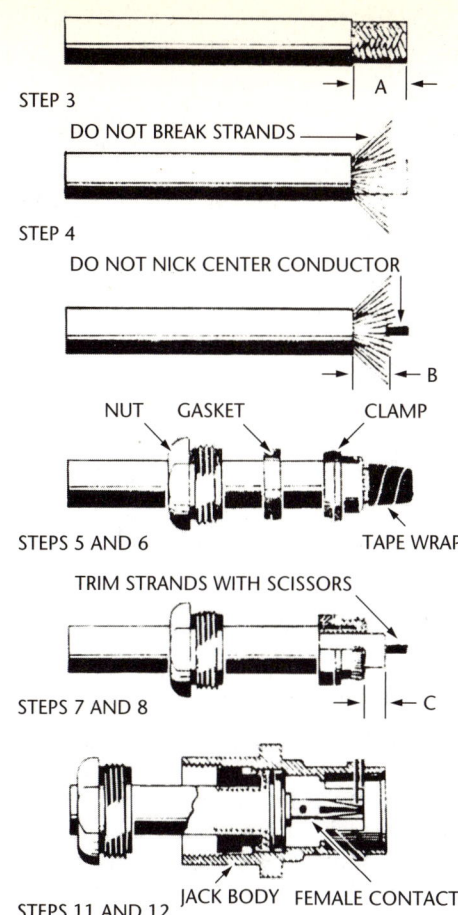

Figure 3-20. Attaching C connectors to coaxial cable

12. Push cable assembly into connector body as far as it will go. Slide gasket into connector body. Make sure gasket is properly seated, with sharp edge of sleeve clamp entering groove in gasket. Then slide nut into connector body and fasten body in vise. See Figure 3-13. Start nut by hand and tighten with end wrench until moderately tight. Gasket should be cut in half during tightening, In plugs, the end of the contact should be flush with insulator. In jacks there should be a clearance of 0.010 inch between end of contact and insulator.

Section 4
HN Series Connectors

HN Connector Types

There are two versions of HN connectors, differing in the method of attaching coaxial cable to the connector body. See Figure 3-2 for typical examples of HN connectors. Table 3-4 lists the more common connectors in the HN series and shows the coaxial cables associated with each.

NOTE: *The HN series of RF connectors are used for replacement purposes only.*

Improved version (See Figure 3-21). Consists of a plug or jack body assembled to coaxial cable with nut, gasket and braid clamp. Plug UG-59E/U and Jack UG-60E/U are typical of this version.

Captivated contact version (see Figure 3-22, next page). Consists of a plug or jack body

Connectors	Dimensions		
	A	B	C
UG-570/U jack	5/16	5/32	3/64
UG-571/U jack	5/16	5/32	3/64
UG-572/U jack	5/16	5/32	3/64
UG-573/U plug	5/16	5/32	3/64
UG-626/U plug	5/16	5/32	3/64
UG-629/U jack	5/16	5/32	3/64
UG-630/U jack	5/16	5/32	3/64
UG-633/U jack	5/16	5/32	3/64
UG-707/U plug	3/8	13/64	11/64
UG-710/U plug	5/16	5/32	3/64

Table 3-3 Assembly dimensions for C series connectors

Plug	Jack	Panel Jack	For Use With Cable Types
*UG-59E/U **UG-1213/U	*UG-60E/U **UG-1214/U	*UG-61E/U **UG-1215/U	RG8/U, 9/U, 213/U and 214/U
*Improved Version; **Captivated Contact Version			

Table 3-4. HN Series connectors with associated cables

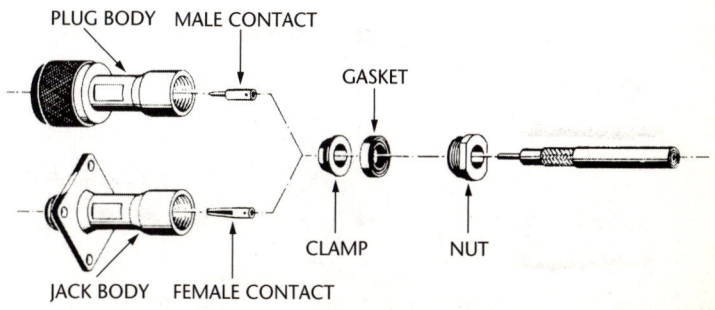

Figure 3-21. Improved HN connectors-exploded view

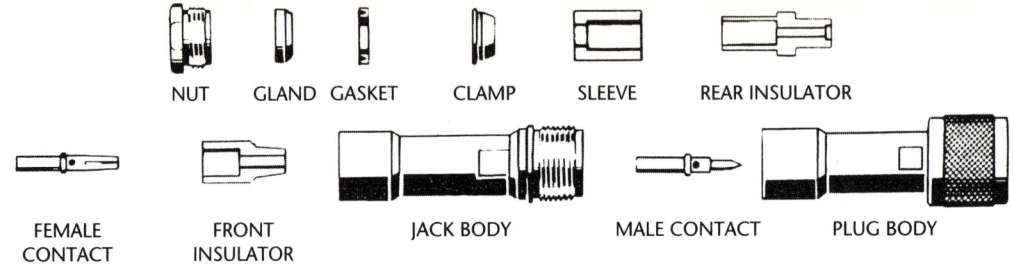

Figure 3-22. HN connectors with captivated contacts-exploded view

assembled to coaxial cable with nut, gland, gasket, clamp, sleeve and front and rear insulators. Plug UG-1213/U and jack UG1214/U are typical of this version.

Assembly Precautions and Procedures

Attaching improved HN connectors to coaxial cable. When attaching improved HN connectors to coaxial cable (see Figure 3-23), follow this procedure:

NOTE: *While attaching connector, observe all general precautions and procedures listed on page 3-2.*

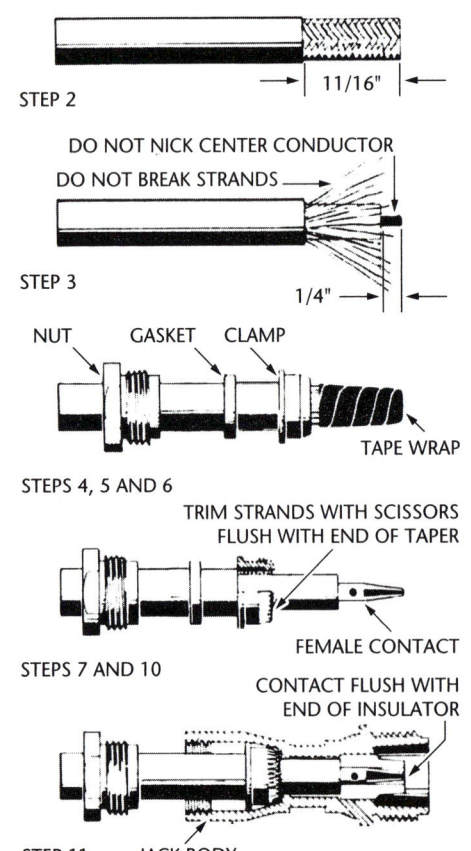

Figure 3-23. Attaching improved HN connectors to coaxial cable

1. Disassemble nut, grooved gasket and braid clamp from plug or jack body. See Figure 3-21.

2. Remove 11/16 inch from outer jacket, exposing shield. See Figure 3-8.

CAUTION: *Do not nick shield.*

3. Comb out shield and strip dielectric 1/4 inch.

CAUTION: *Do not nick center conductor.*

4. Taper shield toward center conductor and wrap a piece of thin pressure tape, wide enough to cover the combed-out shield (one layer is sufficient) around the shielding, forming a cone with the narrow end toward the conductor.

5. Slide nut and gasket in that order over tapered shield onto jacket. Make sure that groove in gasket faces away from the nut.

6. Slide clamp over shield until inside shoulder of clamp butts flush against cut end of jacket.

7. Remove tape from shield, and fold shield strands back over clamp without overlaps. Trim strands with scissors, so that all strands end at end of clamp taper.

8. Tin center conductor as shown in Figure 3-10.

9. Tin inside of contact (male or female) as shown in Figure 3-11.

10. Slip contact over center conductor so that contact butts flush against dielectric. Solder, using a clean, well tinned, soldering iron; contact must still be flush against dielectric after solder has cooled; if it is not, remake the joint. See Figure 3-12.

CAUTION: *Make sure that correct contact is used; a male contact always goes into a plug body, and a female contact always goes into a jack body.*

11. Push cable assembly into connector body as far as it will go. Slide gasket into connector body; make sure gasket is properly seated with sharp edge of braid clamp entering groove in gasket. Slide nut into

connector body and fasten body in vise (see Figure 3-13.) Start nut by hand and tighten with end wrench until moderately tight.

NOTE: *Gasket should be cut in half during tightening.*

Attaching HN connectors with captivated contacts to coaxial cable. When attaching HN connectors with captivated contacts to coaxial cable (see Figure 3-24), follow this procedure:

NOTE: *While attaching connector, observe a general precautions and procedures listed on page 3-2.*

1. Disassemble nut, gland, gasket, clamp, sleeve and front and rear insulators from plug or jack body. See Figure 3-22.

2. Remove 1-5/8 inches from outer jacket, exposing shield. See Figure 3-8.

   CAUTION: *Do not nick shield.*

3. Comb out shield and cut off dielectric 29/32 inch from end of jacket.

   CAUTION: *Do not nick center conductor.*

4. Taper braid wires forward and wrap with tape as described in 3-39d. Slide nut and gland onto jacket. Make sure that sharp edge of gland is toward end of cable; then slide gasket onto jacket with "V" groove toward gland.

5. Slide clamp over the braid until inside shoulder of clamp butts flush against end of jacket.

6. Remove tape and fold shield strands back over clamp without overlaps. Trim strands with scissors so that all strands end at end of clamp taper.

7. Tin center conductor as shown in Figure 3-10, using minimum, amount of heat.

8. Slide sleeve and tear insulator over cable dielectric. Slip contact over center conductor. Rear insulator must seat against cable dielectric, and contact shoulder must be flush with insulator face. Solder contact to center conductor. See Figure 3-12.

9. For jacks only, install front insulator.

   CAUTION: *Make sure that correct contact is used; a male contact always goes into a plug body, and a female contact always goes into a jack body.*

10. Push cable assembly carefully into connector body as far as it will go. Make sure that sharp edge of gland remains in the gasket groove. Tighten nut with wrench, holding body stationary. See Figure 3-13.

    NOTE: *Gasket should be cut in half during tightening.*

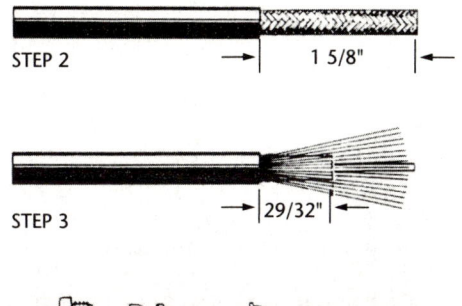

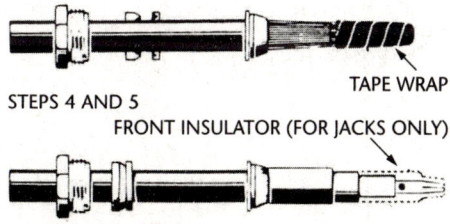

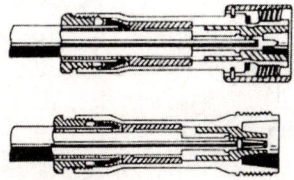

Figure 3-24. Attaching HN connectors with captivated contacts to coaxial cable

Section 5
N Series Connectors

N Connector Types

There are two versions of the N connector, differing in the method of attaching coaxial cable to the connector body. See Figure 3-3 for typical examples of N connectors. Table 3-5 lists the

Plug	Jack	Panel Jack	For Use With Cable Types
Improved Version:			
UG-18D/U	UG-20D/U	UG-19D/U	RG-5/U, 6/U, 21/U and 212/U
UG-21E/U UG-594A/U	UG-23E/U	UG-23E/U UG-160D/U	RG-8/U, 9/U, 213/U and 214/U
UG-536B/U	—	—	RG-55/U and 58/U
Captivated Contact Version:			
UG-1185/U	UG-1186/U	UG-1187/U	RG-8/U, 9/U, 213/U and 214/U

Table 3-5. N series connectors with associated cables

more common connectors in the N series and shows the coaxial cables associated with each.

Improved version (see Figure 3-25). Consists of a body assembled to coaxial cable with nut, grooved gasket and clamp. Plug UG-18D/U and Jack UG-20D/U are typical of this version. The sleeve clamp has a sharp rear face, which cuts into the grooved gasket when the nut is tightened.

Captivated contact version (see Figure 3-26). Consists of a body assembled to coaxial cable with nut, grooved gasket, clamp, washer and front and rear insulators. The plug body assembly omits the rear insulator. Plug UG-1185/U and Jack UG-1186/U are typical of this version.

Assembly Precautions and Procedures

Attaching improved N connectors to coaxial cable. When attaching improved N connectors to coaxial cable (see Figure 3-27), follow this procedure:

> **NOTE:** *While attaching connector, observe all general precautions and procedures listed on page 3-2.*

1. Remove 9/32 inch of outer jacket, exposing shield. See Figure 3-8.

   CAUTION: *Do not nick shield.*

2. Comb out shield and strip dielectric to 1/8 inch from end of jacket, exposing 5/32 inch of center conductor.

   CAUTION: *Do not nick center conductor.*

3. Disassemble nut, gasket and sleeve clamp from plug or jack body. See Figure 3-25.

4. Taper shield toward center conductor and wrap a piece of thin pressure tape, wide enough to cover the combed-out shield (one layer is sufficient) around the shielding, forming a cone with the narrow end toward the conductor.

5. Slide nut and gasket in that order over taped shielding onto jacket. Make sure that grooved side of gasket faces away from the nut. Slide clamp over shield until inside shoulder of clamp butts flush against cut end of packet.

6. Remove tape and fold shield strands back over sleeve clamp taper without overlaps. Trim shield with scissors, so that strands end at end of clamp taper.

7. Check that exposed dielectric is .045 inch beyond shield.

8. Tin center conductor as shown in Figure 3-10.

9. Tin inside of contact (male or female) as shown in Figure 3-11.

10. Slip contact over center conductor so that contact butts flush against dielec-

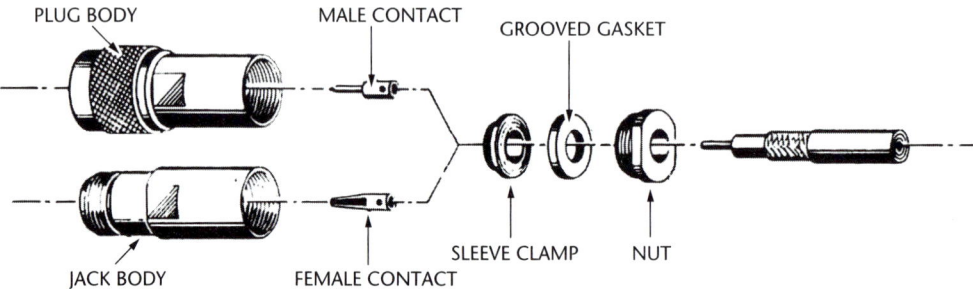

Figure 3-25. Improved N connectors-exploded view

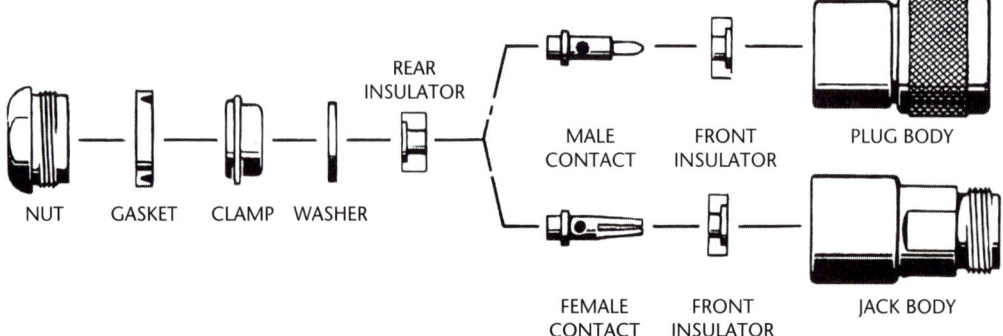

Figure 3-26. N connectors with captivated contacts-exploded view

RF Connectors and Cabling | 3-13

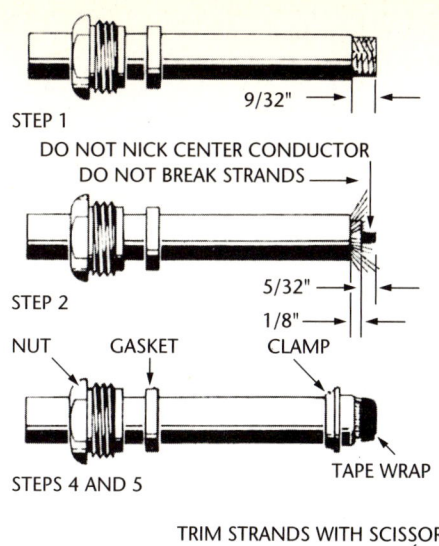

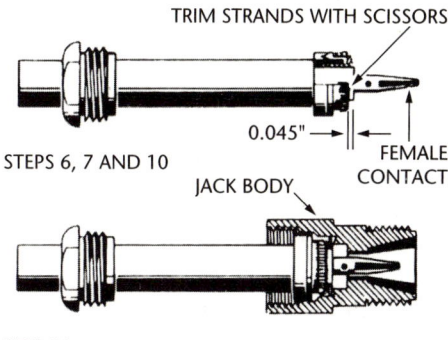

Figure 3-27. Attaching improved N connectors to coaxial cable

1. Remove 23/64 inch of outer jacket, exposing shield. See Figure 3-8.

   CAUTION: *Do not nick shield.*

2. Comb out shield, and cut off cable dielectric 1/8 inch from end of jacket.

   CAUTION: *Do not nick center conductor.*

3. Disassemble nut, gasket, clamp, washer and insulator(s) from plug or jack body.

4. Taper shield toward center conductor, and wrap with tape as described in Figure 3-28. Slide nut and gasket in that order, over tapered shield onto jacket. Make sure grooved side of gasket faces away from nut. Then slide clamp over tapered shield and push back against cable jacket.

tric. Solder, using a clean, well tinned soldering iron; contact must still be flush against dielectric after solder has cooled; if it is not, remake the joint. See Figure 3-12.

CAUTION: *Be sure that correct contact is used; a male contact always goes into a plug body, and a female contact always goes into a jack body.*

11. Push cable assembly into connector body as far as it will go. Slide gasket into connector body. Be sure knife edge of sleeve clamp seats into groove of gasket. Then slide nut into connector body and fasten body in vise. See Figure 3-13. Start nut by hand; tighten with end wrench until moderately tight. Gasket should be cut in half during tightening.

Attaching N connectors with captivated contacts to coaxial cable. When attaching captivated-contact N connectors to coaxial cable (see Figure 3-28) follow this procedure:

NOTE: *While attaching connector, observe all general precautions and procedures listed on page 3-2.*

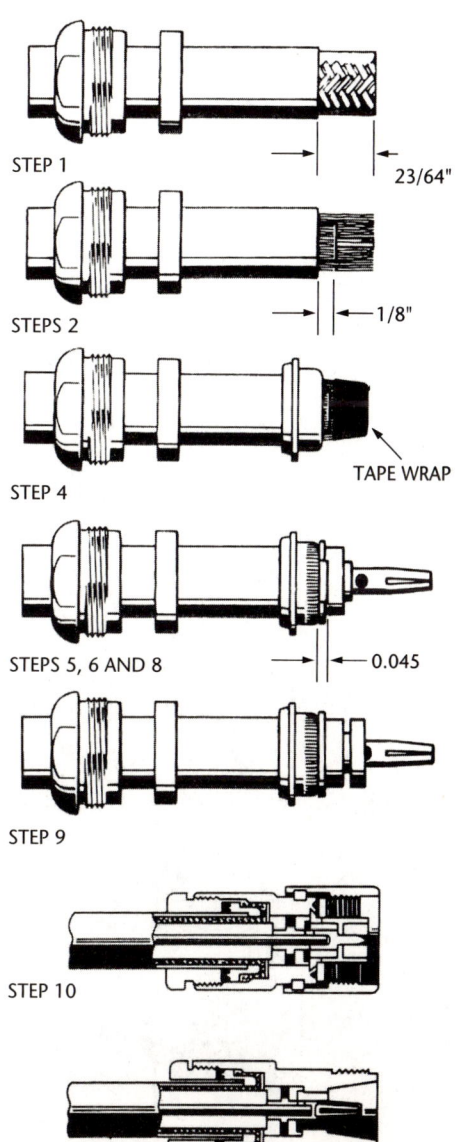

Figure 3-28. Attaching N connectors with captivated contacts to coaxial cable

Plug	Jack	Panel Jack	For Use With Cable Types
Ceramic Insert:			
UG-34/U UG-174/U	—	—	RG-25/U RG-28/U
Rubber Insert:			
UG-180A/U	UG-182A/U UG-1086/U	UG-181A/U	RG-25/U, 64/U, 77/U, 78/U and 88/U

Table 3-6. Pulse series connectors with associated cables

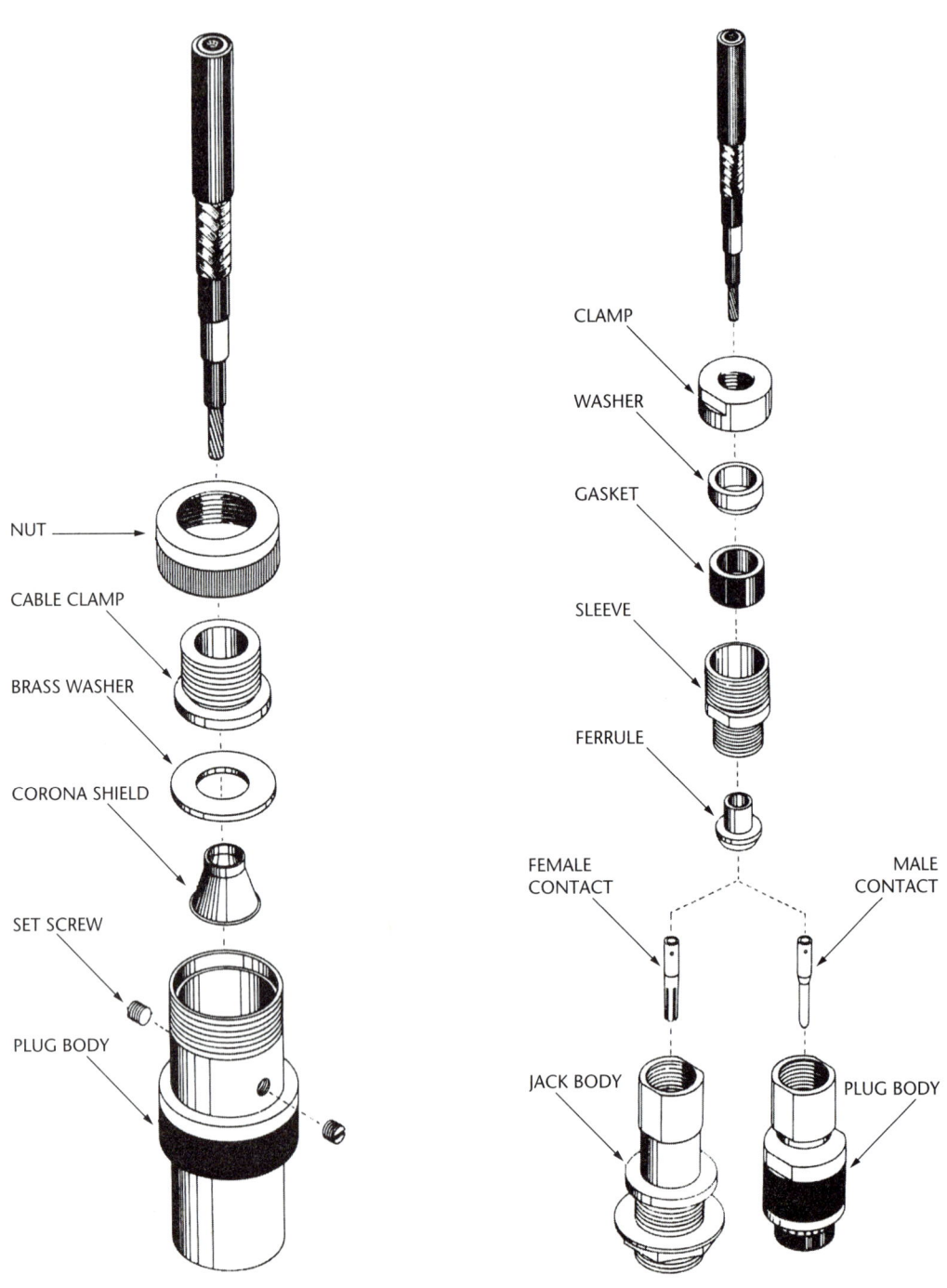

Figure 3-29. Pulse connector-ceramic insert

Figure 3-30. Pulse connector-rubber insert

5. Remove tape, and fold shield strands back over clamp taper without overlaps. Trim shield with scissors so that strands end at end of clamp taper.

6. Check that exposed dielectric is .045 inch beyond shield.

7. Tin center conductor as shown in Figure 3-10, using minimum amount of heat.

8. Slide on washer, rear insulator and contact, so that the counterbored end of the rear insulator butts flush against the dielectric, and the contact shoulder butts flush against the rear insulator. Solder the contact to center conductor. See Figure 3-12.

CAUTION: *Make sure that the correct contact is used; a male contact always goes into a plug body and a female contact always goes into a jack body.*

9. Slide front insulator over contact; make sure the counterbored end of the insulator is toward the mating end of the contact.

10. Push the cable assembly into the connector body. Make sure that the sharp edge of the clamp seats properly in the gasket. Tighten the nut, holding the body stationary. See Figure 3-13.

Section 6
Pulse Series Connectors

Pulse Connector Types

There are two versions of pulse connectors. These versions differ in the material of the inserts, and in the method of attaching the coaxial cable to the connector body. See Figure 3-5 for typical examples of pulse connectors. Table 3-6 lists the more common connectors in the pulse series and shows the coaxial cables associated with each.

Ceramic insert version (see Figure 3-29). Consists of a plug or jack body assembled to coaxial cable with nut, cable clamp, washer and corona shield. Plug UG-174/U is typical of this version.

Rubber insert version (see Figure 3-30). Consists of a plug or jack body assembled to coaxial cable with clamp, washer, gasket, sleeve and ferrule. Plug UG-180A/U and Jack UG-182A/U are typical of this version.

Attaching ceramic insert pulse connectors to coaxial cable. When attaching ceramic insert pulse connectors to coaxial cable (see Figure 3-31, next page) follow this procedure:

NOTE: *The following procedure is for assembling UC-174/U plug to RG-28/U cable, and UG-34/U plug to RG-25/U cable. The two assemblies differ in dimensions as indicated in the procedure steps. Both cables have a double shield.*

NOTE: *While attaching connector, observe all general precautions and procedures listed on page 3-2.*

1. Disassemble nut, cable clamp, washer and corona shield from plug or jack body. See Figure 3-29.

2. Slide nut and cable clamp in that order onto cable jacket. Remove 3-5/8 inches of outer jacket of RG-28/U cable and 2-3/4 inches of RG-25/U cable exposing first shield, (see note at beginning of paragraph). See Figure 3-8.

CAUTION: *Do not nick shield.*

3. Remove first shield to 5/16 inch from cut edge of outer jacket exposing insulating tape.

4. Comb out shield and bend at right angles, as shown. Remove insulating tape even with cut edge of outer jacket, exposing second shield.

CAUTION: *Do not nick shield.*

Slide cable clamp forward against fanned-out first shield. Trim shield strands 1/16 inch below diameter of cable clamp flange.

5. Slide brass washer carefully over second shield against fanned-out shield. Remove second shield to 3/16 inch from brass washer for RG-28/U cable, and 1/8 inch for RG-25/U cable, exposing conducting rubber.

6. Remove layer of conducting rubber to 3/8 inch from face of brass washer for RG-28/U cable and 3/16 inch for RG-25/U cable by making small slit at end of cable core, and removing conducting rubber with dull knife. Scrape insulating rubber underneath to remove any traces of conducting rubber.

CAUTION: *Do not damage insulating rubber.*

7. Slide corona shield over conducting rubber and under second shield until straight part of corona shield enters hole in brass washer approximately 1/16 inch.

8. Solder second shield to brass washer and to corona shield. Remove excess flux. Remove insulating rubber and conduct-

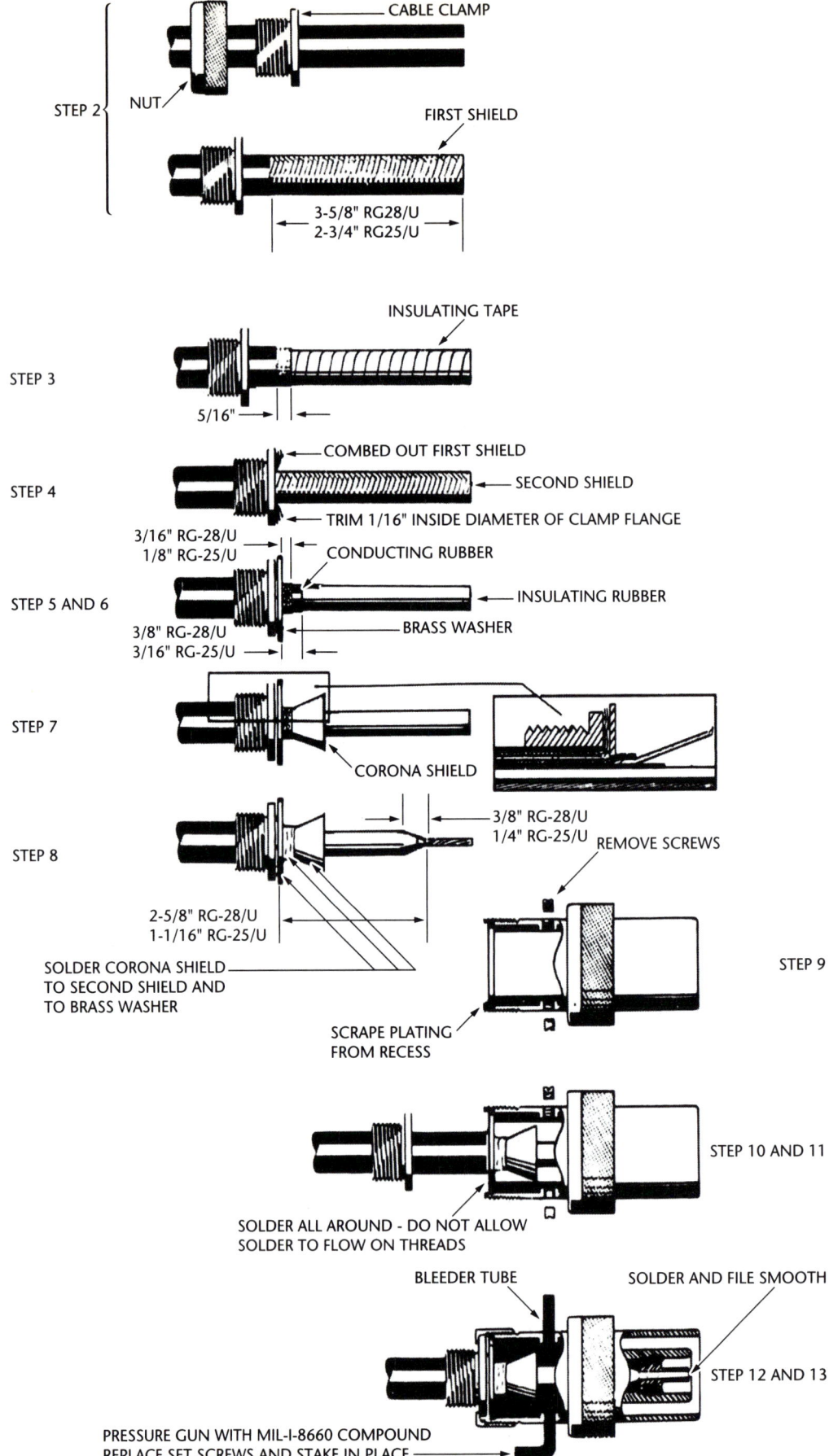

Figure 3-31. Assembly of ceramic insert pulse connector

ing rubber underneath it to 2-5/8 inches from face of brass washer, for RG-28/U cable and 1-1/16 inches for RG-25/U cable exposing center conductor. Taper rubber down to conductor 3/8 inch for RG-28/U or 1/4 inch for RG-25/U. Tin center conductor. Remove excess flux.

9. Scrape nickel plating from recess of plug into which brass washer fits. Remove set screws.

10. Slide cable assembly into plug body, allowing cable clamp to slide back on cable. Solder brass washer to recess in plug by flowing solder into space between washer and groove. Remove excess flux.

11. Slide cable clamp against washer, and nut onto plug body. Start nut by hand and tighten with spanner wrench. Hold plug with strap wrench to prevent it from turning.

12. Cut off excess conductor protruding beyond contact pin. Solder conductor to contact by flowing solder down into hole. Leave drop of solder on end of contact and file smooth. Remove excess flux.

13. Insert bleeder tube in top hole, so it is vertical, insert pressure gun in lower hole and fill plug cavity with Mil-Spec MIL-J-8660 compound until material oozes from bleeder tube. Replace set screws and stake with prick punch.

Attaching rubber insert pulse connectors to coaxial cable. When attaching rubber insert pulse connectors to coaxial cable (see Figure 3-32) follow this procedure:

NOTE: *While attaching connector, observe all general precautions and procedures listed on page 3-2.*

1. Disassemble nut, washer, gasket, sleeve and ferrule from plug or jack body. See Figure 3-30.

2. Slide nut, washer, gasket, and sleeve in that order onto cable jacket. Remove 2-5/8 inches of outer jacket exposing shield. See Figure 3-8.

   CAUTION: *Do not nick shield.*

3. Cut shield(s) to 3/8 inch from cut edge of outer jacket, exposing cable core.

   CAUTION: *Do not nick or cut cable core.*

4. Push ferrule over cable core and under shield(s).

5. Solder shield(s) carefully to ferrule all around its circumference. Be sure solder flows through to all shields. If it is necessary to solder shields separately, fold back outer shield. Solder inner shield, then bring forward outer shield and solder separately on top of inner shield. After solder has cooled, grasp cable in left hand, ferrule in right hand, and give several quick pulls to remove any slack in shield(s). Remove excess flux. Remove cable core with sharp square cut, leaving 1-1/8 inch from ferrule for connection to UG-180A/U, and one inch from ferrule for connection to UG-181A or 182A/U. Trim center conductor to 3/16 inch and tin.

   NOTE: *Cable RG-25A/U, 64A/U, 78/U and 88A/U have a thin layer of red insulating rubber over the cable core. Do not remove this layer. Cables RG-25/U and RG-64/U have a thin layer of black conducting rubber over the cable core. Remove this layer to 1/16 inch from ferrule very carefully with a sharp knife.*

   CAUTION: *Do not nick or cut cable core.*

6. Tin inside of contact (male or female) as shown in Figure 3-11. Slip contact over center conductor so that contact

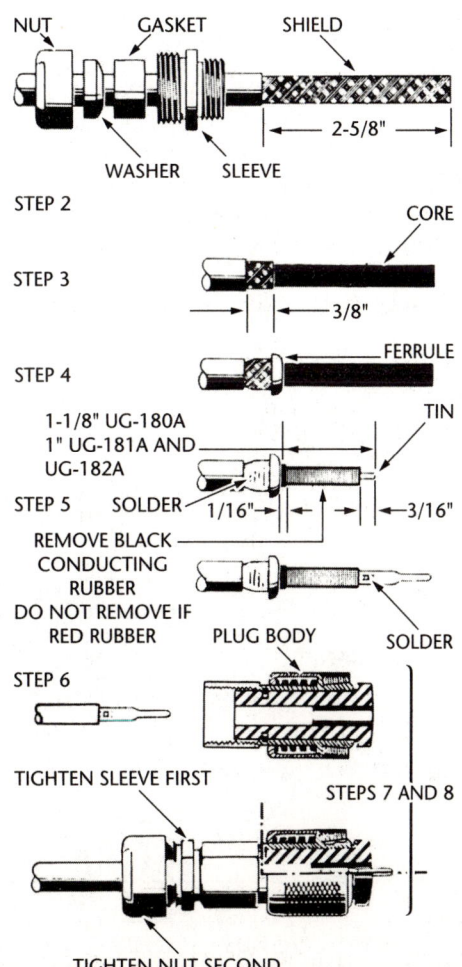

Figure 3-32. Assembly of rubber insert pulse connector

butts flush against cable core. Solder, using a clean, well tinned soldering iron; contact must still be flush against cable core after solder has cooled; if it is not, remake the joint. See Figure 3-12. Remove excess flux.

CAUTION: *Be sure that correct contact is used; a male contact always goes into a plug body, and a female contact always goes into a jack body.*

7. Push cable core into plug or jack body as far as it will go. Insert sleeve and tighten as far as it will go against ferrule, holding body with wrench so it will not turn. See Figure 3-13.

8. Insert gasket, then washer into sleeve. Install nut on sleeve and tighten until gasket deforms around cable to hold it securely.

Section 7
TNC Series Connectors

TNC Connector Types

There is only one version of TNC connector. It consists of a body assembled to coaxial cable by means of a clamp nut, gasket and braid clamp. See Figure 3-33. Table 3-7 lists the more common connectors in the TNC series and shows the coaxial cables associated with each.

Attaching TNC Connectors to Coaxial Cable

When attaching TNC connectors to coaxial cable, follow this procedure: See Figure 3-34.

NOTE: *While attaching connector, observe all general precautions and procedures listed on page 3-2.*

1. Remove 5/16 inch of outer jacket, exposing shield. See Figure 3-8.

CAUTION: *Do not nick shield.*

2. Comb out shield. Take care to prevent breaking shield strands.

CAUTION: *Do not nick center conductor.*

3. Disassemble nut, grooved gasket and sleeve clamp from plug or jack body. See Figure 3-33.

4. Taper shield toward center conductor, and slide nut and grooved gasket in that order over tapered shield onto jacket. Then slide sleeve clamp over tapered shield until inside shoulder of clamp butts flush against cut end of jacket.

5. Comb shield back smoothly over clamp and trim with scissors even with tapered part of clamp.

6. Trim dielectric to 1/8 inch from shield, and cut off center conductor to 7/64 inch from edge of dielectric.

7. Tin center conductor as shown in Figure 3-10

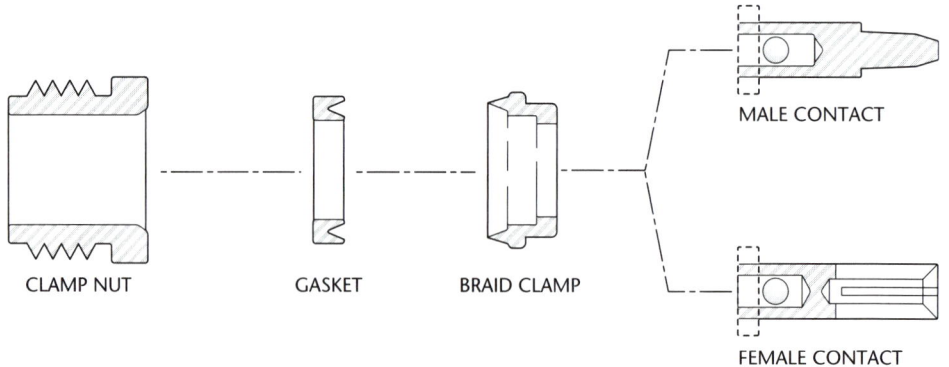

Figure 3-33. TNC connectors-exploded view

Plug		Jack		Panel Jack		Cable RG-
IPC	Plug King	IPC	King	IPC	King	
79875	KA51-03	79600	KA31-02	79425	KA11-04	55/U, 58/U
79525	KA51-02	79500	KA31-03	79925	KA11-03	59/U

Table 3-7. TNC series connectors with associated cables

8. Slide contact over center conductor until contact butts flush against dielectric. Solder contact to center conductor. See Figure 3-12.

CAUTION: *Make sure that correct contact is used; a male contact always goes into a plug body, and a female contact always goes into a jack body.*

9. Push cable assembly into connector body as far as it will go. Make sure gasket is properly seated, with sharp edge of sleeve entering gasket groove. Tighten nut holding the body stationary. See Figure 3-13.

Section 8
RF Connectors

MB Miniature Connector Series

These are small, lightweight, bayonet type, quick connect/disconnect connectors, used with small RF cables where peak voltage is not more than 500 volts. No soldering is required in the assembly of plugs to solid center conductors, such as RG58/U, 59/U, 62/U, 71/U and 141/U. All jacks require soldering. Table 3-8 lists the more common connectors in the MB series and shows the coaxial cables associated with each. These connectors consist of a plug or jack body assembled to coaxial cable with clamp nut, braid clamp and insulator bushing. See Figure 3-35.

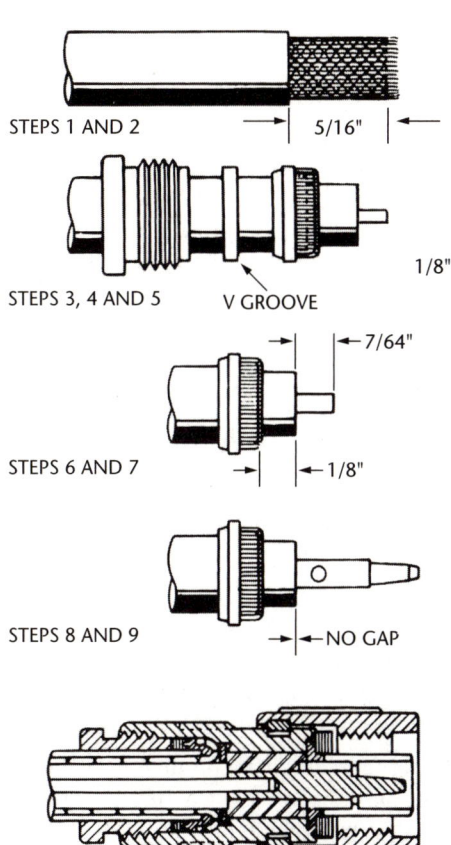

Figure 3-34. Attaching TNC connectors to coaxial cable

Plug (IPC)	Angle Plug (IPC)	Jack (IPC)	Panel Jack (IPC)	Cable RG-
45000 45050	53000	46700	46300	58/U, 141/U
45025 45025	53500	46775	46325	59/U, 62/U,

Table 3-8. MB series connectors with associated cables

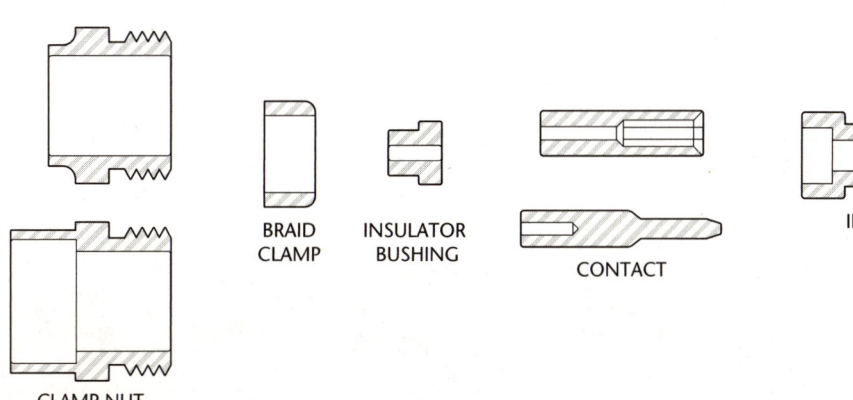

Figure 3-35. MB connectors-exploded view

Attaching MB Connectors to Coaxial Cable

The assembly procedure differs according to the cable used. For assembly to cables RG58/U and RG141/U, the procedure is as follows: See Figure 3-36.

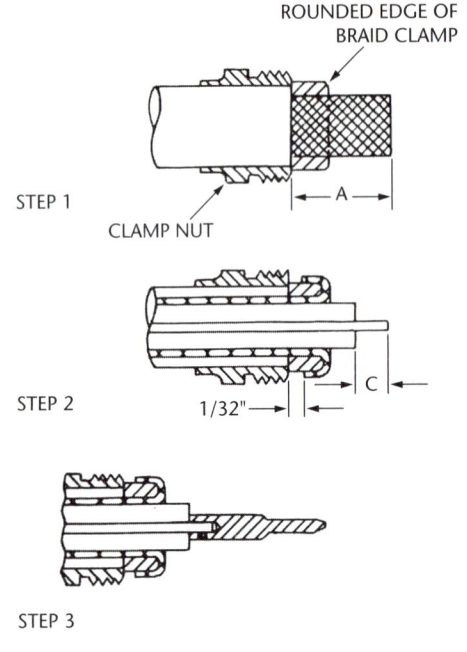

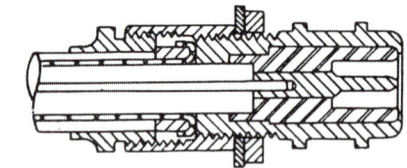

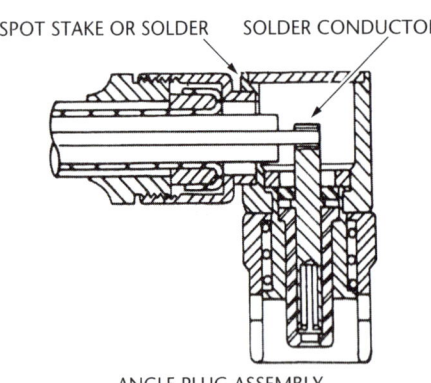

Figure 3-36. Attaching MB connectors to coaxial cable

1. Remove cable jacket to dimension A given in Table 3-9. Insert clamp nut over cable jacket and braid clamp over braid wire.

2. Comb out braid wire, form back over braid clamp and trim to length. Cut off cable dielectric to dimension C in Table 3-9, and tin exposed conductor. If solderless contact is used, omit tinning.

3. Insert contact over conductor. The end of the solderless contact with the shortest slot is inserted over the conductor. If solder contact is used, solder it to the conductor, and remove excess solder from the outside of the contact.

4. Insert assembly minus clamp nut into body and rotate slightly to make sure braid clamp is seated. When assembling straight plugs insert insulator over contact before assembly into body. Thread clamp nut into body and tighten nut, holding body stationary.

For assembly to cables RG59/U, and 62/U the procedure is as follows:

1. Remove cable jacket to dimension A in Table 3-9. Insert clamp nut over cable so that internal shoulder seats against end of cable jacket. Insert braid clamp over wire.

2. Comb out braid wires, form back over braid clamp, and trim to length. Cut off cable dielectric to dimension C in Table 3-9, and tin exposed conductor. If solderless contact is used, omit tinning.

3. Insert contact over conductor. The end of the solderless contact with the shortest slot is inserted over the conductor. Insert insulator bushing over contact if cable RG62/U is being used. If solder contact is used, solder it to the conductor, and remove excess solder from the outside of the contact.

4. Insert insulator over contact. Insert assembly minus clamp nut into body and rotate slightly to make sure braid clamp is seated. Thread clamp nut into body and tighten nut, holding body stationary.

5. Right angle jacks or plugs: Strip cable jacket and dielectric, install parts and form braid as instructed in steps 1 and 2.

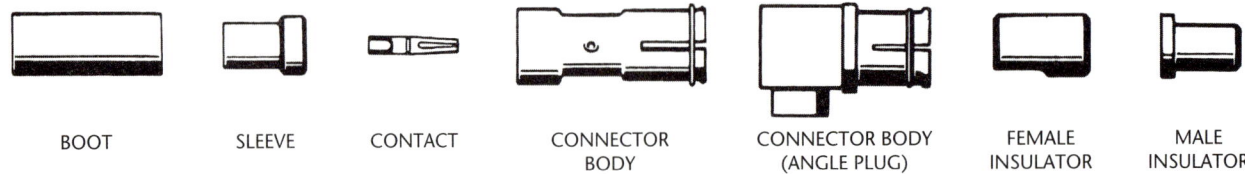

Figure 3-37. Subminiature RF connector-exploded view

6. Tin conductor, and insert assembly into body.

7. With cap removed, solder the conductor in slot of angle plug contact.

8. Insert cap, and spot solder or spot stake.

Subminiature RF Connectors

Subminiature RF connectors (Amphenol #27 Series). These connectors are very small, lightweight connectors designed for use with RG-174/U miniaturized coaxial cable, where peak voltage does not exceed 500 volts. Coupling is either of the screw thread type, or the push-on type. The connectors consist of a plug or jack body assembled to coaxial cable with a sleeve and an insulator. See Figure 3-37. The assembly is crimped into the body, and a vinyl boot shrunk on for cable strain relief. Table 3-10 lists types of #27 series connectors commonly used in aircraft.

Assembly Precautions and Procedures

Attaching subminiature RF connectors to coaxial cable. When attaching subminiature RF connectors to coaxial cable, follow this procedure: (See Figure 3-38).

1. Dilate the boot and slip it over the cable. The boot will remain dilated for approximately five minutes.

2. Trim jacket to dimension A in Table 3-10.

   CAUTION: *Do not nick braid.*

3. Slip sleeve over braid against cable jacket.

Connector Part (IPC)	RG-/U Cable	Stripping Dimensions	
		A	C
Plugs 45000	58, 141	1/4	5/64
45025	59, 62	7/16	5/64
45050	58, 141	1/4	5/64
45550	59, 62	7/16	5/64
Jacks 46300	58, 141	1/4	3/32
46325	59, 62	1/2	3/32
46700	58, 141	1/4	3/32
46775	59, 62	1/2	3/32
Angle 53000	58, 141	3/8	7/64
Plugs 53500	59, 62	3/8	7/64

Table 3-9. Stripping dimensions for coaxial cable assembled to MB connectors

Body Type	A Inches (+0 -1/64)	B Inches (Max)	Position in Tool 27-900 Figure No
Straight Plug	23/64	1/16	3-39 a
Jack, Push-on	3/8	1/16	3-39 b
Jack, Screw-on	3/8	.082	3-39 c
Bulkhead Jack	3/8	.082	3-39 c
Angle Plug	25/32	7/64	3-39 d

Table 3-10. Stripping dimensions and crimping tool positions for subminiature RF connectors

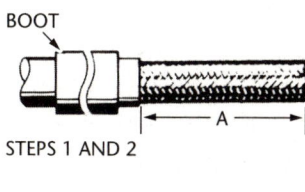
STEPS 1 AND 2

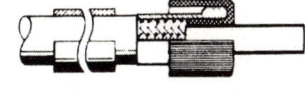
STEP 3

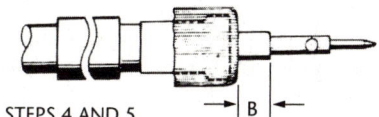
STEPS 4 AND 5

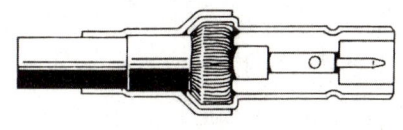

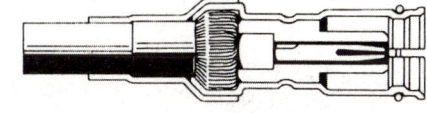
STEP 6

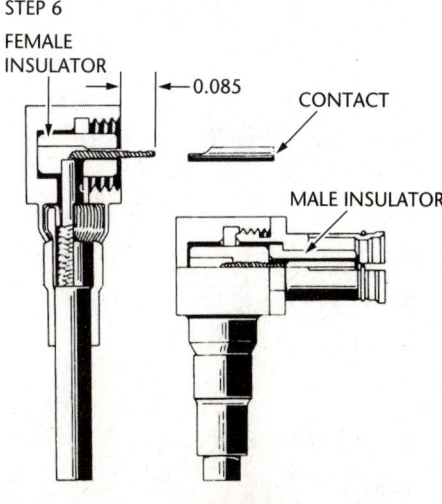
STEP 7   STEPS 8 AND 9

Figure 3-38. Attaching subminiature RF connectors to coaxial cable

Fold braid back over the sleeve and comb out so it lays even without overlapping.

4. Trim dielectric to dimension B in Table 3-10. Tin exposed center conductor, and clean off excess solder.

5. Straight plugs and jacks: Slip contact over center conductor so that it butts flush against cut end of dielectric. Solder contact to conductor, and remove excess solder from outside of contact.

CAUTION: *Avoid overheating during soldering, so as not to deform dielectric.*

6. Slip cable assembly into body and trim off excess braid protruding beyond body end. Crimp the assembly securely (see next procedure for detailed crimping instructions), and pull boot over body as shown.

7. Angle plugs: Unscrew front part of body. Follow procedure of steps 1 through 4 above. Then thread cable through back part of body as shown and crimp. Thread female insulator over conductor and insert into body. Holding the female insulator in place in the body, pull the cable as far forward as possible to remove all slack. Trim conductor to .085 inch maximum.

8. Slip contact over conductor and hurt contact flush against the female insulator. Solder contact to conductor, and remove excess solder.

9. Place male insulator over the contact, and screw the front body part into back body and tighten with end wrench.

Crimping procedure for subminiature connectors. Crimp subminiature connector bodies as follows:

1. Open jaws of tool 27-900 by loosening nut and pulling down lock screws (see Figure 3-39). Place the connector assembly in jaws, and set optimum distance between jaws for each assembly by means of the travel limit screw. Refer to connector placement Figure 3-39 a, b and c for straight plugs and jacks, and Figure 3-39d for angle plugs.

2. Lock jaws by pulling lock screws up, and tighten nut.

3. Squeeze handles to crimp.

4. Release handles to open jaws, and remove crimped assembly. Trim off any excess braid protruding beyond end of body.

5. Slip boot over body end.

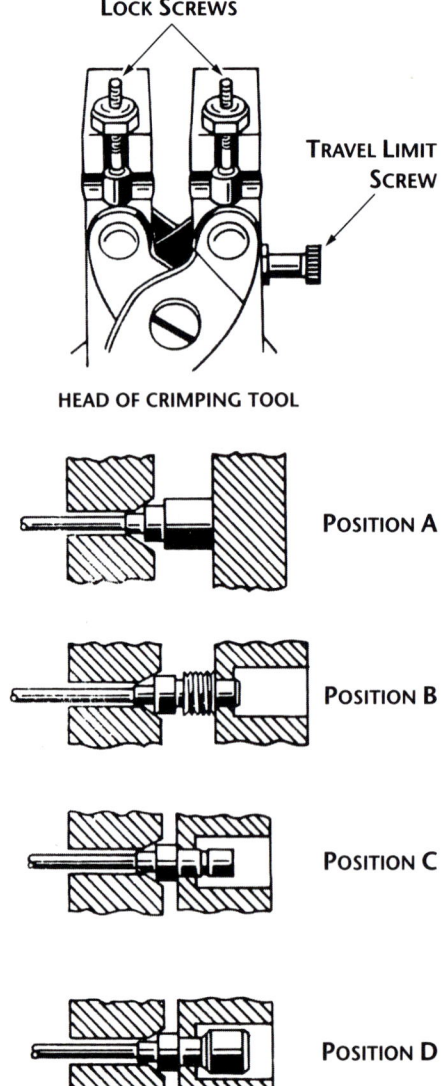

Figure 3-39. Crimping subminiature RF connectors

RF Connectors Used in Fuel Quantity Indicating Systems

Because of their transmission line efficiency, RF connectors are often used in aircraft fuel quantity indicating systems. The connectors most commonly used for this purpose are of two types. One is similar to the standard BNC connector; typical of these are the 163 series made by Avien, and the Liquidometer 9100 series. The second type is miniature RF connectors; the Nu-Line 1200 series (MIL-C-25516), and Liquidometer S62 and S63 are typical. These connectors are designed to be used with coaxial cable, but they are also frequently used with standard shielded or unshielded wire.

Assembling BNC Type Fuel Quantity Indicating Connectors

When assembling BNC type RF connectors, use the following procedure:

Assembly of Avien 163-088 and 163-089 connectors to RG-58A/U coaxial cable. This procedure is as follows (see Figure 3-40):

1. Slide nut, washer and gasket onto cable. Strip outer jacket 7/16 inch, taking care not to nick the braid. Slide the clamp over the braid so it rests flush against the cut end of the jacket.

2. Comb out braid and fold it back over the clamp, and trim the braid even with edge of clamp.

3. Strip the dielectric 3/32 inch from edge of clamp.

4. Cut exposed conductor to 7/64 inch.

5. Slide contact onto conductor, and check that there is no exposed conductor between the insulation and the contact solder hole, check that the distance from the braid to the end of the contact is 1/2 inch as shown in Figure 3-40.

6. Tin the conductor and the solder hole, and solder the contact to the conductor. Remove any excess solder.

7. Push assembly into connector body, screw nut into body and tighten.

Assembly of Avien 163-07 and 163-027 connectors to AN No. 20 unshielded wire. This procedure is as follows (see Figure 3-41):

1. Strip wire to expose 1/4 inch of conductor, and tin the stripped wire. Tin inside of contact.

2. Slide nut, washer and gasket over the wire. Install TFE clamp over the wire, so that the inside shoulder of the clamp butts against the cut insulation.

3. Slide contact over the wire so that it butts against end of the clamp. While doing this, hold the insulation firmly in place against the clamp shoulder.

4. Solder the contact to the conductor, and remove any excess solder.

5. Insert the assembly into the connector body. The contact end should be flush with the end of the TFE clamp, but a recess of 1/32 inch maximum is acceptable.

6. Tighten nut, holding the body stationary.

Assembly of **Liquidometer 9100 series** connectors to RG-58/U coaxial cable is as follows (see Figure 3-42, next page):

1. Slide nut, washer and gasket over the cable; strip outer jacket 3/8 inch, taking care not to nick the braid. Slide the large eyelet over the braid so that it butts against the cut end of the jacket.

2. Comb out the braid up to the large eyelet, and slide the smaller eyelet over the dielectric.

3. Clamp the braid between the two eyelets with special slotted pliers as shown, (see Figure 3-42). Trim off excess braid.

4. Cut dielectric 1/8 inch from the small eyelet, and cut off the exposed conductor to 1/8 inch.

5. Tin the exposed conductor and the inside hole of the contact.

6. Slide the contact onto the conductor and solder. Remove any excess solder.

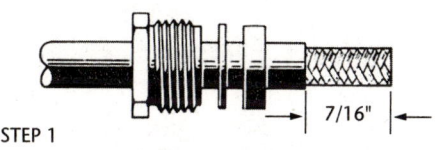
STEP 1

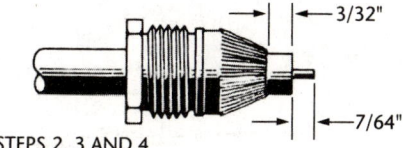
STEPS 2, 3 AND 4

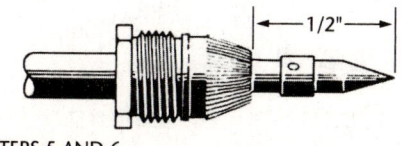
STEPS 5 AND 6

Figure 3-40. Attaching Avien 163-088 and 163-089 connectors to coaxial cable

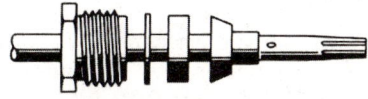
STEPS 1 THROUGH 4

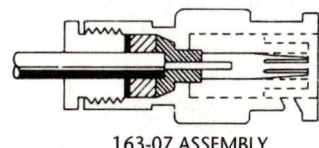
163-07 ASSEMBLY
STEPS 5 AND 6

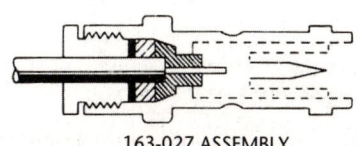
163-027 ASSEMBLY
STEPS 5 AND 6

Figure 3-41. Attaching Avien 163-07 and 163-027 connectors to unshielded wire

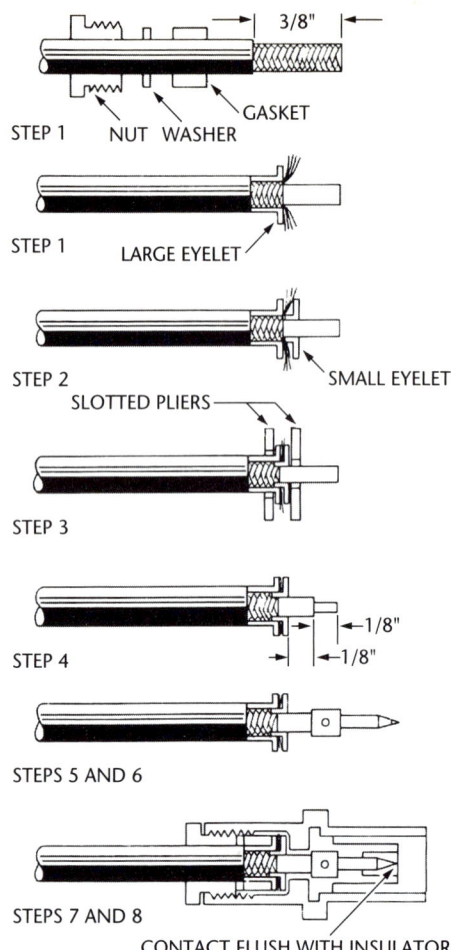

Figure 3-42. Attaching Liquidometer 9100 series connectors to coaxial cable

5. Slide braid clamp sleeve (large ID toward gasket) over braid, and trim off excess braid with scissors. Install O-ring as shown.

6. Place contact retainer firmly over dielectric, and strip off dielectric flush with contact retainer, exposing center conductor.

7. Cutoff center conductor .08 inch from cut end of dielectric.

8. Slide contact, male or female, over center conductor, so that contact butts flush against dielectric.

9. Solder contact to center conductor; contact must still be flush against dielectric after soldering.

10. Slide cable assembly into connector body as far as it will go. Torque nut to approximately five inch pounds.

Assembly of **Liquidometer S62 and S63 series connectors** to RG-58/U coaxial cable is as follows (see Figure 3-44):

1. Slide nut and bushing back over cable. Cut outer jacket to dimension A in Table 3-11, being careful not to nick braid.

7. Seat contact and eyelets into connector body and install the gasket firmly against the eyelets. Insert washer firmly against the gasket.

8. Screw nut into connector body and tighten, holding connector body stationary. Tighten to 30 to 35 inch pounds torque.

Assembling miniature RF fuel quantity indicating connectors. When assembling miniature RF connectors to wire or cable, use the following procedure:

a. Assembly of Nu-Line 1200 Series coaxial connectors to coaxial cable: (See Figure 3-43).

1. Remove 1/2 inch of outer jacket, exposing shield.

2. Slide nut, washer and gasket, in that order, onto outer jacket.

3. Screw threaded braid clamp over jacket as shown.

4. Comb out braid, and fold braid back over braid clamp, without overlap.

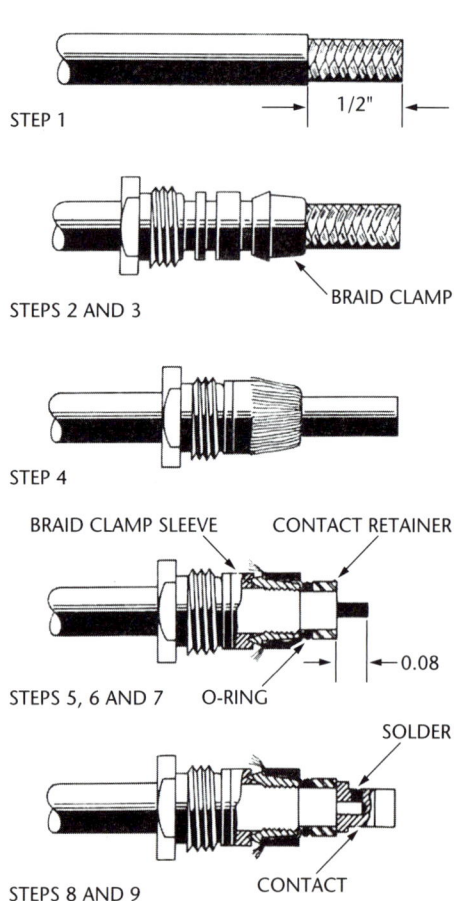

Figure 3-43. Attaching Nu-Line 1200 series connectors to coaxial cable

Connector (Liquidometer Number)	Stripping Dimension A
S62-1 and -4	11/32 inch
S62-2 and -3	5/16 inch
S63-1 and -4	5/16 inch
S63-2 and -3	5/32 inch

Table 3-11. Stripping dimensions for coaxial cable assembled to Liquidometer S62 and S63 series connectors

2. Place the two halves of the cable clamp over the cable, lining up front (large) end of clamp with cut end of outer jacket. Slide bushing over tapered end of clamp. Compress bushing over cable clamp with special pliers as far as it will go.

NOTE: *Special pliers No. TJF-107 are available from connector manufacturer.*

3. Comb braid back over clamp and trim braid around edge of clamp.

NOTE: *Omit next step (4) when using connectors S62-4 and S63-4.*

4. Slide washer, gasket and second washer over dielectric and push up to cable clamp. Cut off dielectric to expose 3/32 inch of conductor, and tin conductor.

5. Slide contact over conductor and solder. Remove any excess solder.

6. Slide assembly into plug or jack body until contact shoulder seats on insulator. Screw clamp nut into body, using a torque of 15 inch pounds.

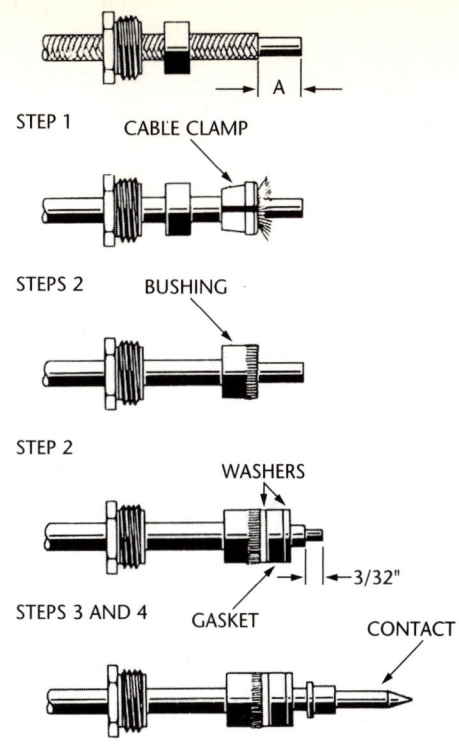

Figure 3-44. Attaching Liquidometer S62 and S63 connectors to coaxial cable

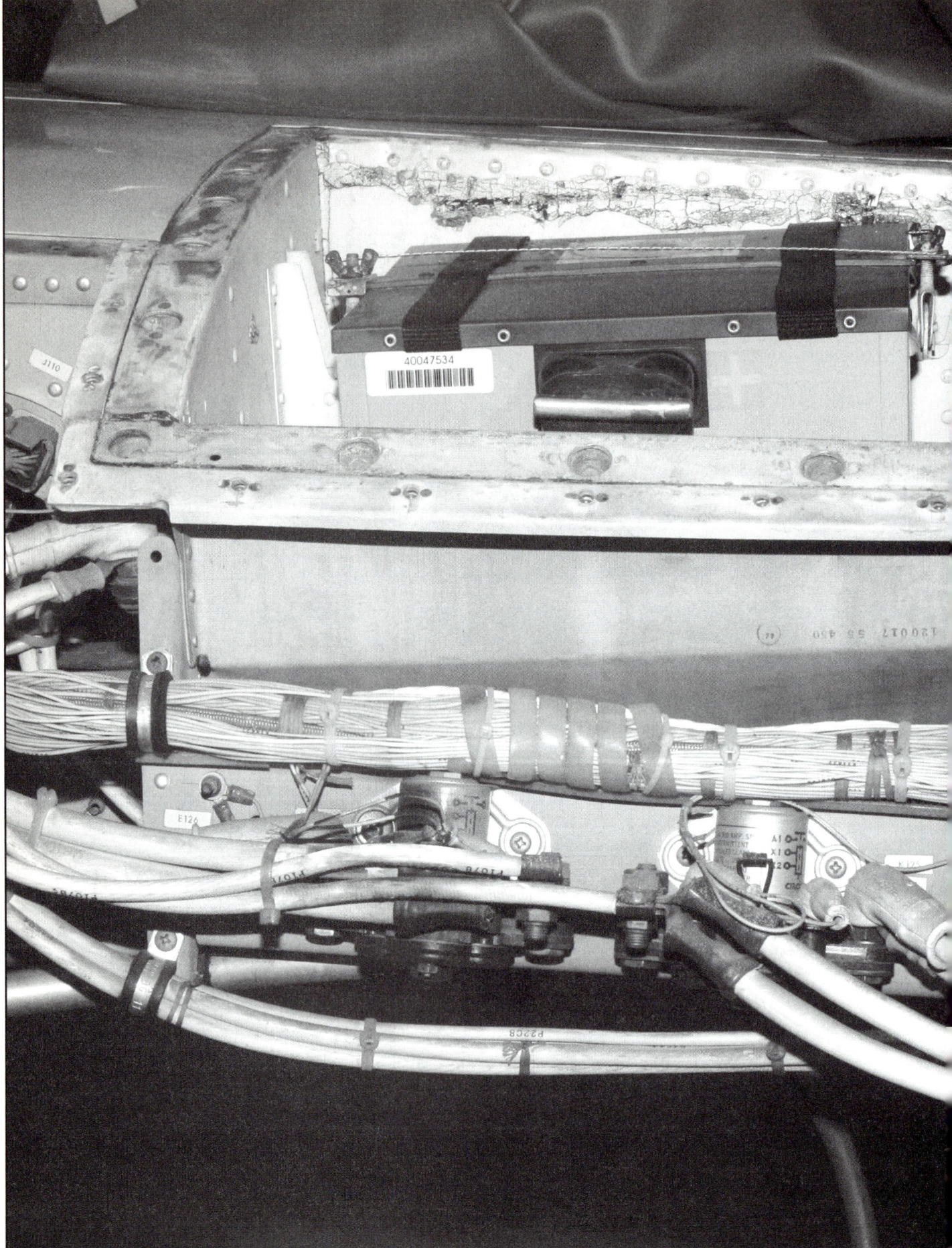

Left. Solderless terminal lugs permit easy and efficient connection to and disconnection from terminal boards, bus bars and other electrical equipment.

Chapter 4
SOLDERLESS terminations and splices

Section 1
Identification

Electric wires are terminated with solderless terminal lugs to permit easy and efficient connection to and disconnection from terminal boards, bus bars, and other electrical equipment. Solderless splices join electric wires to form permanent continuous runs.

This chapter describes recommended methods for terminating copper and aluminum wires, using solderless terminal lugs. It also describes recommended methods for permanently joining (splicing) wires, using solderless splices. Termination of thermocouple wires is covered in Chapter 5.

Description

Solderless terminal lugs and splices are copper or aluminum, and are preinsulated or uninsulated depending on the application. Terminal lugs and splices for high temperature applications are silver-or nickel-plated copper, and are insulated with TFE or a similar material.

> **NOTE:** *Use copper terminations only on copper wire. Use aluminum terminations only on aluminum wire.*

Terminal lugs are available in four styles: straight, 90 degree upright, angle and flag, for use under different space conditions. Figure 4-1 (next page) shows typical terminal lugs and splices. Terminal lugs and splices are crimped to wires by means of hand or power crimping tools.

Power tools are portable or stationary (bench-mounted). Typical crimping tools are illustrated

Learning Objectives:
- Identification
- Terminating Small Copper Wires
- Terminating Large Copper Wires
- High Temperature Terminal Lugs
- Terminating Aluminum Wire
- Splicing

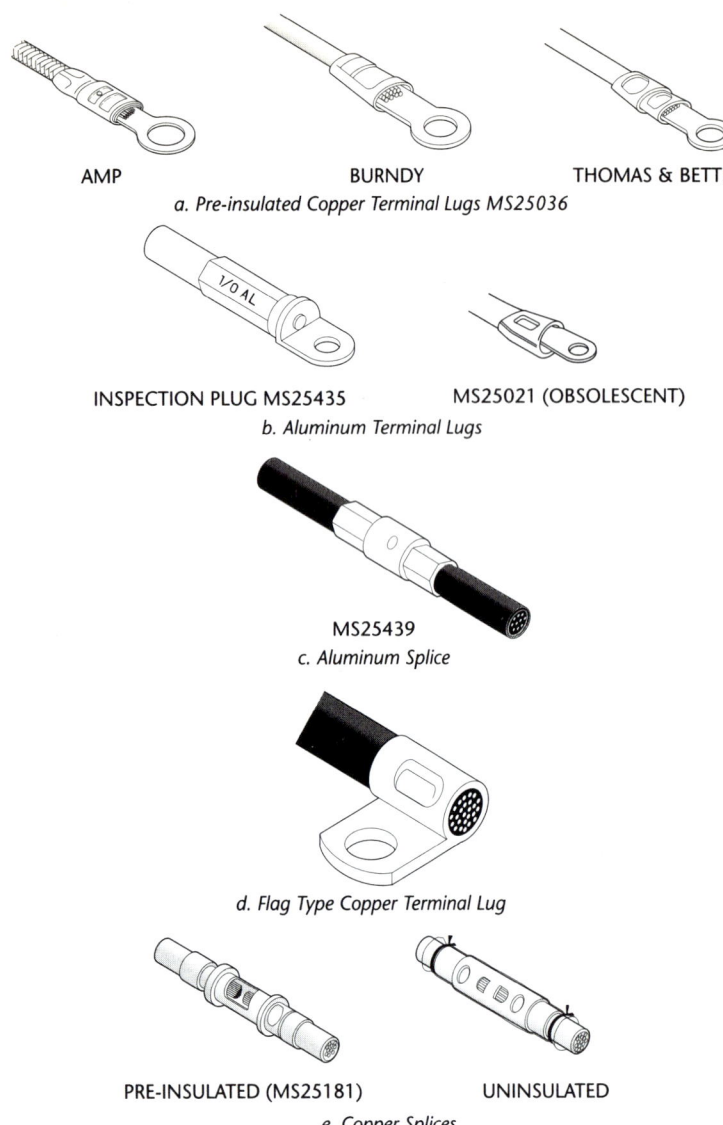

Figure 4-1. Solderless terminal lug and splices

Color of Terminal Lug Insulation	To Be Used On Wire Sizes
Yellow	#26-#24
Red	#22-#20, #18
Blue	#16-#14
Yellow	#12-#10

Table 4-1. Color coding of copper terminal lug insulation

where they are mentioned in this chapter. Solderless terminal lugs and splices most commonly used are made by AMP, Burndy and T&B (Thomas and Betts); this chapter is therefore limited to these items.

Terminal lugs and splices, and the tools used to install them on wires, are divided into two classes, as follows:

Class 1 lugs and splices. These meet all the requirements of the applicable mil-spec and Standard when installed with the specified crimping tools. Class 1 tools are those that meet all the requirements of the applicable mil-spec and Standard.

Class 2 lugs and splices. These are replaceable by Class 1 terminals, and which meet the performance requirements of the applicable mil-spec when installed with a tool recommended by the terminal manufacturer. Class 2 tools are those that will crimp terminals to meet the performance requirements of the applicable mil-spec.

> **NOTE:** *Class 1 tools, splices and lugs shall be used in all cases where designated.*

Section 2

Terminating Small Copper Wires (Sizes No. 26 through No. 10)

Preinsulated terminal lugs. Small copper wires, (sizes No. 26 through No. 10) are terminated with solderless preinsulated straight copper terminal lugs conforming to Mil-spec MIL-T-7928 and drawing MS25036. As shown in Figure 4-2, the insulation is part of the terminal lug and extends beyond its barrel, so that it will cover a portion of the wire insulation; this makes the use of an insulation sleeve unnecessary. In addition, preinsulated terminal lugs have an insulation-grip (a metal reinforcing sleeve) beneath the insulation, for extra gripping strength on the wire insulation.

Preinsulated terminals accommodate more than one size of wire; the insulation is color-coded, as shown in Table 4-1, to identify the wire sizes that can be terminated with each of the terminal lug sizes.

Tools

Crimping tools. Hand, portable power and stationary power tools are available for crimping terminal lugs. These tools crimp the barrel to the conductor, and simultaneously crimp the insulation grip to the wire insulation. Use the standard (Class 1) tool MS25037 to crimp the standard (Class 1) copper terminal lugs MS25036. Use the crimping tools of each manufacturer for the non-standard (Class 2) terminal lugs of the same manufacturer. Crimping tools most commonly used are listed in Table 4-2.

Hand crimping tool. MS25037 hand crimping tool has a self-locking ratchet that prevents the tool from opening until crimp is complete. This mechanism must never be disassembled since it insures proper crimping pressure. This tool has nests identified by color-coded arrows. (See Figure 4-3).

Hand tool inspection. The standard tool MS 25037 is checked by means of a gage for proper adjustment of crimping jaws. For good crimping results, this is done before each series of crimping operations. Return hand tools that are out of tolerance for repair. Check tools as follows:

(MS25430 Gage is used to check this tool.)

1. Place the lower stop bar of MS25430-1 in the yellow nest and gage the red-blue upper nest with MS25430-3 "GO-NO GO" gage.
2. Place the upper stop bar of MS25430-1 in the red-blue nest and gage the yellow lower nest with MS25430-2 "GO-NO GO" gage.

Power tools. Power crimping tools operate on air pressure. Power trigger must be depressed until crimp is complete. As indicated in Table 4-2 power tools use specific inserts, called "Heads", "Dies" or "Die Sets" for each terminal lug size. Use correct insert for each terminal

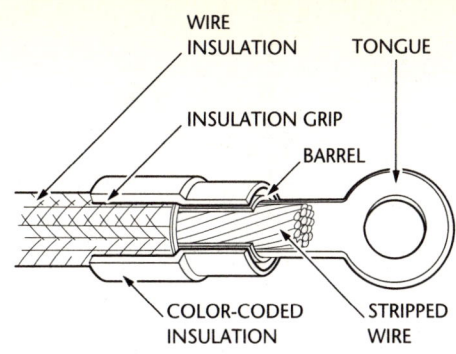

Figure 4-2. Pre-Insulated terminal lug-cut-away

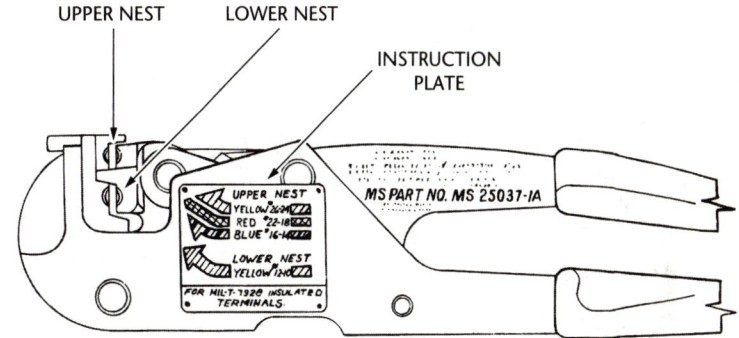

Figure 4-3. Tool MS25037 hand crimping - pre-insulated copper terminal lugs

Wire Size Range and Color Code	Hand Tool	Portable Power Tools		Stationary Power Tools	
		Tool Number	Insert Name and Number	Tool Number	Insert Name and Number
MS25036					
#26 - #24, Yellow #22 - #18, Red #16 - #14, Blue #12 - #10, Yellow	Standard Tool MS25037	—	—	—	—
AMP					
#26 - #24, Yellow	—	69005 69100	Head 47469 Head 46224	69011	Die 45155
#22 - #18, Red	—	69005 69100	Head 47516 Head 47806	69011	Die 47498
#16 - #14, Blue	—	69005 69100	Head 47517 Head 47807	69011	Die 47499
#12 - #10, Yellow	—	69010 69100	Head 47518 Head 47808	69012	Die 47500
Burndy					
#26 - #24, Yellow	—	M8ND Y8ND	Die Set N22HET-1	Y10NCP	Die Set R220HET-2
#22 - #18, Red #16 - #14, Blue #12 - #10, Yellow	—	M8ND Y8ND	Die Set N10ET-9	Y10NCP	Die Set R10ET-3
T&B					
#26 - #24, Yellow	—	—	—	—	—
#22 - #18, Red #16 - #14, Blue #12 - #10, Yellow	—	—	—	21728 & 21729	

Table 4-2. Copper terminal lugs (small) and crimping tools

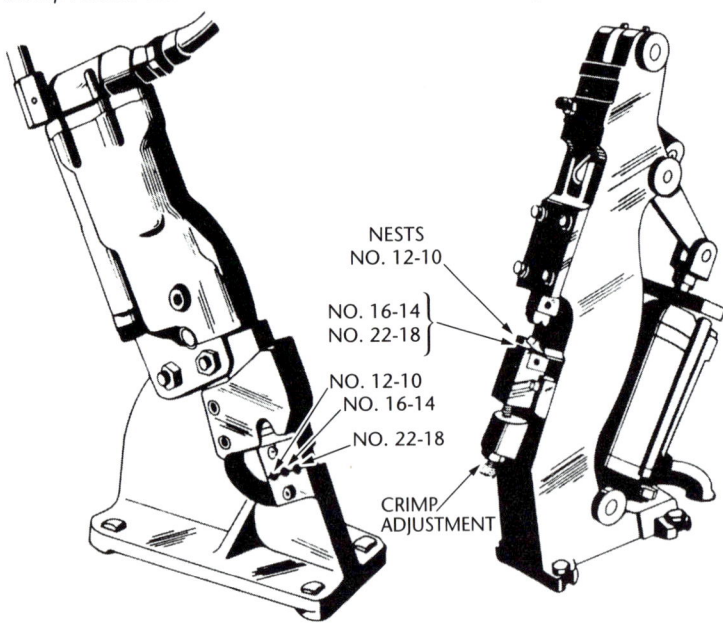

Figure 4-4. Tools-power crimping-pre-insulated copper terminal lugs

a. Amp Portable Tool 69005 or 69010
b. Burndy Portable Tool Y8ND
c. Burndy Stationary Tool Y10NCP
d. Thomas & Betts Stationary Tool 27129

lug being crimped. See Figure 4-4 and Table 4-2 for details.

WARNING: *ALWAYS disconnect power tool from its air pressure source, BEFORE installing or removing insert.*

Power tool inspection and adjustment. AMP, Burndy and T&B power tools are checked by means of gages for proper adjustment. For good crimping results, this is done before each series of crimping operations. When the tool is adjustable, make proper correction; otherwise, return tool to manufacturer for repair. Tools are checked as follows:

- AMP power tools are checked with the tool fully bottomed. The gap between the barrel crimping jaws and the gap between the insulation crimping jaws shall meet the requirements of Table 4-3. Note that the "GO" gages shall be able to enter the jaws and the "NO GO" gages shall not be able to enter. AMP power tools are not adjustable in the field.

- Check Burndy tool Y10NCP, with die set R10ET-3 closed. The gap in the 22-18 (red) nest shall accept an .087 diameter rod and shall not accept a .105 diameter rod. Replace dies that are out of tolerance.

- Check Burndy tool Y8ND, with die set N10ET-9 fully closed. The gap in the 12-10 (yellow) nest shall accept a .140 diameter rod and shall not accept a .146 diameter rod. Replace dies that are out of tolerance.

- T&B power tool No. 21728 is checked by closing the die over a No. 38 (.101) drill inserted in the smaller nest (for Red and Blue terminal lugs). The crimp adjustment screw is then tightened until the drill cannot be removed. Checking is required each time the insert is replaced.

Crimping Procedures

Crimping procedure for MS25037 standard hand tool. Hand crimp preinsulated copper terminal lugs in the No. 26-No. 10 wire size range with MS25037 standard hand tool as follows:

1. Strip wire insulation (lengths are given in Table 4-4). Use one of recommended stripping procedures in Chapter 1, page 1-14.

2. Check tool for correct adjustment in accordance with the procedure outlines on page 4-3. Tools out of adjustment must be returned to manufacturer for repairs.

3. Insert terminal lug, tongue first, into hand tool barrel crimping jaws, until terminal lug barrel butts flush against tool stop. See Figure 4-5 for correct insertion method.

Tool Wire Size Range	Tool	Gaging Dimension (Inches)			
		For Barrel Crimping Jaws		*For Crimping Insulation Jaws	
		GO	NO GO	GO	NO GO
22-16	47386	.098	.104	.030	.090
	59250	.109	.11	—	—
16-14	47387	.108	.114	.040	.100
	59250	.115	.125	—	—
12-10	59239	.139	.145	.064	.139

*When tools have adjustment pins GO gaging is done with pins in Position No. 1; and NO GO gaging with pins in Position No. 3.

Table 4-3. Gaging dimensions for AMP tools

4. Squeeze tool handles slowly until tool jaws hold terminal lug barrel firmly in place, but without denting it.

5. Insert stripped wire into terminal lug barrel until wire insulation butts flush against end of barrel.

6. Squeeze tool handles until ratchet releases.

7. Remove completed assembly and examine it for proper crimp, in accordance with the instructions on page 4-20.

Crimping procedure for AMP power tools. Power-crimp AMP preinsulated copper terminal lugs with AMP power tools as follows:

1. Select, from Table 4-2, power tool and the correct insert for the terminal lug size being crimped.

2. Install insert in tool.

3. Check tool for proper adjustment in accordance with the procedures on page 4-3.

4. Check that removable stop plate is present on insert. When using tools No. 69005 and 69010 set insulation adjustment pins as follows:

   a. Place both adjustment pins in the #3 position.

   b. Place terminal in the tool crimping jaws and insert the unstripped wire into the terminal insulation grip only. Crimp the terminal.

   c. Hold terminal firmly and bend the wire back and forth once. The terminal insulation grip should hold the wire.

   d. If wire pulls out, set adjustment pins in the #2 position. Repeat test until insulation grip holds the wire firmly.

5. Strip wire insulation, using recommended stripping practices described in Chapter 1, page 1-14; stripping lengths are given in Table 4-4.

6. Insert stripped wire into terminal lug barrel until wire insulation butts flush against end of barrel.

7. Insert wire and terminal lug assembly in tool, until terminal lug barrel butts flush against tool stop (when tool has no stop, center terminal lug barrel under indentor).

8. Squeeze tool trigger, or actuate foot treadle.

9. Remove completed assembly and examine it for proper crimp in accordance with the procedure on page 4-20.

Crimping procedure for Burndy power tools. Power-crimp Burndy preinsulated copper terminal lugs with Burndy power tools as follows:

1. Strip wire insulation, using recommended stripping practices described in Chapter 1 on page 1-14. Stripping lengths are given in Table 4-4.

2. Select from Table 4-2 power tool that fits crimping conditions best, and the correct die set.

3. When using tool No. Y10NCP, check that removable stop-plate is present on die.

WARNING: *ALWAYS disconnect power tool from its pressure source, BEFORE installing or removing dies.*

4. Install insert in tool.

5. Insert stripped wire into terminal lug barrel until wire insulation butts flush against end of barrel.

6. Insert wire and terminal lug assembly in tool, until terminal lug barrel butts flush against tool stop (when tool has no stop, center terminal lug barrel under indentor).

7. Squeeze tool trigger, or actuate foot treadle.

8. Remove completed assembly and examine it for proper crimp in accordance with the procedure on page 4-20.

Wire Size	Stripping Length (in inches)
#26 & #24	5/32
#22 through #14	3/16
#12 & #10	9/32

Table 4-4. Wire stripping lengths terminal for small copper lugs

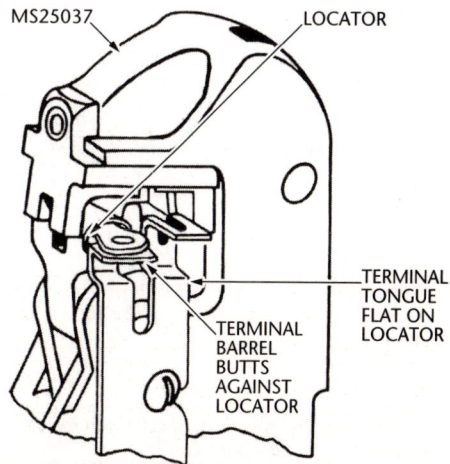

Figure 4-5. Inserting terminal lug into hand tool

Crimping procedure for T&B power tools.
Power-crimp T&B preinsulated copper terminal lugs with T&B power tools, as follows:

1. Strip wire insulation, using recommended stripping practices described in Chapter 1, on page 1-14; stripping lengths are given in Table 4-4.

2. Select from Table 4-2 the power tool that fits crimping condition best. T&B power tools listed include dies for wire sizes No. 22-No. 10.

   WARNING: *ALWAYS disconnect power tool from its pressure source, BEFORE installing or removing dies.*

3. Check power tools for proper adjustment in accordance with the procedure on page 4-3.

4. Insert stripped wire into terminal lug barrel until wire insulation butts flush against end of barrel.

5. Insert wire and terminal lug assembly in tool, until terminal lug barrel butts flush against tool stop. When tool has no stop, center terminal lug barrel under indentor.

6. Squeeze tool trigger, or actuate foot treadle.

7. Remove completed assembly and examine it for proper crimp in accordance with the procedure on page 4-20.

Section 3
Terminating Large Copper Wires

Terminal Lugs. Copper terminal lugs of all three styles, straight, angle and flag, are used to terminate copper wires of sizes No. 8 through No. 4/0. The style to be used depends on existing space conditions. As indicated in Table 4-5, these terminal lugs are available uninsulated in all types, and preinsulated in the straight and angle types, though not necessarily in all wire sizes, or from all manufacturers. Straight preinsulated terminal lugs conform to mil-spec MIL-T-7928 and Standard Drawing MS25036. As shown in Figure 4-2, preinsulated terminal lugs have the insulation extending beyond the barrel, so that it will cover a portion of the wire insulation. This makes the use of a separate insulating sleeve unnecessary. Straight uninsulated terminal lugs conform to mil-spec MIL-T-7928 and Standard Drawing MS20659.

Insulating sleeves. Uninsulated straight and right-angle type terminal lugs are insulated, after assembly to wire, by heat-shrink tubing or by lengths of transparent tubing, called sleeves. These methods of insulation provide

	Crimping Tools and Size Range		
Terminal Lugs	Band	Portable Power	Stationary Power
1. Straight, Pre-insulated			
MS25036	—	MS25494, with MS25441-1 head, MS25441-3 hose & MS23002 dies (8-4/0)	MS25441, with MS23002 dies (8-4/0)
2. Straight, Uninsulated			
MS20659	AN3427(8-4/0)	—	—
Burndy	—	Y29B (8-4/0)	Y29NSC (8-2/0)
T&B	—	13986, with head 13642M (8-4/0)	13581, with head 13642M (8-4/0) 21073 (8-4/0)
3. Right-angle, Pre-insulated			
AMP	—	69061 (8-2)	69068 (8-2) (8-4/0)**
Burndy	MY28-6(8-2/0)	Y29B (8-2/0)	Y29BUC (8-2/0)
4. Right-angle, Uninsulated			
Burndy	MY28	Y29B (8-2/0)	—
T&B	WT-115 (8-4) WT-127 (2-4/0)	—	21073 (8-4/0)
5. Hag, Uninsulated			
MS25189	AN3427		
Burndy	—	Y29B (8-2/0)	Y29NSC (8-1/0)
T&B	—	13586, with head 13642M (8-4/0)	13581, with head 13642M (8-4/0)

* Using Head No. 69051 with proper dies for each wire size
** Using Head No. 69000 with proper dies for each wire size

Table 4-5 Copper terminal lugs (large) and crimping tools

electrical and mechanical protection at the connection. When the size of sleeving used is such that it will fit tightly over the terminal lug, the sleeving need not be tied; otherwise, it is tied with lacing cord. (See Figure 4-6). Tight fitting sleeves are expanded in methyl-ethyl-ketone solvent before installation. When the solvent evaporates, the sleeve will shrink tightly over the terminal lug.

WARNING: *Methyl-ethyl-ketone is highly inflammable.*

Tools

Crimping tools. Hand, portable power and stationary power tools are available for crimping large copper terminal lugs. The crimping tools used on the preinsulated terminal lugs crimp the barrel to the conductor, and simultaneously crimp the terminal lug insulation to the wire insulation. For best results Class 1 tools (those designated as either AN or MS tools) should be used. If Class 2 tools must be used, the crimping tools of each manufacturer should be used for the terminal lugs of the same manufacturer. The crimping tools most commonly used are listed in this manner in Table 4-5. The numbers in parentheses after the tool numbers indicate the range of terminal lug sizes the tool can crimp; do not use any tools for larger terminal lug sizes than listed as this will result in a poor connection.

NOTE: *Use power tools for large terminal lugs whenever possible.*

Hand tools. Hand crimping tools are available for MS20659, Burndy and T&B large copper uninsulated terminal lugs. These tools are illustrated in Figure 4-7. The AN3427 and the Burndy hand tools, No. MY28 and MY28-6, each have a single nest and indentor; the nest is adjustable by means of a screw to the proper position for each terminal lug size. Note the shape of the indentor in Figure 4-7a. It is used for crimping copper terminal lugs; do not confuse it with the indentor used for aluminum terminals as shown in Figure 4-13. T&B tool No. WT 115 has a single indentor and a nest wheel containing a specific nest for each terminal lug size; the nest is identified by means of letters, which correspond to terminal lug sizes, as listed on the tool itself. T&B tool No. WT 127 requires a different nest for each terminal lug size to be crimped; these nests are listed in Table 4-6.

Hand tool adjustment. The hand tool AN3427 has two positioning plates as shown in Figure 4-7a. One plate is for installing one-piece terminal lugs and the other is for installing two piece and flag type terminal lugs. Adjust the position of each plate as follows:

1. Adjust one-piece terminal position plate to indicator guide line of No. 8 and No. 6 when an .090 diameter pin is just held by indentor and nest.
2. Adjust two-piece and flag type terminal position plate to indicator guide line of No. 8 and No. 6 when a .120 diameter pin is just held by indentor and nest.

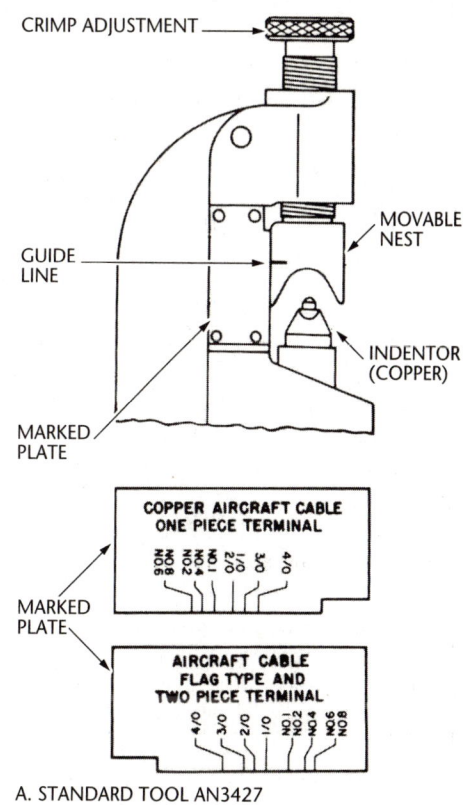

A. STANDARD TOOL AN3427

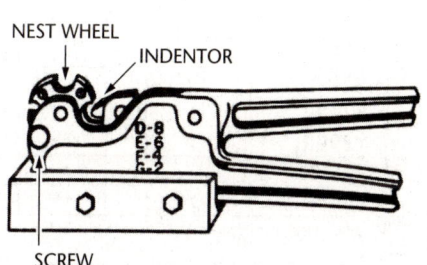

B. THOMAS & BETTS TOOL WT115

Figure 4-7. Tools - hand crimping large copper terminal lugs

Terminal Lug Size	For Use on Uninsulated Lugs	
	Nest Number	Letter
2	21651	G
1	21652	H
1/0	21653	J
2/0	21654	K
3/0	21655	L
4/0	21656	M

Table 4-6. Nests for T&B hand tool WT 127

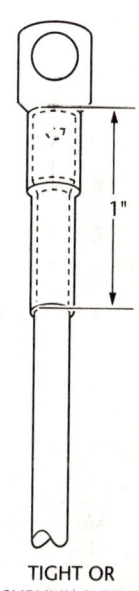

TIGHT OR SHRUNK SLEEVE

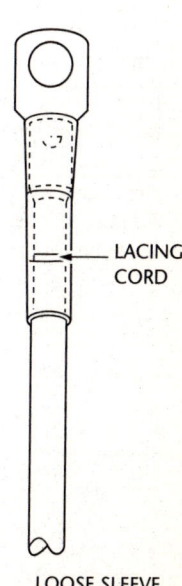

LOOSE SLEEVE

Figure 4-6. Insulating sleeves

Solderless Terminations and Splices

Table 4-7 Dies and gages for MS25441 and MS25494 power tools

Terminal Lug Size	MS Die Part Number (with Head MS25441-1)	MS Gage Part No.
8	MS23002-8	MS23003-8
6	MS23002-6	MS23003-6
4	MS23002-4	MS23003-4
2	MS23002-2	MS23003-2
1	MS23002-1	MS23003-1
1/0	MS23002-01	MS23003-01
2/0	MS23002-02	MS23003-02
3/0	MS23002-03	MS23003-03
4/0	MS23002-04	MS23003-04

Table 4-8. Dies for AMP power tools 69061 and 69068

Terminal Lug Size	For Head No. 69061		For Head No. 69066	
	Vinyl Ins.	Nylon Ins.	Vinyl Ins.	Nylon Ins.
8	48752	47820	48858	
6	48753	47821	48859	
4	48754	47822	48860	
2	48755	47823	48861	
1/0			48756	47824
2/0			48757	47825
3/0			48758	47915
4/0			48759	47918

Table 4-9 Die sets for Burndy power tools Y29B & Y29NSC

Terminal Lug Size	For Use on Straight & Right-Angle Type Uninsulated Terminal Lugs		For Use on Flag Type Uninsulated Terminal Lugs	
	Nest No.	Indentor No.	Nest No.	Indentor No.
8	DV8L	Y29PL	DV8B	Y29PBL
6	DV6L	Y29PL	DV6L	Y29PBL
4	DV4L	Y29PL	DV4BL	Y29PL
2	DV2L	Y29PL	DV2BL	Y29PBL
1	DV1L	Y29PL	DV1BL	Y29PL
1/0	DV25L	Y29PR	DV25BL	Y29PR
2/0	DV26L	Y29PR	*DV28L	*Y29PR
3/0	*DV27L	*Y29PR	—	—
4/0	*DV28L	*Y29PR	*DV28L	*Y29PR

*Use these nests & indentors on tool No. Y29B only, since tool No. Y29NSC is not powerful enough to be used on terminal lugs of this size.

The Burndy hand tool MY-28 is similar to Standard Tool AN 3427 and is adjusted the same way.

Thomas & Betts hand tool WT 115 is not adjustable and must be returned to the manufacturer if the following check reveals any defects.

1. Nest D should accept a 5/32 inch (.156) diameter pin when tool is fully closed.
2. Nest D should not accept a .166 diameter pin when tool is fully closed.

Power tools. Power crimping tools operate on either hydraulic power or air pressure. The power tools MS25441 and MS25494, all AMP tools, and T&B tools No. 13586 and 13581 are hydraulically powered. All Burndy tools, and T&B tools other than those previously noted operate on air pressure. See Figure 4-8 for illustrations of power tools.

Dies for the Class 1 tools MS25441 and MS25494 are listed in Table 4-7. Each terminal lug size requires a special die.

AMP power tool dies are listed in Table 4-8. Each terminal lug size requires a special die.

Each Burndy power tool, except tool No. Y29BUC, requires a different die set, (nest and indentor combination), for each size and type of terminal lug. These die sets are listed in Table 4-9 for uninsulated terminal lugs, and in Table 4-10 for preinsulated terminal lugs. Burndy tool No. Y29BUC, uses two rack and indentor combinations, as listed in Table 4-10.

T&B power tools numbers 13586 and 13581 require separate indentor/nest combinations for each terminal size, as listed in Table 4-11. T&B power tool No. 21073 includes dies.

Power tools adjustment. The MS series, AMP, Burndy and T&B power tools can be checked for proper adjustment. For good crimping results, this must be done before each series of crimping operations. When tool is adjustable, proper correction must be made;

Table 4-10. Die sets for Burndy power tools Y29B & Y29BUC for PRE-INSULATED COPPER terminal lugs

Terminal Lug Size	Burndy Die Set Number for Right Angle Terminal Lugs			
	For Tool No. Y29B		For Tool No. Y29BUC	
	Nest No.	Indentor No.	Rack No.	Indentor No.
8	DEV8L	Y29PLE-1	Y29BUR-7	Y29PU5
6	DEV6L	Y29PLE-1	Y29BUR-7	Y29PU5
4	DEV4L	Y29PLE-1	Y29BUR-7	Y29PU5
2	DEV2L	Y29PLE	Y29BUR-8	Y29PU1
1	DV26L	Y29PLE	Y29BUR-8	Y29PU1
1/0	DEV25L	Y29PLE	Y29BUR-8	Y29PU1
2/0	DEV26L	Y29PLE	Y29BUR-8	Y29PU1

Terminal Lug Size	Straight Type Terminals		Flag Type Terminals	
	Indentor	Nest	Indentor	Nest
8	13649M	13651	21731M	21733
6	13649M	13652	21731M	21734
4	13649M	13653	21731M	21735
2	13650M	13654	21732M	21736
1	13650M	13655	21732M	21737
1/0	13650M	13656	21732M	21738
2/0	13650M	13657	21732M	21739
3/0	13650M	13658	21732M	21740
4/0	13650M	13659	21732M	21741

Table 4-11. Die sets for T&B power tools 13586 and 13581

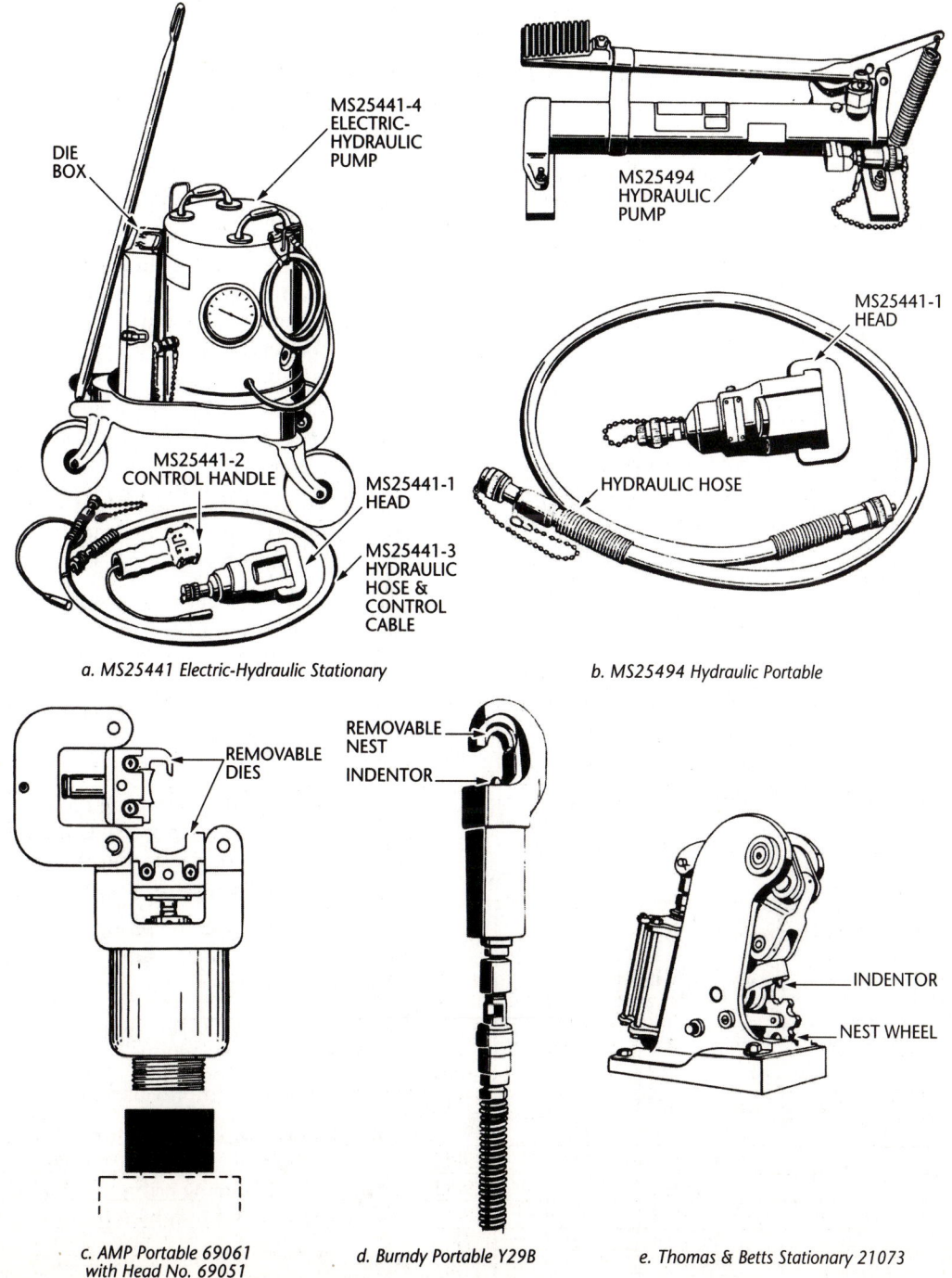

a. MS25441 Electric-Hydraulic Stationary
b. MS25494 Hydraulic Portable
c. AMP Portable 69061 with Head No. 69051
d. Burndy Portable Y29B
e. Thomas & Betts Stationary 21073

Figure 4-8. Tools-power crimping-large copper terminal lugs

otherwise, tool must be returned to manufacturer for repair. Check tools as follows:

  a. Gage the dies of the tool in the closed position with the appropriate GO/NO GO gages listed in Table 4-7.

  b. AMP tools are checked with tool fully bottomed. Gap between barrel crimping jaws of the dies shall meet the requirements of Table 4-12. Note that "GO" gages shall be able to enter between jaws, and that "NO GO" gages shall be unable to enter.

  c. Burndy tools Y29B and Y29BUC when equipped with nests and indentors for preinsulated terminal lugs (Table 4-10) are checked to "G" dimensions listed in Table 4-13. The "G" dimension is the clearance between nest and indentor when the tool is fully bottomed.

  d. T&B power tool No. 21073, is checked when tool is fully bottomed. An 11/32 inch (.344) drill shall enter between nest K and the indentor, and a 23/64 inch (.360) drill shall not enter. Dies for T&B tools 13586 and 13581 (similar to the MS series power tools listed in Table 4-7) are gaged in the same way as the MS dies.

AN Wire Size	Gaging Dimensions (in Inches)	
	GO	NO GO
8	.200	.208
6	.236	.244
4	.268	.276
2	.318	.326
1/0	.399	.409
2/0	.445	.455
3/0	.497	.507
4/0	.559	.569

Table 4-12. Gaging dimensions for AMP tools

AN Wire Size	"G" Dimensions (in Inches)			
	Y29B		Y29BUC	
	Max.	Min.	Max.	Min.
8	.121	.094	.147	.132
6	.176	.149	.172	.157
4	.186	.161	.192	.177
2	.216	.184	.210	.190
1	.296	.266	.290	.270
1/0	.325	.295	.320	.300
2/0	—	—	.340	.320

Table 4-13. "G" dimensions for Burndy power tools

Crimping Procedure for Hand Tools

AN3427 Standard Hand Tools (Burndy Tool #MY-28). Hand crimping of large copper terminal lugs is done as follows with Standard Tool AN3427 (Burndy #MY-28):

1. Strip wire insulation as described in Chapter 1 on page 1-14. Wire stripping length for both uninsulated and preinsulated large copper terminals shall be conductor barrel length plus 1/16 inch.

2. Set tool to proper crimping position by use of adjustment screw. (See Figure 4-7a.)

3. Slide insulating sleeve (see page 4-6) over wire insulation, well clear of crimping area. Do not use insulating sleeves with flag type terminal lugs.

4. Insert stripped wire into terminal lug barrel until wire insulation butts flush against end of barrel.

5. Insert wire and terminal lug assembly in tool nest. Center barrel under tool indentor. Close handles all the way (until movable handle reaches fixed stop).

NOTE: *Tongue of flag type terminal lug must rest flat against upper level of nest as shown in Figure 4-9.*

6. Remove completed assembly and examine it for proper crimp, in accordance with the instructions on page 4-20.

7. Slide insulating sleeve over terminal lug barrel and secure it in accordance with the instructions on page 4-6 and Figure 4-6.

Crimping procedure for T&B hand tools. Hand-crimp large T&B copper terminal lugs as follows:

1. Strip off wire insulation, using recommended practices described in Chapter 1 on page 1-14. See previous number 1 for instructions on stripping length.

2. Select from Table 4-5 the proper hand tool for terminal lug type and size being crimped.

3. Check tool for correct adjustment in accordance with the instructions on page 4-8.

4. Prepare tool for crimping as follows:

  a. Tool No. WT 115: loosen thumbscrew, turn nestwheel until proper nest for terminal lug size being crimped is opposite indentor and retighten thumbscrew.

  b. Tool No. WT 127: insert proper nest for terminal lug size being crimped, in accordance with Table 4-6.

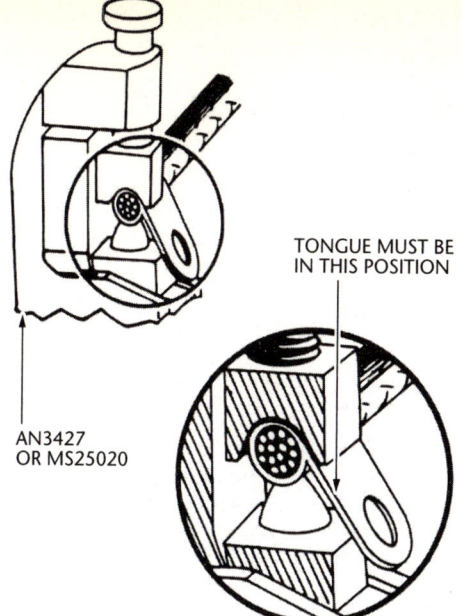

Figure 4-9. Positioning flag type terminal lugs

5. Slide insulating sleeve if required (see page 4-6) over wire insulation well clear of crimping area. Do not use insulating sleeve with preinsulated terminal lugs, or with flag type terminal lugs.

6. Insert stripped wire into terminal lug barrel until wire insulation butts flush against end of barrel.

7. Insert wire and terminal lug assembly in tool nest. Center barrel under tool indentor.

8. Close tool handles all the way (until movable handle reaches fixed stop).

9. Remove completed assembly and examine it for proper crimp, in accordance with the instructions on page 4-20.

10. Slide insulating sleeve over terminal lug barrel and secure it, in accordance with the instructions on page 4-7 and Figure 4-6.

Crimping Procedure for Power Tools

Crimping procedure for MS series power tools. Power-crimp large copper terminals MS25036, as follows:

1. Select proper die for wire size from Table 4-7, and install die in tool.

2. Strip wire insulation, using recommended practices described in Chapter 1 on page 1-14; stripping lengths are given on previous page.

3. Insert stripped wire into terminal barrel until wire insulation butts flush against end of barrel.

4. Insert wire and terminal lug assembly into die.

5. Actuate the power tool (press button on handle for electrically operated tool; actuate handle for manual hydraulically operated tool).

6. Remove the crimped assembly, and examine it for proper crimp, in accordance with the instructions on page 4-20.

Crimping procedure for AMP power tools. Power-crimp large AMP copper preinsulated terminal lugs as follows:

1. Strip wire insulation, using recommended stripping practices in Chapter 1 on page 1-14; stripping lengths are given on previous page.

2. Select the proper insert for the terminal lug size being crimped from Table 4-8.

   WARNING: *ALWAYS disconnect a power tool from its pressure source, whether air or hydraulic, BEFORE installing or removing its insert.*

3. Install insert in tool.

4. Check tool for proper adjustment in accordance with the procedure on previous page.

5. Insert stripped wire into terminal lug barrel until wire insulation butts flush against end of barrel.

6. Insert wire and terminal lug assembly in tool nest, so that terminal lug barrel butts flush against tool stop.

7. Squeeze tool trigger or actuate foot treadle.

8. Remove completed assembly and examine it for proper crimp in accordance with the procedure on page 4-20.

Crimping procedure for Burndy power tools. Power-crimp large Burndy copper terminal lugs as follows:

1. Strip wire insulation, using recommended stripping practices described in Chapter 1 on page 1-14; stripping lengths are given on page 4-10.

2. Select from Table 4-5 a power tool for terminal lug style and size being crimped. Pick tool that fits crimping conditions best.

3. Select proper insert for tool from Tables 4-9 or 4-10, depending on terminal lug style.

   WARNING: *ALWAYS disconnect power tool from its pressure source, BEFORE installing or removing its insert.*

When using tool No. Y29BUC, ALWAYS remove nest rack BEFORE installing or removing indentor.

4. Install insert in tool.

5. Slide insulating sleeve over wires to be connected to uninsulated terminal lugs (see page 4-6). Do not use insulating sleeves for flag type terminals.

6. Insert stripped wire into terminal lug barrel until wire insulation butts flush against end of barrel.

7. Insert wire and terminal lug assembly in tool nest. Center barrel under indentor.

8. Squeeze tool trigger or actuate foot treadle.

9. Remove completed assembly and examine it for proper crimp in accordance with the procedure on page 4-20.

10. Where used, slide insulating sleeve over terminal lug barrel and secure, in accordance with the procedure on page 4-6 and Figure 4-6.

Crimping procedure for T&B power tools. Power-crimp large T&B copper terminal lugs as follows:

1. Strip wire insulation, using recommended stripping practices described in Chapter 1 on page 1-14; stripping lengths are given on page 4-10.

2. Select from Table 4-5 a power tool for terminal lug style and size being crimped. Pick tool that fits crimping conditions best.

3. Set tool No. 21073 by rotating nest wheel to index proper nest opposite indentor. Install correct nest and indentors in other tools, from Table 4-11.

4. Slide insulating sleeve over wires to be connected to uninsulated terminal lugs (see page 4-6). Do not use insulating sleeves for flag type terminals.

5. Insert stripped wire into terminal lug barrel until wire insulation butts flush against end of barrel.

6. Insert wire and terminal lug assembly in tool nest. Center barrel under indentor.

7. Squeeze tool trigger or actuate foot treadle.

8. Remove completed assembly and examine it for proper crimp in accordance with the procedure on page 4-20.

9. Where used, slide insulating sleeve over terminal lug barrel and secure, in accordance with the procedure on page 4-6 and Figure 4-6.

Section 4
High Temperature Terminal Lugs

In areas where the temperature is expected to exceed 300°F, special terminal lugs are used. These are copper, either nickel or silver plated, and insulated with a high temperature material such as TFE or Kel-F. These terminal lugs can be used where the continuous operating temperature does not exceed 500°F. Terminal lugs made by Burndy and Thomas & Betts are preinsulated; those made by AMP are provided with a TFE sleeve having a metal crimping ring which is crimped into the terminal lug after the wires have been inserted.

Tools

Crimping tools. Hand, portable power and stationary power tools are used to install high temperature terminal lugs on electrical wire. These tools are listed in Table 4-14.

Hand tools. Hand tools are of the self-locking ratchet type. AMP hand tools are used on one terminal size only; the tool handle is color coded to match the insulation color of the terminal. Burndy and T&B tools are furnished with dies for the wire ranges; the opening for each wire size range is marked.

Power tools. Power crimping tools operate on air pressure. AMP tools require an insert or die for each wire size range; Burndy tools require a separate die set for each wire size range. (See Table 4-14).

Crimping Procedure

Crimping procedure for AMP hand tools. Hand crimp AMP high temperature terminal lugs as follows: (See Figure 4-10).

1. Strip wire length of terminal barrel, plus 1/32 inch.

2. Open tool jaws by squeezing tool handles until ratchet releases.

3. Place terminal in crimping nest and center the terminal barrel in the nest.

4. Place insulating sleeve on the wire, well back from the crimping area.

5. Insert stripped wire into terminal, and crimp by closing tool handles until ratchet releases. Remove the crimped terminal and install the insulating sleeve over the barrel as shown.

Wire Size Range	Color Code	Hand Tool	Crimping Tools	
			Portable Power	Stationary Power
AMP				
#22 - #20	Silver	46467	69100, with die 46464	—
#18 - #16	Red	46468	69100, with die 46465	—
#14	Blue	46469	69100, with die 46466	—
#12 - #10	Yellow	46470	—	—
Burndy				
#22 - #16	Red	MR8-36T	M8ND, Y8ND, with die set N10JT-9	Y10NCP, with die set R10HJT-1T
#16 - #14	Blue	MR8-36T	M8ND, Y8ND, with die set N10JT-9	Y10NCP, with die set R10HJT-1T
#12 - #10	Yellow	MR8-36T	M8ND, Y8ND, with die set N10JT-9	Y10NCP, with die set R10HJT-1T
#8	—	MY4-25	Y29NC, with die set DJV8C-HT	
T&B				
#22 - #20	Green Stripe	WT-187	—	—
#20 - #18	Red Stripe	WT-188	—	—
#16 - #14	Yellow Stripe	WT-188	—	—

Table 4-14. High temperature terminal lugs and crimping tools

6. Center the sleeve ring in the crimping nest of tool; close handles only far enough to hold the sleeve firmly in place. Wire maybe moved to position terminal sleeve, so that it extends 1/32 inch beyond end of terminal barrel.

7. Complete crimp by squeezing handles until ratchet releases.

Crimping procedure for Burndy hand tools. The procedure for crimping Burndy high temperature terminal lugs is as follows:

1. Strip wire to length of terminal barrel, plus 1/32 inch.
2. Insert stripped wire into terminal.
3. Open jaws of tool by squeezing handles together as far as they will go, and then releasing them.
4. Place terminal and wire assembly into correct groove of tool, so that terminal barrel rests against the stop.
5. Crimp by closing handles all the way.

Crimping procedure for T&B hand tools. The procedure for crimping T&B high temperature terminal lugs is similar to the procedure described on page 4-4 for small copper terminal lugs.

Crimping procedure for AMP power tools. Power-crimp AMP high temperature terminal lugs as follows:

1. Strip wire to length of terminal barrel, plus 1/32 inch.
2. Open die and insert terminal barrel so it centers in stationary die nest. Brazed seam on terminal barrel should face upward.
3. Close jaws just enough to hold terminal in place. Make sure terminal barrel remains centered in die.
4. Place insulating sleeve on wire back from stripped end.
5. Insert stripped wire in terminal barrel, and crimp, holding wire firmly in place.
6. Remove assembly, and pull sleeve into place over terminal barrel.
7. Place assembly into die so that sleeve ring is centered in nest of stationary die.
8. Close jaws just enough to hold assembly, and make sure sleeve ring remains centered. Complete crimp and remove assembly.

Crimping procedure for Burndy power tools. The procedure for crimping Burndy high temperature terminal lugs is similar to the procedure for crimping small copper terminals to wire. See page 4-5.

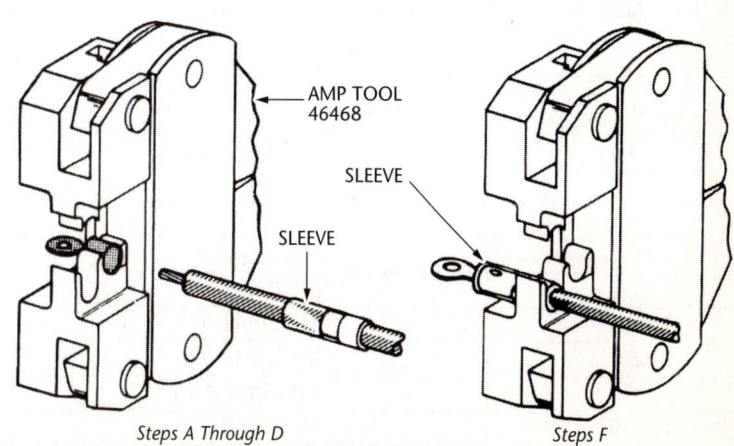

Figure 4-10. Crimping high temperature terminal lugs

Section 5
Terminating Aluminum Wire

Aluminum wire is used in aircraft because of its weight advantage over copper. Aluminum, however, has the disadvantage of being softer than copper. Further, bending aluminum wire will cause "work hardening" of the metal that makes it more brittle. This will result in failure or breakage of strands much sooner than in copper wire. Aluminum also forms a high resistance oxide film immediately upon exposure to air. For these reasons, aluminum wire is no longer used in new construction. Where possible, aluminum wiring should be replaced with copper during repair procedures.

Aluminum wire may still be found in many older aircraft. It is important to follow carefully the recommended installation procedures in order to compensate for the disadvantages of aluminum wire.

WARNING: *Do not use any aluminum wire that has nicked or broken strands. Damaged strands will fail in service.*

Hand Tool	Hand Tool	Power Crimping Tools and Wire Size Range	
		Portable Power	Stationary Power
MS25435, 25436, 25437 and 25438	—	MS25494, with Head MS25441 (8-4/0)	MS25441-1 (8-4/0)
*Burndy: *Straight *MS25021	*MS25020	Y29B (8-4/0)	Y29NSC (8-4/0)
*Burndy: *Flag *MS25022	*MS25020	Y29NC (8-1/0)	—
*Items are obsolete			

Table 4-15. Aluminum terminal lugs and crimping tools

Terminal Size	Dies for MS Tools Head #MS25441-1
8	MS25442-8A
6	MS25442-6A
4	MS25442-4A
2	MS25442-2A
1	MS25442-1A
1/0	MS25442-01A
2/0	MS25442-02A
3/0	MS25442-03A
4/0	MS25442-04A

Table 4-16. Dies used on power tool for crimping aluminum terminal lugs

Aluminum terminal lugs. Aluminum terminal lugs are used only to terminate aluminum wires. See Figure 4-1 for typical connections. Aluminum terminal lugs are available in three types, straight, 90 degree upright, and angle (left or right). The barrels of aluminum terminal lugs are filled with a petroleum-base abrasive compound. This compound, by a grinding process during the crimping operation removes the oxide film from the aluminum. The compound also prevents oxide from re-forming in the completed connection. All aluminum terminals have an inspection hole to allow checking the depth of wire insertion (see Figure 4-1). This inspection hole is sealed with a removable plastic plug, which also serves to retain the oxide-inhibiting compound. Each aluminum terminal lug is marked with the letters "AL" indicating it is for use with aluminum wire, and also with the wire size it will accommodate.

CAUTION: *Do not remove the inspection plug until the crimp has been completed, and the wire insertion is to be inspected. Replace plug after inspection.*

Insulating sleeves. Aluminum terminal lugs are not preinsulated, therefore it is necessary to insulate them after assembly, by lengths of transparent flexible tubing, called sleeves. The sleeve provides mechanical and electrical protection at the connection. (See Figure 4-6).

Tools

Crimping tools. Power tools, either stationary or portable, are recommended to install aluminum terminal lugs. Use tools MS25441 and MS25494 to install MS aluminum terminal lugs. Install other aluminum terminal lugs with the tools listed in Table 4-15. The numbers in parentheses listed after the tool numbers indicate the range of wire sizes that can be safely crimped with that tool.

CAUTION: *Do not use any crimping tool beyond its rated capacity.*

Hand tools. The hand tool MS25020 shown in Figure 4-13 is obsolete for use on aluminum terminal lugs. This tool was formerly used to crimp MS25021 and MS25022 aluminum terminal lugs, which have now been superseded by the MS terminals listed in Table 4-15.

CAUTION: *Do not use the MS25020 tool on Terminals MS25435, 25436, 25437 or 25438. Use this tool only for such obsolete aluminum MS terminals as may be still stocked.*

Power tools. See Figure 4-8. MS 25441 is a complete electric-hydraulic tool, consisting of an electric pump, hydraulic head, hydraulic hose, and control handle. A foot operated pump MS25494 can be used to supply hydraulic pressure for MS25441-1 head. All power tools require a specific insert die for each terminal lug size. These dies are listed in Table 4-16, and 4-17. Burndy power tools may be used only to install the obsolete terminal lugs MS25021 and 25022.

CAUTION: *Do not use Burndy power tools on terminal lugs MS25435, 25436, 25437 or 25438.*

Power tool inspection. Dies for the tools MS25441 and MS25494 are checked with MS25472 gages. Check the dies in the closed position with the appropriate GO/NO GO gages listed in Table 4-18.

Crimping Procedures

Crimping procedure for power tools. Power-crimp aluminum terminal lugs (except MS25021 and 25022) as follows:

1. Select power tool from Table 4-15. Use tool that fits crimping conditions best.

2. Select proper die for wire size from Table 4-16. Die is stamped with the wire size on both upper and lower faces and with the letters AL. Install die in tool head.

   WARNING: *ALWAYS disconnect power tool from its pressure source BEFORE installing or removing die.*

3. Strip wire insulation carefully, using recommended stripping practices for aluminum wire described in Chapter 1 on page 1-14; stripping lengths are listed in Table 4-19.

4. Install insulating sleeve over wire insulation, well back from crimping area.

5. Inspect to see that inner barrel is well coated with compound.

6. Insert wire into terminal barrel.

   CAUTION: *Do not remove the inspection plug as this keeps the compound in the barrel. When the wire is inserted to the full depth of the barrel the compound is forced between and around the conductor strands.*

7. Wipe off any excess compound squeezed out of terminal lug barrel with a clean soft cloth.

8. Insert assembly into the die correctly positioned as shown in Figure 4-11.

9. Actuate the power tool. When using MS25441, press the lower button on the control handle. Do not release the button until the dies open automatically. When using MS25494, close the release knob

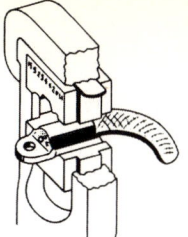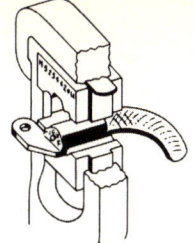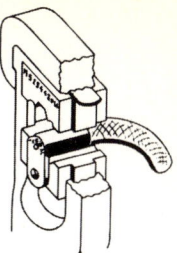

Figure 4-11. Positioning aluminum terminal lugs in die nests

Lug Size	Die Sets			
	For Straight Type Terminal Lugs		For Flag Type Terminal Lugs	
	Nest	Indentor	Nest	Indentor
8	DV8L-1	Y29PL	DV8B	Y29PBL
6	DV6L-1	Y29PL	DV6L	Y29PBL
4	DV4L	Y29PLA	DV4BL	Y29PBL
2	DV2L	Y29PLA	DV2BL	Y29PLA
1	DV25L	Y29PA	DV1BL	Y29PA
1/0	DV26L	Y29PA	DV25BL	Y29PA
2/0	*DV27L	*Y29PA	*DV28L	*Y29PA
3/0	*DV28L	*Y29PA	*DV28L	*Y29PA
4/0	*DV28L	*Y29PA	*DV28L	*Y29PA

*For use on tool No. Y29B only, since tools Y29NC & Y29NSC are not powerful enough for these terminal lug sizes.

Table 4-17. Die sets for Burndy for crimping MS25021 power tools Y29B, Y29NC & Y29NSC, and 25022 aluminum terminal lugs ONLY

Die Dash Number	Gage MS25472 Dash Number
-8A	-1
-6A	-2
-4A	-3
-2A	-4
-1A	-5
-01A	-6
-02A	-7
-03A	-8
-04A	-9

Table 4-18. Gages for MS25442 crimping dies

Wire Size	Stripping Length (in inches)		
	MS25435, 25436, 25437, & 25438	MS25021	MS25022
8	11/16	7/16	1/2
6	13/16	1/2	1/2
4	27/32	5/8	9/16
2	1-1/32	5/8	**5/8
1	1-1/32	3/4	*3/4
1/0	1-1/32	13/16	**13/16
2/0	1-7/32	15/16	15/16
3/0	1-9/32	15/16	15/16
4/0	1-7/16	1	1

Table 4-19. Stripping lengths for aluminum wire

and actuate the pump handle. When the dies have been closed to a pre-determined pressure the terminal can be released by opening the release valve.

NOTE: *Wire sizes No. 8 to No. 2/0 require only one crimp. Wire sizes No. 3/0 and No. 4/0 require two crimps. Locate the second crimp centrally on the portion of the barrel remaining after the first crimp. See Figure 4-12.*

10. Check visually to see that the correct wire size is imprinted on the barrel.

11. Remove the inspection plug and check visually or with the aid of a probe to see that wire is fully inserted. Replace the plug after inspection.

12. Slide insulating sleeve over the terminal lug barrel and secure in accordance with the procedure on page 4-7 and Figure 4-6.

Crimping procedure for MS25020 hand tool. This tool shall be used only to crimp obsolete aluminum terminal lugs MS25021 and 25022. The procedure is as follows:

1. Set tool to proper crimping position for terminal lug size being crimped. This is done by moving nest by means of adjustment screw until guide line on nest is centered over line on marked plate corresponding to terminal lug size being crimped. See Figure 4-13.

2. Carefully strip wire insulation, using recommended stripping practices for aluminum wire described in Chapter 1 on page 1-14; stripping lengths are listed in Table 4-19.

3. When crimping straight or right-angle type terminal lugs, slip insulating sleeve over wire insulation, well clear of crimping area.

4. Remove protective cover from terminal lug barrel. See Figure 4-14.

5. Check that terminal lug barrel is at least half-full of petrolatum-zinc dust compound. If not, add additional compound.

6. Hold terminal lug in one hand, with finger fully blocking inspection hole to prevent loss of compound, and pick up wire in other hand.

7. Insert stripped wire into terminal lug barrel. Do not twist wire during insertion.

8. Hold wire and terminal lug assembly firmly together and wipe off any excess compound squeezed out of terminal lug barrel with a soft dry cloth.

9. Check inspection hole, visually or with a wire probed, and make sure that wire end reaches all the way into terminal lug barrel.

10. Insert wire and terminal lug assembly into tool, so that terminal lug barrel is centered over tool indentor. Tongue of flag type terminal lug must rest flat against side of tool nest. See Figure 4-9.

11. Close tool handles all the way or until movable handle meets stop.

12. Remove completed assembly from tool.

13. Wipe off excess compound squeezed out of terminal lug barrel with a soft dry cloth.

14. Examine assembly for proper crimp in accordance with the procedure on page 4-20.

15. Slide insulating sleeve over terminal lug barrel and secure it in accordance with the procedure on page 4-6 and Figure 4-6.

Crimping procedure for Burndy power tools. The Burndy power tools listed in Table 4-15 are used only to crimp obsolete aluminum terminal lugs MS25021 and MS25022. The following procedure is used:

1. Select the correct insert from Table 4-17.

2. Install insert in tool.

WARNING: *ALWAYS disconnect power tool from its pressure source BEFORE installing or removing insert.*

3. Carefully strip wire insulation to length given in Table 4-19 using recommended stripping practices for aluminum wire described in Chapter 1 on page 1-14.

4. When crimping straight type terminal lugs, slip insulating sleeve over wire insulation, well clear of crimping area.

5. Remove protective cover from terminal lug barrel. (See figure 4-14).

6. Check that terminal lug barrel is at least half-full of petrolatum-zinc dust compound. If not, add additional compound.

7. Hold finger over inspection hole to prevent loss of compound. Insert stripped wire

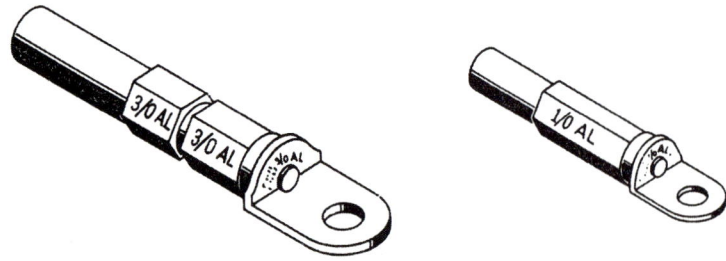

Figure 4-12. Single and double crimp on aluminum terminal lugs

into terminal lug barrel. Do not twist wire.

8. Hold wire and terminal lug firmly together and wipe off any excess compound squeezed out of terminal lug barrel with a soft dry cloth.

9. Check inspection hole, visually or with a wire probe, and make sure that wire end reaches all the way into terminal lug barrel.

10. Insert wire and terminal lug assembly into tool, so that terminal lug barrel is centered over tool indentor.

11. Actuate power tool.

12. Remove completed assembly from tool.

13. Examine assembly for proper crimp in accordance with the procedure on page 4-20.

14. Slide insulating sleeve (if used) over terminal lug assembly and secure it in accordance with the procedure on page 4-6 and Figure 4-6.

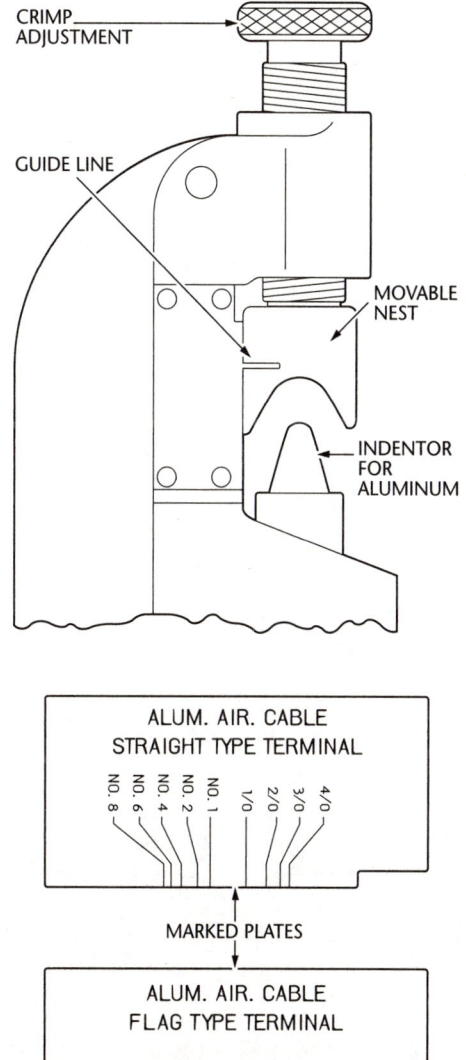

Figure 4-13. Tool-hand crimping-aluminum terminal lugs

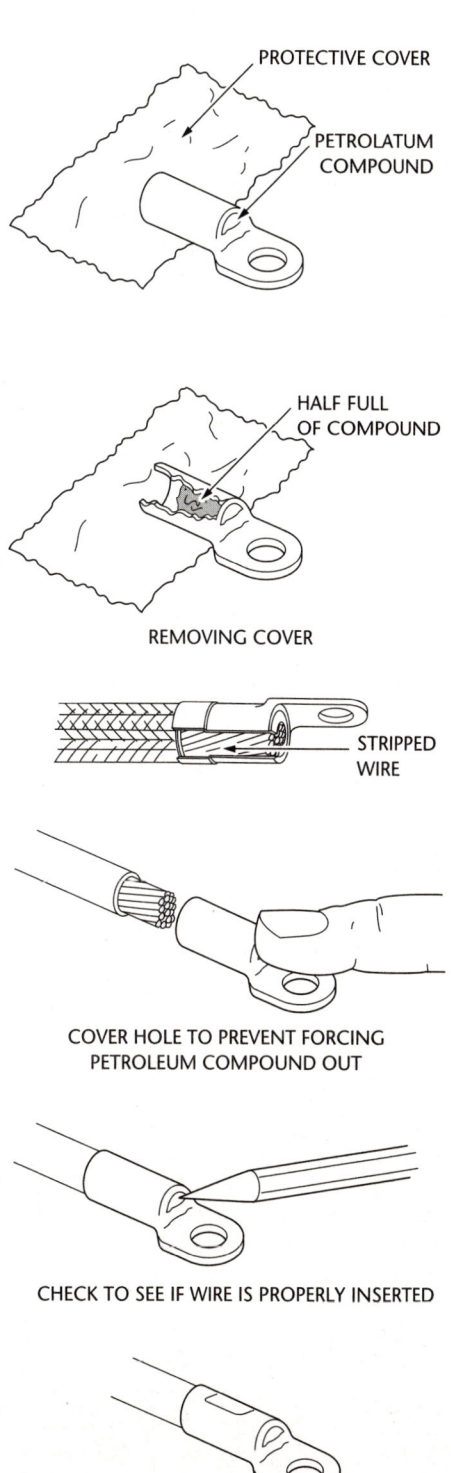

Figure 4-14. Inserting aluminum wire into MS25021 aluminum terminal lugs

Section 6

Splicing

Splicing Small Copper Wires

Preinsulated splices. Preinsulated permanent copper splices are used to join small copper wires of AN wire sizes No. 26 through No. 10. Typical splices are shown in Figure 4-1. Note that splice preinsulation extends over the wire insulation. Each splice size can be used for more than one wire size. Splices are color coded in the same manner as preinsulated small copper terminal lugs; refer to page 4-3 and Table 4-1 for details. Splices are also used when authorized to reduce wire sizes, as shown in Figure 4-15, for special applications (see Figure 4-16), and for multiple wire applications.

Crimping tools. The tool MS25037 is used to crimp the small copper splices MS25181. See page 4-4 for tool description and inspection.

Crimping procedure for MS25037 standard hand tool. Crimp small preinsulated copper splices in the No. 26 No. 10 wire size range as follows:

1. Swing the stop plate used for installing small insulated terminal lugs out of the way. (See Figure 4-17)
2. Strip wire to length given in Table 4-20, following the procedures described in Chapter 1 on page 1-14.
3. Position the splice in the correct die nest as indicated by instruction plate, so that side of splice to be crimped is on the

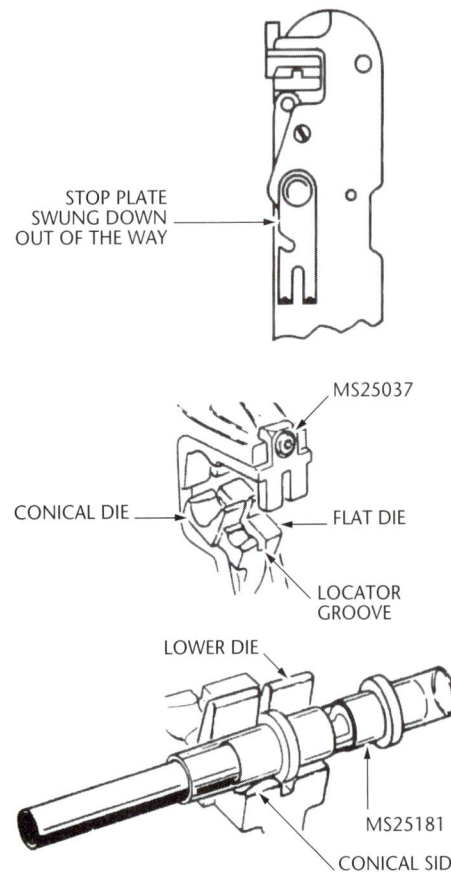

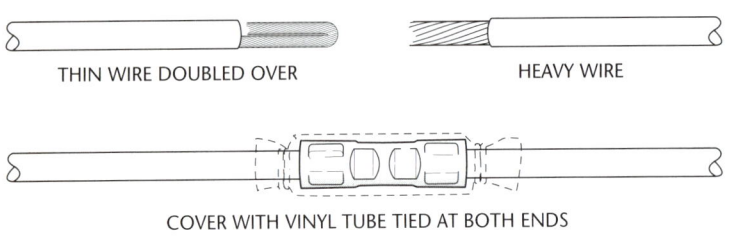

Figure 4-15. Reducing wire size with permanent splice

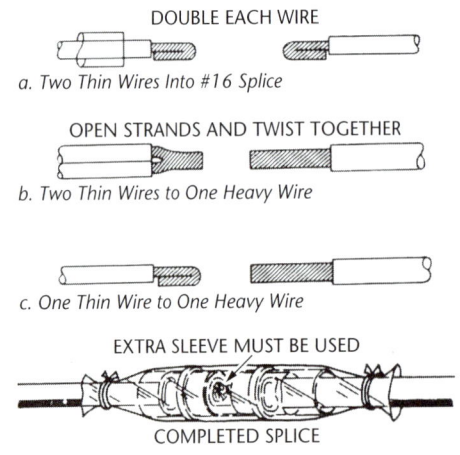

Figure 4-16. Special splices

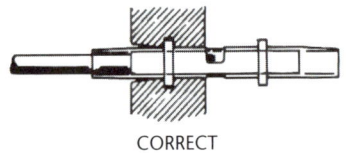

CORRECT

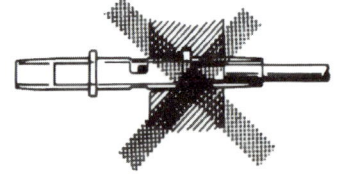

INCORRECT

Figure 4-17. Locating splice in crimping tool

Wire Size	Stripping Length (inches)
26-24	5/32
22-14	7/32
12-10	5/16

Table 4-20. Stripping length for small copper splices

conical die side of the locator groove, as shown in Figure 4-16.

4. Insert wire into the "wire in" side of splice so that the stripped conductor butts against the stop in the center of the splice. This can be seen through the splice inspection window.

5. Crimp by closing tool handles; tool will not open until full crimping pressure has been applied.

6. After crimping, check that wire end is still visible through inspection window.

7. Reverse position of splice, and repeat steps 2 through 6 to crimp other side of splice.

Splicing Large Copper Wires

Splices. Uninsulated splices are used to join large copper wires of sizes No. 8 through No. 4/0. There is a different splice for each wire size. Uninsulated splices are insulated after assembly with either heat-shrink tubing or transparent flexible sleeving to provide electrical and mechanical protection. If flexible sleeving method is used, cut sleeve to length in accordance with Table 4-21. Center sleeve over splice and tie at both ends with cord as shown in Figure 4-18.

Crimping Tools. The crimping tools for Burndy large copper splices are AN3427 (standard tool) and MY28, Y29B and Y29NC. AN3427 and MY28 are hand tools: Y29B and Y29NC are portable air powered tools which require a different die set for each splice size: these die sets are listed in Table 4-22. The portable hydraulic crimping tool head for T&B large copper splices is 13642; it requires a different die set for each splice size, as listed in Table 4-23.

NOTE: *Use power tools, if available for large copper splices.*

Crimping procedures. Crimping procedures for hand or power tools are the same as those described on page 4-10 for large copper terminal lugs, except that wire is stripped to dimensions given in Table 4-24, and crimping operation is done twice, at both ends of splice.

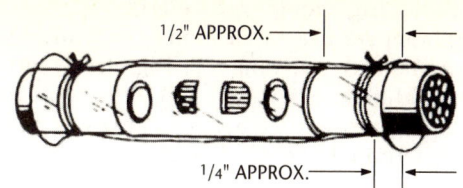

Figure 4-18. Insulating sleeves for splices size 8 and large

Splice Wire Size	Insulating Sleeve Length (in inches)
8	1-15/16
6	2-1/8
4	2-1/8
2	2-13/32
1	2-13/32
1/0	2-17/32
2/0	2-25/32
3/0	2-13/16
4/0	2-15/16

Table 4-21. Length of insulating sleeves

Splice Wire Size	Die Set Number	
	Nest	Indentor
8	DV8L	Y29PL
6	DV6L	Y29PL
4	DV4L	Y29PL
2	DV2L	Y29PL
1	DV1L	Y29PL
1/0	DV25L	Y29PR
2/0	DV26L	Y29PR
3/0	*DV27L	*Y29PR
4/0	*DV28L	*Y29PR

*For tool No. Y29B only, since tool No. Y29NC is not powerful enough for these sizes.

Table 4-22 Die sets used on Burndy tools Y29B & Y29NC for crimping large copper splices

Splice Wire Size	Indentor	Nest
8	13649M	13651
6	13649M	13652
4	13649M	13653
2	13650M	13654
1	13650M	13655
1/0	13650M	13656
2/0	13650M	13657
3/0	13650M	13658
4/0	13650M	13659

Table 4-23 Die sets used for crimping on T&B head large copper 13642 splices

Splicing high temperature wires. Splices for high temperature applications are available in the same wire size ranges as high temperature terminal lugs, (see Table 4-14.) The tools and crimping procedures are the same for splices as for terminal lugs. Crimp splice at both ends.

Splicing aluminum wires. Splice large aluminum wires sizes No. 8 through No. 4/0 with splice MS25439. Use the MS series power tools listed in Table 4-15, with the correct dies from Table 4-16. Follow the same procedure as for aluminum terminal lugs outlined on page 4-15, positioning the splice in the tool as shown in Figure 4-19. Crimp-splice at both ends.

AN Wire Size	Strip Length (inches)
8	7/16
6	1/2
4	1/2
2	5/8
1	5/8
1/0	11/16
2/0	13/16
3/0	13/16
4/0	7/8

Table 4-24. Wire stripping lengths for large copper splices

Multi-splicing. Multi-splicing is the crimping together of three or more wires in a single splice. This is a special application and may be used only when called for on the applicable engineering drawing. See Table 4-25 for acceptable combinations of wire sizes. For multi-splicing applications using standard MS25181 splices, the MS21980-128 ferrule is used to hold the wire together during the crimping operation.

Crimping multi-splice connections. To terminate multiple conductors in the MS25181 splice to make a multisplice, use MS21980-128 ferrule with MS17776 and MS25037 crimping tools, with the following procedure:

1. Strip wire ends 1/2 inch.
2. Insert wire in MS21980-128 ferrule to within 3/32 inch, plus or minus 1/32 inch from the insulation of all conductors.
3. Crimp ferrule, using the MS17776 crimping tool.
4. Round ferrule in rounding die of this tool.
5. Trim off excess wire strands protruding through open end of ferrule to extend not more than 1/32 inch beyond ferrule.
6. Insert this assembly into the MS25181-3 splice and crimp with MS25037 crimping tool.
7. Inspect ferrule assembly to insure that it is visible through window, or must not be more than 1/64 inch back under the edge of the splice window.

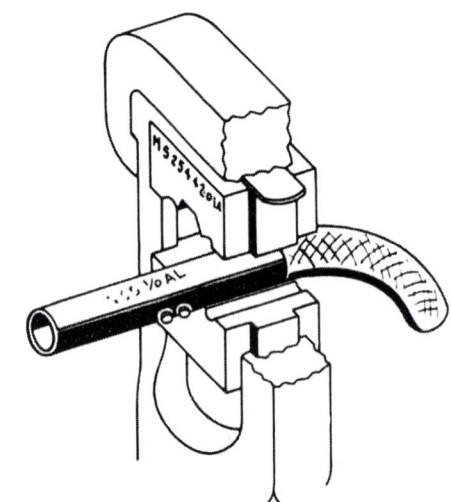

Figure 4-19. Positioning aluminum splice in die nest

Inspection of Crimped Connections

Visual inspection. Examine the crimped connection carefully for the following:

1. Indent centered on terminal lug barrel or splice barrels.
2. Indent in line with barrel; not cracked.
3. Terminal lug or splice barrel not cracked.
4. Terminal lug or splice insulation not cracked.
5. Insulation grip crimped.

CAUTION: *Do not use any connection that is found defective as a result of the visual inspection. Cutoff defective connec-*

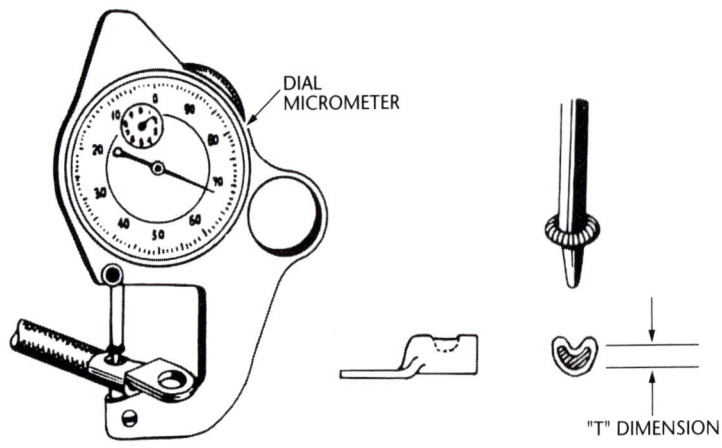

Figure 4-20. Indent inspection

tion and remake using a new terminal lug or splice.

Checking depth of indent. Sizes 8 through 4/0 uninsulated terminal lugs and splices crimped with the MS and Burndy tools are checked for depth of indent as follows:

1. Use micrometer or depth indicator with point adapter to measure "T" dimension as shown in Figure 4-20.

No. of Wires	Wire Size	Combined With:		No. of Wires	Wire Size	Combined With:	
		No. of Wires	Wire Size			No. of Wires	Wire Size
2	#22	—	—	2	#16	—	—
3	#22	—	—	3	#16	—	—
4	#22	—	—	1	#16	1	#22
2	#20	—	—	1	#16	2	#22
3	#20	—	—	1	#16	3	#22
4	#20	—	—	1	#16	1	#20
1	#20	1	#22	1	#16	2	#20
1	#20	2	#22	1	#16	3	#20
1	#20	3	#22	1	#16	1	#18
1	#20	4	#22	1	#16	2	#18
2	#20	1	#22	2	#16	1	#22
2	#20	2	#22	2	#16	2	#22
2	#20	3	#22	2	#16	1	#20
3	#20	1	#22	2	#16	1	#18
3	#20	2	#22	1	#16	1 #20 and 1	#22
2	#18	—	—	1	#16	1 #20 and 2	#22
3	#18	—	—	1	#16	1 #18 and 1	#22
4	#18	—	—	1	#16	1 #18 and 1	#20
1	#18	1	#22	1	#16	2 #20 and 1	#22
1	#18	2	#22	1	#16	1 #18 and 2	#22
1	#18	3	#22	1	#16	1 #18, 1 #20 and 1	#22
1	#18	4	#22	2	#14	—	—
1	#18	1	#20	1	#14	1	#22
1	#18	2	#20	1	#14	2	#22
1	#18	3	#20	1	#14	3	#22
2	#18	1	#22	1	#14	1	#20
2	#18	2	#22	1	#14	2	#20
2	#18	3	#22	1	#14	3	#20
2	#18	1	#20	1	#14	1	#18
2	#18	2	#20	1	#14	2	#18
3	#18	1	#22	1	#14	1	#16
3	#18	1	#20	2	#14	1	#22
1	#18	1 #18 and 1	#22	1	#14	1 #20 and 2	#22
1	#18	1 #20 and 1	#22	1	#14	1 #18 and 1	#22
1	#18	2 #20 and 1	#22	1	#14	1 #18 and 2	#22
2	#18	1 #18 and 1	#22	1	#14	2 #20 and 1	#22

Table 4-25. Wire size combinations (per splice end) for multi-splices

	"T" Dimensions (in Inches)			
	Using AN3427		Using Burndy Tools No. Y29NC Y29NSC & Y29B	
AN Wire Size	Min.	Max.	Min.	Max.
8	.090	.120	.096	.150
6	.090	.120	.109	.150
4	.140	.175	.155	.197
2	.140	.180	.195	.236
1	.220	.260	.227	.281
1/0	.280	.320	.252	.306
2/0	.280	.330	.270	.330
3/0	.365	.410	.335	.395
4/0	.410	.455	.320	.380

Table 4-26. "'T" Splices dimensions for Burndy uninsulated copper and straight and right-angle terminal lugs

2. "T" limits for COPPER terminal lugs and splices are listed in Tables 4-26 and 4-27.

3. "T" limits for MS25021 and MS25022 aluminum terminal lugs are listed in Tables 4-28 and 4-29.

NOTE: *Do not attempt to measure "T" dimensions of insulated terminal lugs. The soft plastic insulation will give a false indication.*

When a dial micrometer is not available, a hand micrometer is modified as shown in Figure 4-21. The indicated dimension A must be subtracted from the micrometer reading to obtain the dimension T.

	"T" Dimensions (in Inches)			
	Using AN3427 Tool		Using Burndy Tools No. Y29B & Y29NSC	
AN Wire Size	Min.	Max.	Min.	Max.
8	.130	.170	.180	.205
6	.130	.180	.205	.235
4	.205	.260	.270	.315
2	.265	.318	.270	.315
1/0	.325	.378	*.340	.440
2/0	.390	.450	.405	.465
4/0	.510	.570	.500	.580

*For Tool Y29B these dimensions are .310 - .440

Table 4-27. "T" dimensions for MS25189 copper (flag) terminal lugs

	"T" Dimesnions (in Inches)			
	Using MS25020 Tool		Using Burndy Tools No. Y29B, Y29NC, Y29NSC	
AL Wire Size	Min.	Max.	Min.	Max.
8	.145	.200	.155	.185
6	.205	.245	.190	.220
4	.265	.305	.260	.290
2	.272	.305	.230	.280
1	.338	.373	.255	.305
1/0	.391	.423	.255	.315
2/0	.370	.406	.345	.405

Table 4-29. "T" dimensions for MS25022 aluminum (flag) terminal lugs

	T Dimensions (in Inches)			
	Using MS25020 Tool		Using Burndy Tools No. Y29B, Y29NC, Y29NSC	
AL Wire Size	Min.	Max.	Min.	Max.
8	.098	.123	.098	.168
6	.125	.150	.125	.150
4	.141	.176	.136	.176
2	.259	.294	.171	.240
1	.290	.320	.204	.280
1/0	.340	.400	.232	.336
2/0	.375	.420	.290	.390
3/0	.380	.435	.310	.400
4/0	.410	.485	.380	.479

Table 4-28. "T" dimensions for (straight) terminal MS25021 aluminum lugs

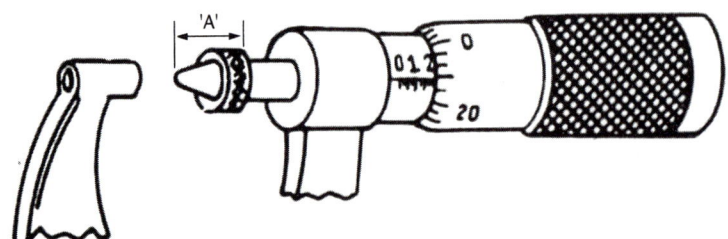

Figure 4-21. Modifying hand micrometer for indent inspection

Chapter 5

THERMOCOUPLE wire soldering and insulation

Section 1

Description and Identification

Thermocouples are used throughout aircraft to detect and measure temperature changes. Thermocouples are prefabricated into spark plug gaskets, bayonets for insertion into oil sumps and into probes for use in exhaust stacks. These thermocouples are supplied with short leads, usually 12 inches long and end in terminals such as AN5548 or AN5539. The installation mechanic fabricates extension leads to carry the voltages generated by the thermocouple to the indicating instruments. The components of a thermocouple system are designed to have a high degree of accuracy. Correct installation by a good mechanic will maintain this accuracy.

This chapter describes and illustrates recommended procedures for fabrication and installation of thermocouple extension leads.

The importance of good workmanship in the fabrication and installation of thermocouple wires cannot be over-emphasized. A good mechanic is careful to be neat and thorough in soldering and installing wires.

Thermocouple wire leads. See Figure 5-1 (next page). Thermocouple extension wires are paired in a braided jacket and color-marked as listed in Table 5-1 (next page). The material for extension leads is the same as the thermocouple material. Iron-constantan extensions are used for iron-constantan thermocouples, chromel-alumel extensions for chromel-alumel thermocouples, and copper-constantan extensions for copper-constantan thermocouples.

Learning Objectives:

- Thermocouple Wire Preparation
- Hard Soldering Thermocouple Wire
- Soft Soldering Thermocouple Wire
- Thermocouple Wiring Installation

Left. Thermocouple selection is based on location within the aircraft and on temperature conditions.

5-2 | Thermocouple Wire Soldering and Insulation

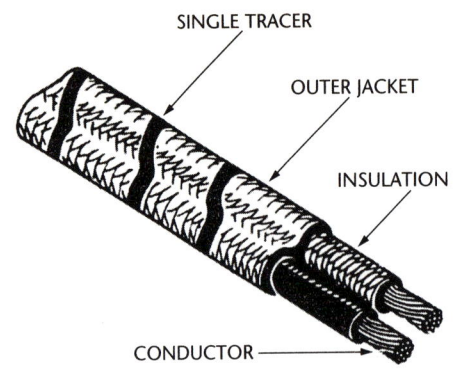

Figure 5-1. Thermocouple wire

Thermocouple terminals and connector. See Figure 5-2. Selection of terminals for thermocouple wiring is governed by AN drawing 10406, and drawings MS33560 and MS33599. The selection is based on location within the airframe, and on temperature conditions. Hot areas are those subject to high temperature, such as engine section, exhaust pipe, etc. Cool areas are those on the side of the firewall away from the engine or other heat producing elements. Where the temperature does not exceed 250°F, use terminals listed in Table 5-2 (derived from drawings AN10406, MS33560 and MS33599).

	A. IRON CONSTANTAN SYSTEM			
	Conductor	Insulation Color	Polarity	
	Iron	Black	Positive (+)	
	Constantan	Yellow	Negative (−)	
	Type II – 8 Ohms per 100 feet		Type III – 8 Ohms per 200 feet	
	Class A	Class B	Class A	Class B
Outer Jacket Base Color:	Light blue	Light blue	Light blue	Light blue
Tracer Color:	None	One red	Two black	Two red
Temperature Limit of Insulation:	120°C (248°F)	230°C (446°F)	120°C (248°F)	230°C (446°F)

	B. CHROMEL-ALUMEL SYSTEM		
	Conductor	Insulation Color	Polarity
	Chromel	White	Positive (+)
	Alumel	Green	Negative (−)
	Type II Class A 7 Ohms per 25 feet	Type III, Class A 7 Ohms per 50 feet	Type IV, Class A 7 Ohms per 100 feet
Outer Jacket Base Color:	White	White	White
Tracer Color:	One green	Two green	Three green
Temperature Limit of Insulation:	315°C (600°F)	315°C (600°F)	315°C (600°F)

	C. COPPER CONSTANTAN SYSTEM		
	Conductor	Insulation Color	Polarity
	Copper	Red	Positive (+)
	Constantan	Yellow	Negative (−)
	Type II : 7 Ohms per 200 feet		
	Class A	Class B	
Outer Jacket Base Color:	Black	Black	
Tracer Color:	One White	Two White	
Temperature Limit of Insulation:	120°C (248°F)	230°C (446°F)	

Table 5-1. Thermocouple system

Dash letters after basic numbers indicate whether terminal is plain or lock type, except for AN5538, where dash number indicates change in size only.

CAUTION: *Do not use solderless terminals on thermocouple wire unless specified in applicable engineering drawing.*

Thermocouple connector AN5537 as shown in Figure 5-3 is used to carry thermocouple connections through firewalls. This is a plug and jack connection, supplied with an insulating plate for attachment to the firewall. Plugs and jacks are supplied in chromel-alumel or iron-constantan combinations. The jack part of the connector is installed on the cool side of the firewall. The pin plug part of the connector is installed on the hot side of the firewall.

Thermocouple contacts in MS connectors. MS type connectors may be supplied with iron-constantan or chromel-alumel contacts in sizes #12, #16 or #20 in some insert arrangements for thermocouple connections. These contacts are coded to identify the material. (See Table 5-3).

Definitions

- **Soft solder.** A mixture of 60% tin and 40% lead, as specified in Federal Specification QQ-S-571. It may be in bar form to be melted for tinning, or in the form of rosin core solder wire for use with soldering iron.

	HOT AREAS (Silver Soldered)	COOL AREAS (Tin-Lead Soldered)
IRON-CONSTANTAN	AN5539	AN5538
CHROMEL-ALUMEL	AN5548	AN5538

Table 5-2. Thermocouple terminals

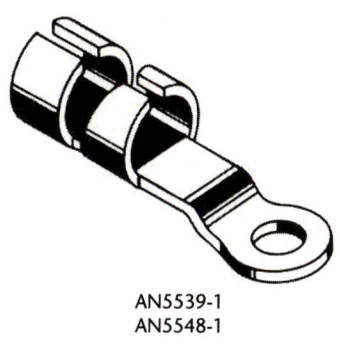

AN5539-1
AN5548-1

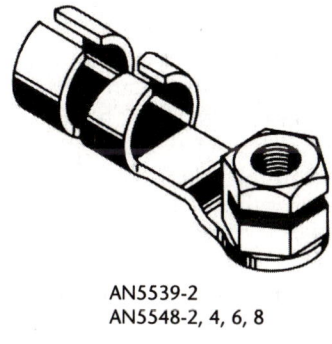

AN5539-2
AN5548-2, 4, 6, 8

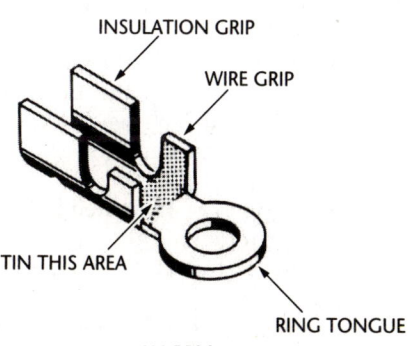

AN 5538

Figure 5-2. Thermocouple terminals

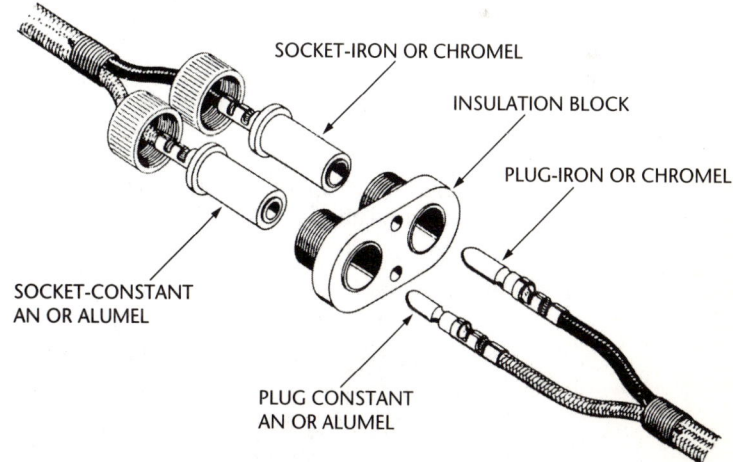

Figure 5-3. Thermocouple connector assembly (AN5537)

Manufacturer	Method of Coding	Code			
		Iron	Constantan	Chromel	Alumel
Amphenol	Color	White	Red	Green	Orange
Bendix	Letters	Ir.	Con.	Ch.	Al.
Cannon	Letters	IR	CO	CH	AL

Table 5-3. Coding for thermocouple contacts in MS connectors

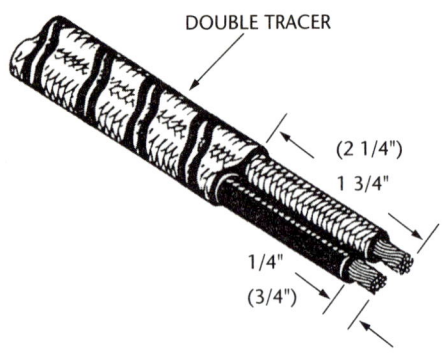

Figure 5-4. Stripping thermocouple wire for terminal and for AN5537 connector installation

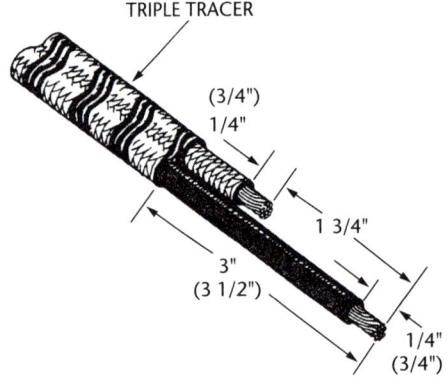

Figure 5-5. Stripping thermocouple wire for splice installation

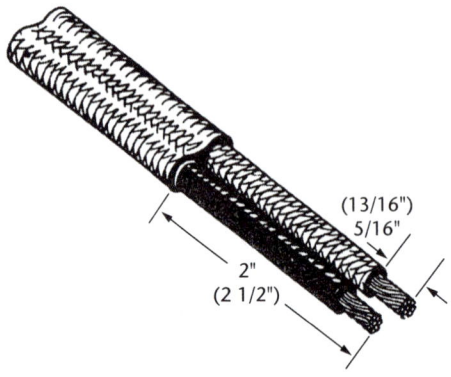

Figure 5-6. Stripping thermocouple wire for MS connector installation

- **Hard solder.** Silver alloy with flow point at approximately 635°C (1175°F), as specified in Federal Specification QQ-S-561.
- **Soft-solder flux.** For use with soft solder, flux is pure water-white rosin (Federal Specification LLL-R-626), if necessary powdered and mixed to a paste-like consistency with denatured alcohol. Other, more active soldering fluxes may be necessary. See page 5-8 for details.
- **Hard solder flux.** For use with hard solder, flux is borax or other similar material (Federal Specification O-F-499) mixed to a paste-like consistency with water.
- **Soldering and brazing.** For purposes of this chapter, the term "soldering" includes soft soldering, silver (hard) soldering and brazing.

Section 2

Thermocouple Wire Preparation

Cutting and identifying thermocouple wire. Cut thermocouple wire with diagonal pliers to length specified in drawing. Cut so that end is clean and square. Identify wire with sleeves as described in Chapter 1 on page 1-9. If outer covering is removed more than three inches from termination, install sleeve just back of serving at branching point (refer to page 5-8).

Stripping thermocouple wire. Remove outer covering of thermocouple wire with a knife by slitting between parallel conductors and trimming the fabric braid with scissors or diagonal pliers. The stripping dimensions for each use are shown in Figures 5-4 through 5-6. Note that longer stripped lengths are required if the wires are to be resistance tinned.

Use a hand stripper, as illustrated in Chapter 1, Figure 2-21, for removing the primary insulation from each conductor.

> **CAUTION:** *Do not cut or nick strands of the conductor.*

Cleaning wire prior to soldering. Clean stripped conductor, if necessary, as follows: Remove grease and dirt by washing with Stoddard's solvent. Rinse in methylene chloride for no longer than five seconds.

> **CAUTION:** *Do not use extra heat and special fluxes as a substitute for clean soldering surfaces.*

Section 3

Hard Soldering Thermocouple Wire

Tinning with Silver Solder

Torch tinning with silver solder. Before wires are soldered to terminals or other connections, they are tinned. The inability to obtain a good tinned surface indicates that the wire was not clean. See Figure 5-7. The procedure for torch tinning is as follows:

1. Dip half of exposed, clean conductor into hard solder flux.

2. Protect wire insulation with notched copper sheet shield, to prevent scorching.

3. Apply flame to wire until flux bubbles. Then feed small amount of silver solder in wire form to fluxed area while flame is kept there. After the silver solder has flowed, remove the flame and allow the wire to cool in the air.

CAUTION: *Silver solder will flow and adhere to conductor at approximately 635°C (1175°F). Avoid greater heat then necessary.*

Excess heat will decompose flux and prevent alloying of silver solder to the wire.

Dip tinning wire with silver solder. Thermocouple wires can be dip tinned in molten silver solder if a solder pot capable of maintaining the required 635°C (1175°F) heat is available. The process is similar to that used in dip tinning copper wire in soft solder as described in Chapter 1 on page 1-17. The procedure for dip tinning with silver solder is as follows (see Figure 5-8):

1. Dip half of exposed, clean conductor into hard solder flux.

2. Dip fluxed conductor into solder pot. Do not dip conductor deeper than one half of exposed area.

NOTE: *Powdered borax sprinkled over top of molten solder will retard oxidation of solder and aid alloying of silver solder to the wire.*

3. After solder has flowed between strands, remove the wire and allow it to cool in air.

Resistance tinning wire with silver solder. Electrical resistance heat is a good method for

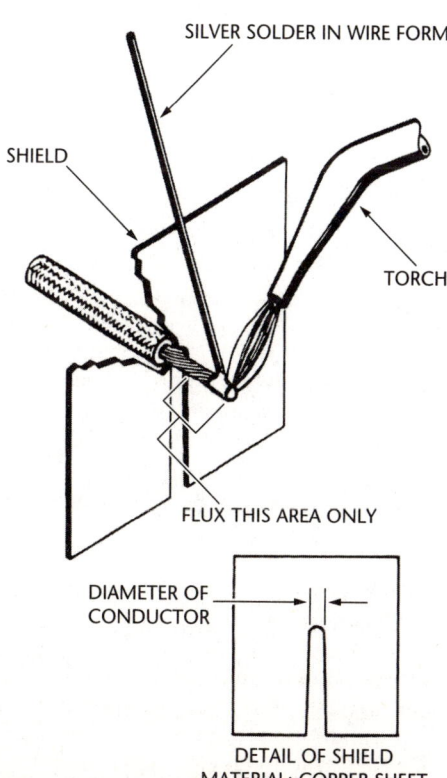

Figure 5-7. Torch tinning thermocouple wire

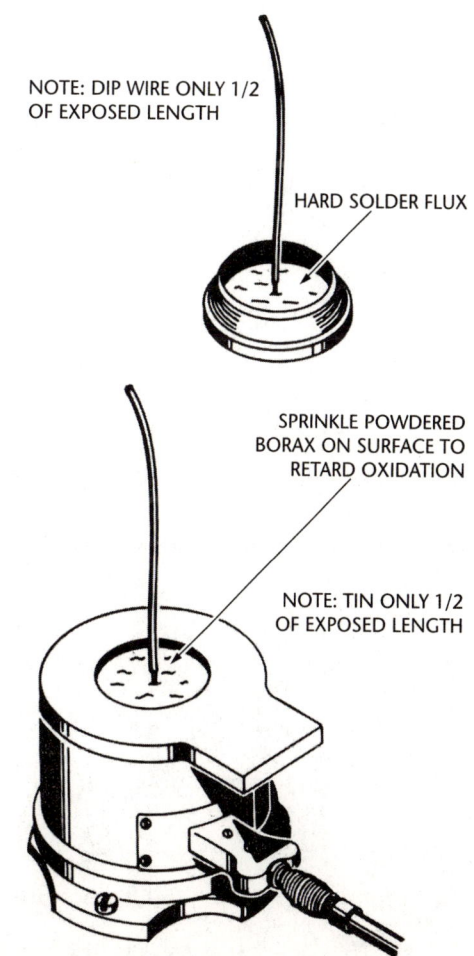

Figure 5-8. Dip tinning thermocouple wire in silver solder

silver soldering thermocouple wires. Use a unit that has a capacity of 1000 watts. See Figure 5-9. Wire that is to be tinned by means of electrical resistance should be stripped 1/2 inch longer than wire that is to be dip tinned or torch tinned. The extra 1/2 inch provides a holding area that is removed after tinning is complete. See Figures 5-4 through 5-6 for stripping dimensions. The procedure for resistance tinning is as follows:

1. Apply hard solder flux to area to be tinned. This is an area about 1/8 inch long as shown in Figure 5-8.
2. Grasp end of wire in resistance heating pliers. Grasp wire only as shown.
3. Apply current for approximately five seconds and then touch silver solder wire to area previously fluxed.
4. After solder has flowed between strands, shut off the current and allow the wire to cool in air.

CAUTION: *Do not overheat the wire by allowing the current to remain on longer than necessary to flow the silver solder.*

5. Trim off the holding area of the exposed conductor. The conductor should be trimmed with diagonal pliers to the point of tinning.

Tinning terminals with silver solder. Tin only section of thermocouple terminals inside wire grip as shown in Figure 5-2. Terminals for silver soldering should not be plated.

CAUTION: *Do not allow any flux or solder to get on the insulation grip or on the ring tongue.*

1. With a brush, apply a small amount of hard solder flux to the area to be tinned.
2. Using a torch or the resistance heating pliers melt a thin coat of silver solder onto inside of wire grip. See Figure 5-10 for use of resistance heating pliers in this operation.
3. Allow terminal to cool in air.

Soldering Methods for Attaching Terminals to Thermocouple Wire

Secure terminal to thermocouple wire as follows:

1. Flux previously tinned areas of terminal and wire.
2. Install terminal on wire so that insulation is flush with or protrudes slightly beyond insulation grip. The tinned portion of the conductor should then be inside the wire grip. See Figure 5-11.
3. Crimp wire grip over conductor using modified crimping tool illustrated in Figure 5-12.

NOTE: *Do not crimp insulation grip until after soldering operation. The heat of soldering may damage insulation if insulation grip is tight during soldering.*

Torch Soldering Terminals to Thermocouple Wire

The procedure for torch soldering terminals to thermocouple wire (see Figure 5-11) is as follows:

1. Use copper shield to protect insulation.
2. Heat joint until flux bubbles and then apply silver solder wire to joint as shown. Keep flame in motion to assure uniform heating.
3. When solder has flowed down into wire grip, remove flame and allow joint to cool

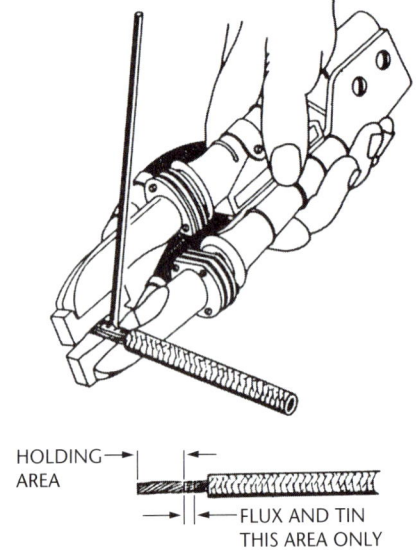

Figure 5-9. Resistance heating to tin wire

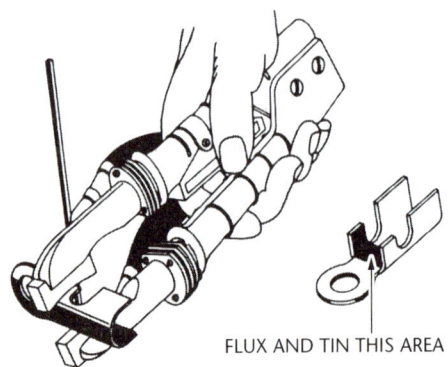

Figure 5-10. Resistance tinning of terminal

without disturbing it. Note that AN5539 terminals require reinforcement with silver solder at indicated areas. See Figure 5-13.

CAUTION: *Do not allow solder to flow onto ring tongue, as this will prevent proper assembly into system.*

Resistance Soldering Terminals to Thermocouple Wire

1. Grasp terminal and wire assembly, prepared in accordance with page 5-6, at wire grip area. The resistance heating pliers are to be in the position shown in Figure 5-10.

2. Apply current until flux bubbles and then apply silver solder wire to connection from conductor end of assembly.

3. Continue to apply heat and watch for flow of solder inside wire grip. When solder is visible at opposite end of wire grip from where it was applied, turn off current.

4. Allow assembly to solidify before removing from pliers.

Cleaning and completing silver soldered terminal connections. After the silver solder has solidified and cooled, the junction must be completed as follows:

1. Remove flux residues with warm water and a bristle brush, and then dry thoroughly.

2. Secure insulation grip on insulation using modified crimping tool shown in Figure 5-12. The final result is shown in Figure 5-11.

NOTE: *Insulation grip ears may be trimmed so they butt.*

3. Examine junction to be sure that silver solder has alloyed to wire and terminal.

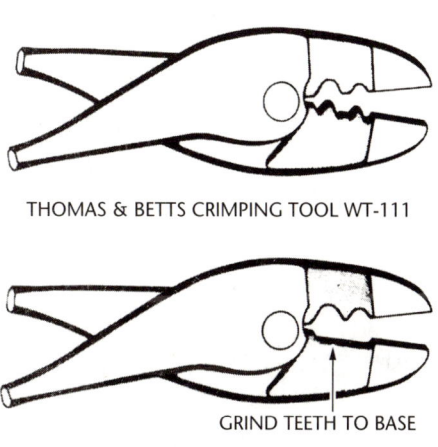

Figure 5-12. Modified crimping tool for thermocouple terminals

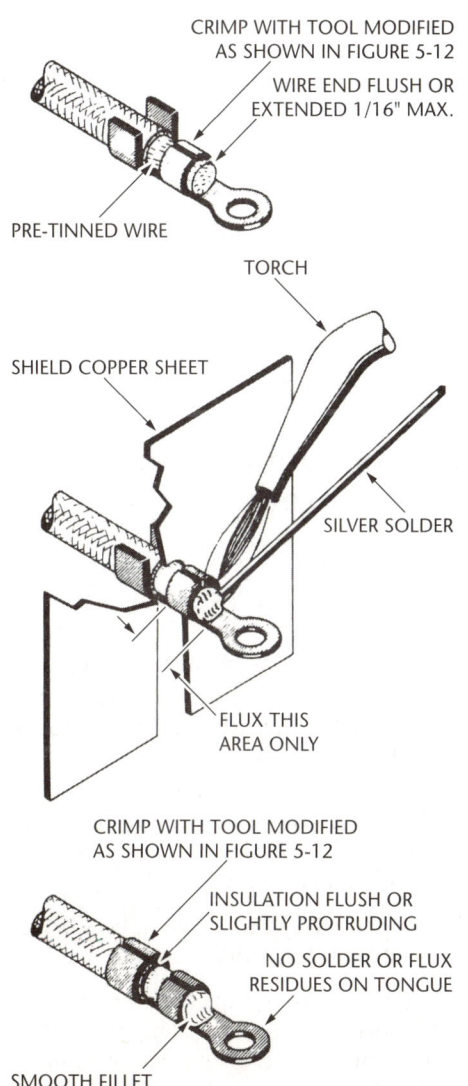

Figure 5-11. Silver soldering thermocouple wire to terminal

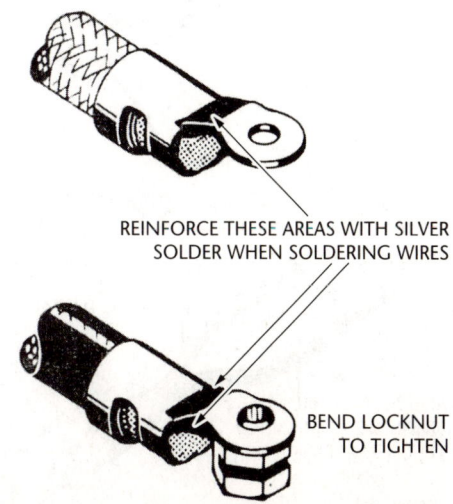

Figure 5-13. Reinforcing solder on an 5539 terminals

Thermocouple Wire Soldering and Insulation

Examine also to be sure that insulation has not been scorched. Rework any connection that is defective.

4. Coat areas indicated in Figure 5-14 with zinc chromate brushing compound.

5. Serve the completed extension lead at branching point as described in the following paragraph, and shown in Figure 5-14.

Serving thermocouple wire. After soldering operation has been completed, and solder has cooled, serve thermocouples at the branching point as shown in Figure 5-14. Use nylon or waxed cotton cord in cool areas, and fiberglass cord in hot areas. Coat the serving with clear lacquer. The serving will prevent unraveling of the outer jacket.

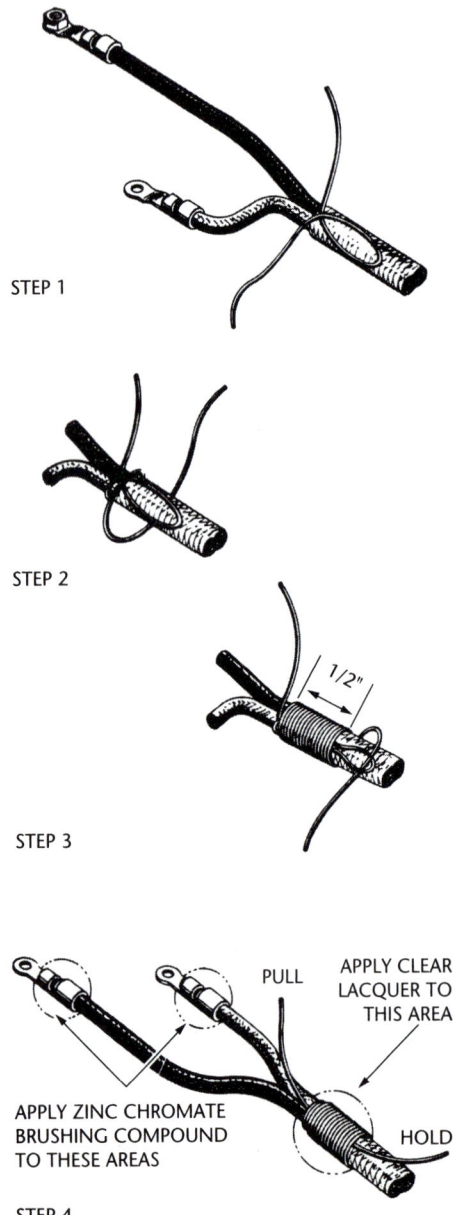

Figure 5-14. Serving thermocouple wire

Section 4
Soft Soldering Thermocouple Wire

Tinning wire for soft soldering. Tin thermocouple wire for soft soldering in the same manner as copper wire as described in Chapter 1 on pages 1-16 through 1-19). Either dip tinning or soldering iron tinning is satisfactory. Occasionally, if wires are oxidized, rosin-alcohol flux may not do a satisfactory job of tinning. If this happens, use the following:

Lactic acid-glycerin-rosin mixture made by mixing (by weight) one part of lactic acid (C. P.) with one part glycerin (U.S.P.), and adding three parts pure, freshly powdered rosin.

> **NOTE:** *When soldering-iron method is used, use a soldering iron of 200 to 250 watts capacity for tinning thermocouple wires.*

Tinning terminals for soft soldering. Tin terminal section inside wire grip, using a 200 to 250 watt soldering iron, and rosin core solder. Do not allow flux or solder to get on the insulation grip or on the ring part of the tongue.

> **NOTE:** *Terminals for use with soft solder should be cadmium plated.*

Procedure for Soft-soldering Wire to Terminals

1. Install terminals on thermocouple wires as described on page 5-6, and illustrated in Figure 5-11.

2. Soft solder, using 200 to 250 watt iron and rosin core solder. Make sure that solder flows inside wire grip and forms a smooth fillet.

> **CAUTION:** *For soft-soldering, do not use any flux other than rosin-alcohol, regardless of flux used for tinning.*

3. Remove excess flux by scrubbing with brush and denatured alcohol.

4. Bend insulation grip ears around insulation using modified crimping tool shown in Figure 5-12. Trim ears so they butt flush around small wires. See Figure 5-11.

5. Coat areas indicated in Figure 5-14 with zinc chromate brushing compound.

Procedure for Soldering Wire to MS Connectors

Thermocouple contacts in MS series connectors are not tinned by the manufacturer. Therefore, it is necessary to properly tin these contacts with soft solder before thermocouple wire is soft soldered into place. MS connector contacts must be removed from inserts for soldering because of the extra heat needed to raise thermocouple wire to solder temperature. Best results are obtained when electrical resistance heating pliers are used to tin the contact and for soldering wire into contact. The procedure for tinning and soldering is as follows:

1. Tin contact by use of resistance heating pliers or torch. Use rosin-alcohol flux and 60/40 tin-lead solder or, if necessary use flux described on page 5-8, with the same 60/40 tin-lead solder.

2. Remove flux residues. Rosin residues are removed by brushing vigorously with Stoddard's solvent or with denatured alcohol. Lactic acid flux is removed by brushing in warm water. Dry each tinned contact thoroughly before proceeding with next step.

3. Check contact coding and wire coding carefully to avoid mismatch of materials. See Tables 6-1 and 6-3.

   CAUTION: *It is important that thermocouple materials match. Make sure that the thermocouple wire is soldered to a contact of the same material.*

4. Insert properly pre-tinned wire into contact and solder using resistance pliers or torch. Use only rosin core solder for this operation. See Figure 5-15.

5. After solder has flowed and alloyed, allow connection to cool without motion. Then remove flux residues with Stoddard's solvent or denatured alcohol.

6. Examine joint to be sure solder has flowed to form smooth fillet, and that no solder is left on outside of solder cup.

7. Reassemble contacts into MS Connector as described in Chapter 2 on pages 2-29 through 2-40. Be careful to reassembly each contact into the hole from which it was removed.

NOTE: *For chromel and alumel contacts, in addition to visual inspection of the contact stamping, further material verification, separation of contacts during connector assembly, and inspection verification after assembly can be made with the aid of a magnet if desired, as a magnet will attract the alumel contact but not the chromel contact.*

Procedure for soldering wire to AN5537 firewall connector. Thermocouple wires are brought through firewalls by means of AN5537 firewall connectors. To preserve the integrity of the system, it is necessary to hard solder wires to the connector on the hot side of the firewall. The cool side of the firewall may be either hard or soft soldered.

   CAUTION: *Be careful to connect wire leads to mating materials of connector. Connector plugs and sockets are coded with letters to indicate materials. Sizes are also different to aid in quick identification. See Table 5-4 for code.*

The procedure for attaching wires is as follows: (See Figure 5-3).

1. Disassemble connector as shown. Slide nuts over the pretinned leads that will be installed on the hot side of the firewall.

2. Tin the wire grips of the socket assemblies using hard solder as described on page 5-6.

3. Assemble and hard-solder wires to socket assemblies as described on pages 5-6 through 5-7.

4. Complete assembly of hot side wires by cleaning, crimping insulation grips and coating with zinc chromate brushing compound as described on page 5-8.

5. Attach plugs to wires on cold side of firewall by using hard or soft solder as

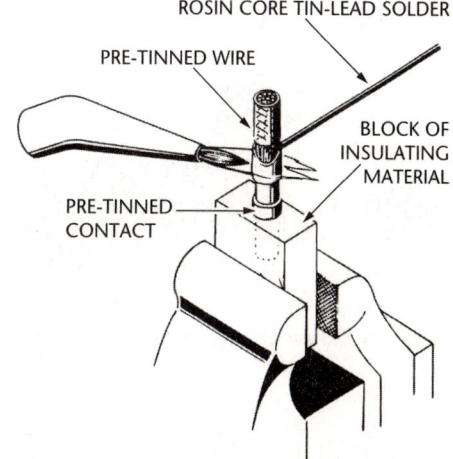

Figure 5-15. Torch soldering thermocouple wire to MS connector contact

Material	Code	Size
Iron	FE	Large
Constantan	CON	Small
Chromel	CR	Large
Alumel	AL	Small

Table 5-4. Code for markings on AN5537

required on applicable drawing for the specific installation. The method of attachment, soldering, cleaning, etc., is the same as that previously described.

Section 5

Thermocouple Wiring Installation

Connecting thermocouple splices. (See Figure 5-16.) Connect thermocouple splices as follows:

1. Slide sleeve over one lead.
2. Bend locknut of lock terminal slightly before assembly to assure tightness.
3. Bring contact areas of two terminals together, and pass screw through plain terminal first and then through locknut of lock terminal.
4. Tighten screw securely.
5. Slide sleeve over terminal and tie securely.

Mounting AN5537 connector assembly. AN5537 firewall connector assemblies are mounted as follows (see Figure 5-3):

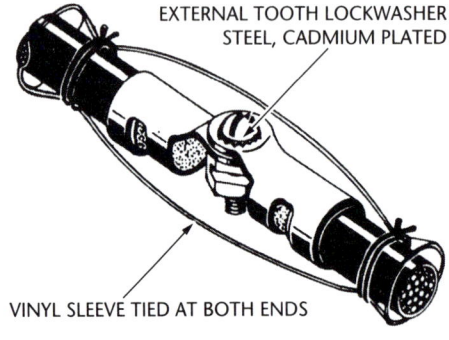

Figure 5-16. Connecting thermocouple splices

1. Attach insulating block to firewall on hot side. Bosses on block should fit into holes in firewall so that block face is flush against wall.
2. Push socket assemblies through holes and lock into place with coupling nuts.
3. Push plugs into socket assemblies from cold side of firewall.

Routing thermocouple wiring. Route thermocouple wiring as described generally in Chapter 10. In addition, observe the following special precautions:

1. Support thermocouple wiring so it will not be exposed to heat producing surfaces such as exhaust pipe or combustion chamber at any point.
2. Do not bend thermocouple leads sharply.
3. Do not splice thermocouple leads except where specifically indicated and then only with approved splices such as shown in Figure 5-16.
4. Protect adjacent wiring against abrasion from thermocouple splices as described on page 5-10.
5. Route thermocouple wiring away from hot spots.

Protection. Insulate thermocouple spliced terminal connections with sleeves to protect the insulation of adjacent wires from abrasion. Use plastic sleeving in cool areas and silicone impregnated rubber or glass sleeving in hot areas. Tie sleeving securely at both ends.

CAUTION: *Do not use sleeving as a substitute for safe routing.*

Slack in thermocouple wiring. (See Figure 5-17.) Thermocouple wire installations require the use of fixed wire lengths to maintain a specified resistance. The slack that results should be distributed by one of the following methods:

1. Distribute excess slack evenly between wire supports, as shown in Figure 5-17a.
2. If sufficient slack is available, take it up, at a support, in the form of a loop of which the diameter is at least 20 times the thickness of the thermocouple wire, as shown in Figure 5-17b.

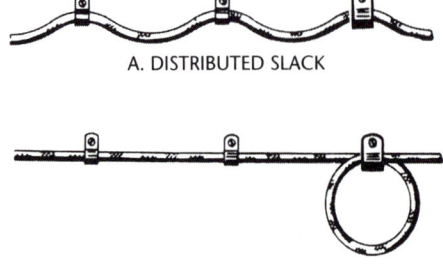

Figure 5-17. Distributing slack in thermocouple wire

CAUTION: *Do not bend thermocouple leads to less than a two-inch radius. When calibration resistors are used in the circuit to adjust for short lengths, do not allow any excess slack, except for approximately three inches at each end for maintenance.*

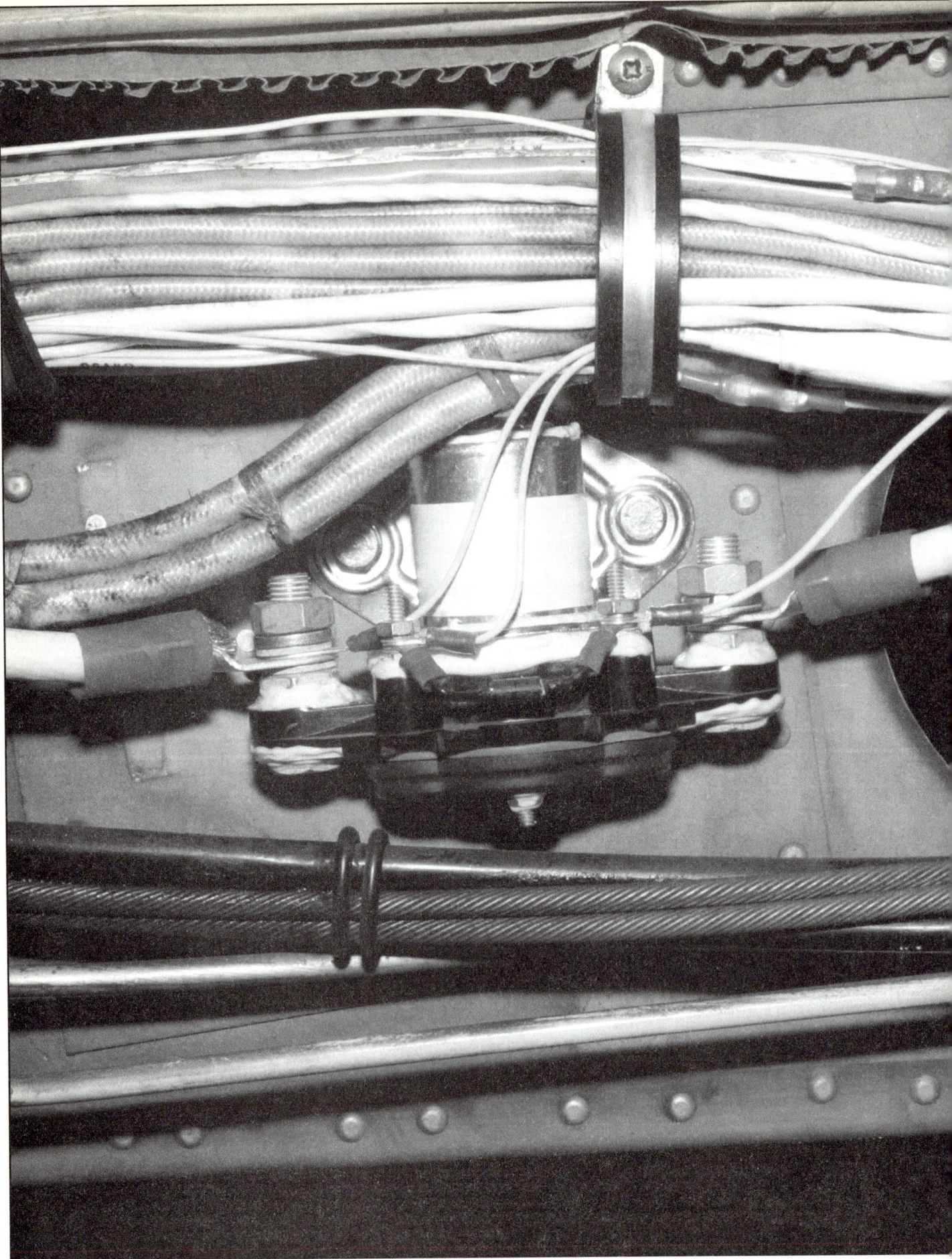

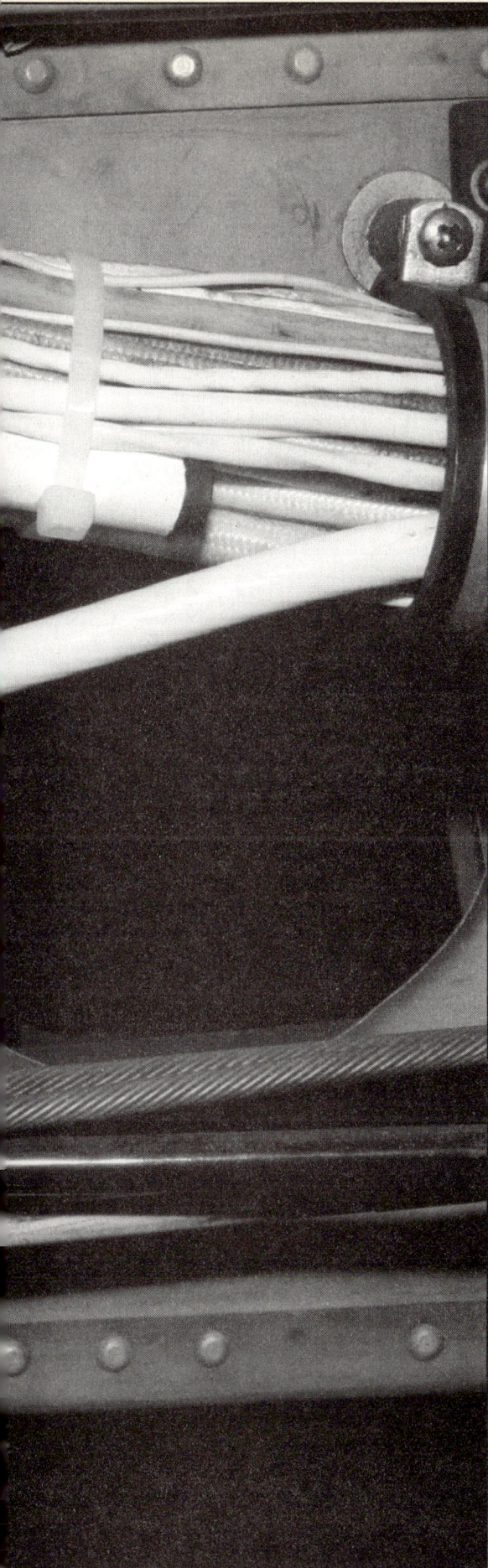

Left. The hardware used to make bonding or grounding connections is selected on the basis of mechanical strength, current to be carried and ease of installation.

Chapter 6

BONDING and *grounding*

Section 1

Description and Identification

Bonding and grounding connections are made in aircraft for the following purposes:

- To protect aircraft and personnel against hazards from lightning discharge.
- To provide power current return paths.
- To prevent development of RF potentials.
- To protect personnel from shock hazard.
- To provide stability and homogeneity of radio transmission and reception.
- To prevent accumulation of static charge.
- To provide fault current return paths.

This section describes and illustrates the recommended procedures to be followed in the preparation and installation of bonding and grounding connections.

Learning Objectives:

- Description & Identification
- Bonding & Grounding Hardware
- Preparation of Bonding or Grounding Surfaces
- Methods of Bonding or Grounding
- Testing Bonds and Grounds
- Refinishing

Definitions

- **Bonding.** The electrical connecting of two or more conducting objects not otherwise adequately connected.
- **Grounding.** The electrical connecting of conducting object to primary structure, for return of current.
- **Primary structure.** The main frame, fuselage and wing structure of the aircraft (commonly referred to as ground).

General Precautions and Procedures. When making bonding or grounding connections in

Structure	Screw or Bolt; Lock-Nut	Plain Nut	Washer A	Washer B	Washer C & D	Lock-Washer E	Lock-Washer F
Aluminum Terminal and Jumper							
Aluminum Alloys	Cadmium Plated Steel	Tin Plated Brass	Aluminum Alloy	Aluminum Alloy	Cadmium Plated Steel or Aluminum	Cadmium Plated Steel	Cadmium Plated Steel
Magnesium Alloys	Cadmium Plated Steel	Cadmium Plated Steel	Magnesium Alloy	Magnesium Alloy	Cadmium Plated Steel or Aluminum	Cadmium Plated Steel	Cadmium Plated Steel
Steel, Cadmium Plated	Cadmium Plated Steel	Cadmium Plated Steel	None	None	Cadmium Plated Steel or Aluminum	Cadmium Plated Steel	Cadmium Plated Steel
Steel, Corrosion Resisting	Corrosion Resisting Steel	Cadmium Plated Steel	None	None	Cadmium Plated Steel or Aluminum	Cor. Resist. Steel	Cadmium Plated Steel
Tinned Copper Terminal and Jumper							
Aluminum Alloys	Cadmium Plated Steel	Cadmium Plated Steel	Aluminum Alloy	Aluminum Alloy	Cadmium Plated Steel	Cadmium Plated Steel	Cadmium Plated Steel or Aluminum
Avoid Connecting Copper to Magnesium							
Magnesium Alloys Steel, Cadmium Plated	Cadmium Plated Steel	Cadmium Plated Steel	None	None	Cadmium Plated Steel	Cadmium Plated Steel	Cadmium Plated Steel
Steel, Corrosion Resisting	Corrosion Resisting Steel	Cor. Res. Steel	None	None	Cadmium Plated Steel	Cor. Resist. Steel	Cor. Resist. Steel

Table 6-1. Hardware for stud bonding or grounding to flat surface (refer to Figure 6-1)

Structure	Screw or Bolt; Nut Plate	Rivet	Lock Washer	Washer A	Washer B
Aluminum Terminal and Jumper					
Aluminum Alloys	Cadmium Plated Steel	Alum. Alloy	Cadmium Plated Steel	Cadmium Plated Steel or Aluminum	None
Magnesium Alloys	Cadmium Plated Steel	Alum. Alloy	Cadmium Plated Steel	Cadmium Plated Steel or Aluminum	None or Mag. Alloy
Steel, Cadmium plated	Cadmium Plated Steel	Cor. Res. Steel	Cadmium Plated Steel	Cadmium Plated Steel or Aluminum	None
Steel, Corrosion Resisting	Corrosion Resist. Steel or Cadmium Plated Steel	Cor. Res. Steel	Cadmium Plated Steel	Cadmium Plated Steel or Aluminum	Cadmium Plated Steel
Tinned Copper Terminal and Jumper					
Aluminum Alloys	Cadmium Plated Steel	Alum. Alloy	Cadmium Plated Steel	Cadmium Plated Steel	Alum. Alloy
Magnesium Alloys	**Avoid Connecting Copper to Magnesium**				
Steel, Cadmium plated	Cadmium Plated Steel	Cor. Res. Steel	Cadmium Plated Steel	Cadmium Plated Steel	None
Steel, Corrosion Resisting	Corrosion Resist. Steel	Cor. Res. Steel	Cadmium Plated Steel	Cadmium Plated Steel	None

Table 6-2. Hardware for plate nut bonding or grounding to flat surface (refer to Figure 6-2)

ALUMINUM TERMINAL AND JUMPER						
STRUCTURE	SCREW OR BOLT; LOCK NUT	LOCK WASHER	WASHER A	WASHER B	WASHER C	
Aluminum Alloy	Cadmium Plated Steel	Cadmium Plated Steel	Cadmium Plated Steel or Aluminum	None	Cadmium Plated Steel or Aluminum	
Magnesium Alloy	Cadmium Plated Steel	Cadmium Plated Steel	Magnesium Alloy	None or Mag. Alloy	Cadmium Plated Steel or Aluminum	
Steel, Cadmium Plated	Cadmium Plated Steel	Cadmium Plated Steel	Cadmium Plated Steel	Cadmium Plated Steel	Cadmium Plated Steel or Aluminum	
Steel, Corrosion Resisting	Corrosion Resisting Steel or Cadmium Plated Steel	Cadmium Plated Steel	Corrosion Resisting Steel	Cadmium Plated Steel	Cadmium Plated Steel or Aluminum	
TINNED COPPER TERMINAL AND JUMPER						
Aluminum Alloy	Cadmium Plated Steel	Cadmium Plated Steel	Cadmium Plated Steel	Alum. Alloy	Cadmium Plated Steel	
Magnesium Alloy	Avoid Connecting Copper to Magnesium				
Steel, Cadmium Plated	Cadmium Plated Steel	Cadmium Plated Steel	Cadmium Plated Steel	None	Cadmium Plated Steel	
Steel, Corrosion Resisting	Corrosion Resisting Steel or Cadmium Plated Steel	Cadmium Plated Steel	Corrosion Resisting Steel	None	Cadmium Plated Steel	

Table 6-3. Hardware for bolt and nut bonding or grounding to flat surface (refer to Figure 6-3)

aircraft, observe the following general precautions and procedures:

- Bond or ground parts to the primary aircraft structure where practical.
- Make bonding or grounding connections in such a way as not to weaken any part of the aircraft structure.
- Bond parts individually wherever possible.
- Make bonding or grounding connections against smooth, clean surfaces.
- Install bonding or grounding connections so that vibration, expansion or contraction, or relative movement incident to normal service use will not break or loosen the connection.
- Locate bonding and grounding connections in protected areas whenever possible; locate connections whenever possible near hand holes, inspection doors or other accessible areas to permit easy inspection and replacement.
- Do not compression-fasten bonding or grounding connections through any non-metallic material.

Section 2
Bonding & Grounding Hardware

Selection of Hardware

Hardware used to make bonding or grounding connections is selected on the basis of mechanical strength, current to be carried and ease of installation. Where connection is made by aluminum or copper jumpers to structure of dissimilar material, a washer of suitable material is installed between the dissimilar metals so that any corrosion which may occur will occur in the washer, which is expendable, rather than in the structure, which is not expendable.

NOTE: *When repairing or replacing existing bonding or grounding connections, be sure to use the same type of hardware as in the original connection. Do not make any changes.*

Hardware material and finish. Select hardware material and finish from Tables 6-1, 6-2 or 6-3, depending on material of structure to which attachment is made, and material of

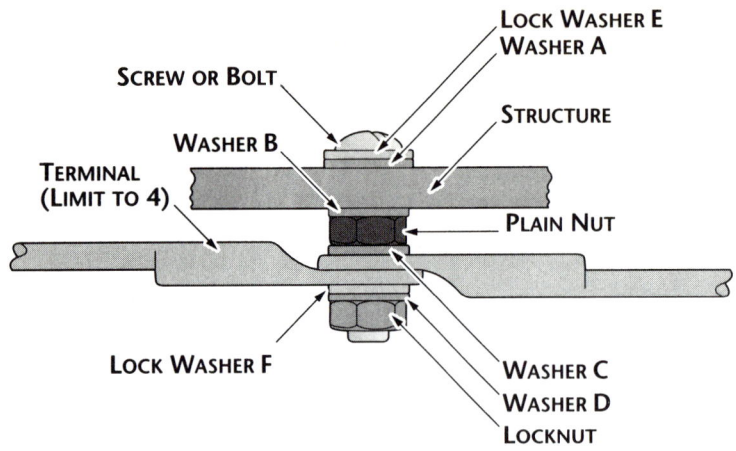

Figure 6-1. Stud bonding or grounding to flat surface

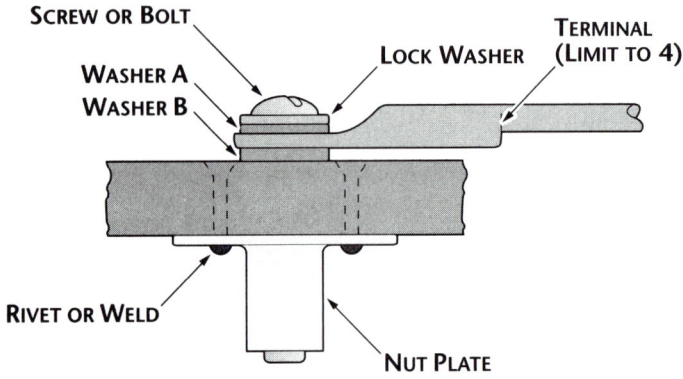

Figure 6-2. Plate nut bonding or grounding to flat surface

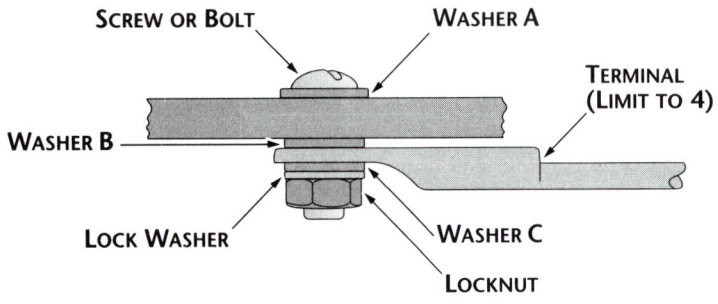

Figure 6-3. Bolt and nut bonding or grounding to flat surface

jumper and terminal specified for the bonding or grounding connection.

Selection of stud. Use either an AN screw or bolt of the proper size for the specified jumper terminal. Length of screw or bolt should be such that when bonding or grounding connection is fully tightened approximately 1/8 inch of screw protrudes beyond top of nut.

Selection of nuts. Use AN nuts, either plain or self-locking where indicated in Figures 6-1 and 6-3. Use an all-metal self-locking nut if practical. Always use an all-metal self-locking nut where current is, or will be present. Where installation conditions require, use an AN nutplate, riveted to structure.

Selection of washers. Use AN plain washers and split lockwashers where indicated in Figures 6-1, 6-2, 6-3. Always use split lockwashers with nuts, either plain or self-locking. With aluminum terminals use a plain washer of at least the diameter of the terminal tongue, next to the aluminum terminal. If an AN washer does not meet this requirement, use a washer of the SAE heavy series, or a special washer made for this application

Selection of cable clamp. For bonding or grounding to cylindrical surfaces use an AN735 clamp. Where an AN735 clamp is not available, or where installation conditions do not allow its use, an uncushioned AN742 clamp may be substituted.

CAUTION: *Do not use cushioned clamps in any bonding or grounding connection.*

Section 3

Preparation of Bonding or Grounding Surfaces

Clean bonding and grounding surfaces thoroughly before making the connection. Remove all oil, grease, paint, anodic film or other non-conducting material from an area slightly larger than the connection.

CAUTION: *Do not use abrasives such as emery cloth, crocus cloth, steel wool, etc. These may leave particles imbedded in the surface or scattered in the area which may cause corrosive action.*

Cleaning procedure for aluminum surfaces. Apply a coating of petrolatum compound to bonding or grounding surface of aluminum structure, and clean surface thoroughly, using steel wire brush with pilot as shown in Figure 6-4. Wipe off the petrolatum compound with a clean dry cloth.

Cleaning procedure for magnesium alloy surfaces. Prepare magnesium alloy surfaces for bonding or grounding as follows:

1. Remove grease and oil from surface with Stoddard's solvent.

2. Remove paint or lacquer, if present, from surface with lacquer thinner (Mil-Spec

MIL-T-6094).

3. Brush area liberally with chrome pickle solution (Mil-Spec MIL-M-3171) for one minute, then rinse within five seconds by brushing with clean water.

4. Dry thoroughly.

Cleaning procedure for steel surfaces. When the surface is corrosion-resisting or plated steel, clean bonding or grounding surface as described in the previous section, in steps 1 and 2.

CAUTION: *Do not remove zinc or cadmium plate from steel surfaces.*

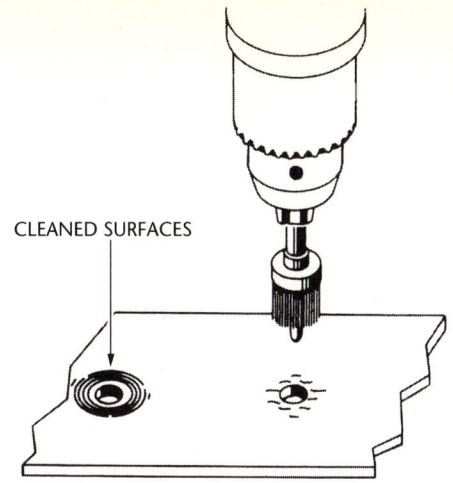

Figure 6-4. Steel wire brush with pilot for cleaning aluminum surfaces

Section 4
Methods of Bonding or Grounding

Bonding or grounding connections are made directly to a flat surface of basic structure, or to a cylindrical surface of a basic structure.

Connection to Flat Surfaces

Bonding and grounding connections are made to flat surfaces by means of through bolts or screws where installation has easy access. The three types of bolted connections are as follows:

Stud connection. See Figure 6-1 and Table 6-1. In this type of connection a bolt or screw is locked securely to structure, thus becoming in effect a stud. Grounding or bonding jumpers can be removed or added to the shank of stud without removing stud from structure.

Nut plate and bolt connection. See Figure 6-2 and Table 6-2. Nut plates are used where access to the nut for repairs may be difficult. Nut plates are riveted or welded to a clean area of the structure. Cleaning of structure is done in accordance with the procedures for cleaning aluminum of magnesium alloy surfaces as applicable.

Nut and bolt connection. See Figure 6-3 and Table 6-3. In this connection the bolt or screw is not attached permanently to structure. When jumpers are to be added or removed, the entire connection is remade.

Tables 6-1, 6-2 and 6-3 list materials and platings that are compatible with the structure to which they are mounted. These materials are selected

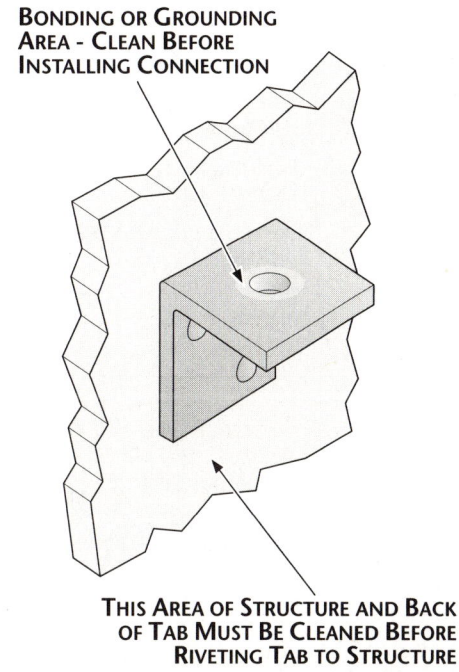

Figure 6-5. Bonding tab riveted to structure

so that corrosion, if it occurs, will occur in the washers, which are expendable, rather than in the structure which is not expendable.

Connection to tab riveted to structure. For bonding leads carrying high current, size AN-4 or larger, do not make the connection directly to the structure, but to a tab of suitable size adequately bonded to the aircraft structure. See Figure 6-5. When a bonding or grounding connection is made to a tab riveted to structure rather than directly to the structure, clean the bonding or grounding surface and make the connection exactly as though the connection were being made to structure. If it is necessary to remove the tab for any reason, replace rivets with one size larger. Make sure

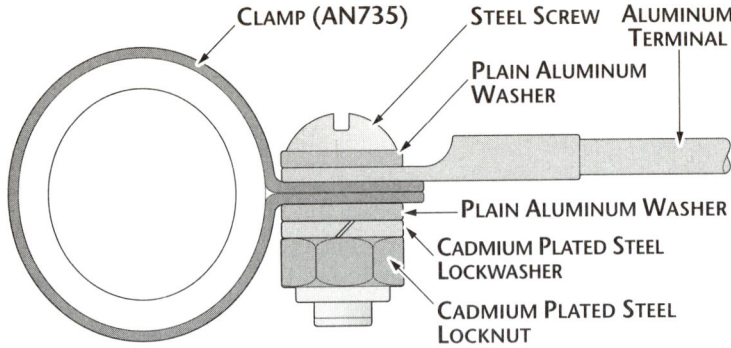

Figure 6-6. Aluminum jumper connection to tubular structure

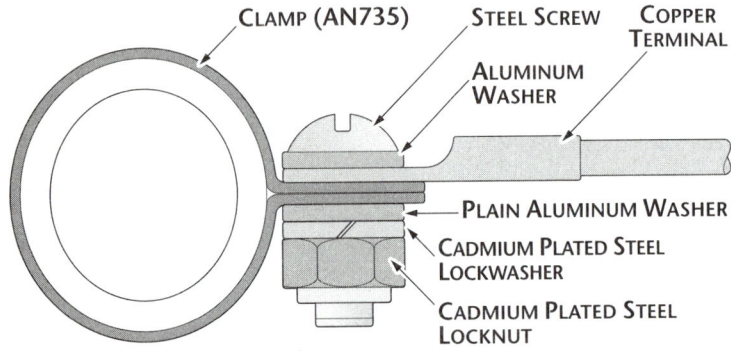

Figure 6-7. Copper jumper connection to tubular structure

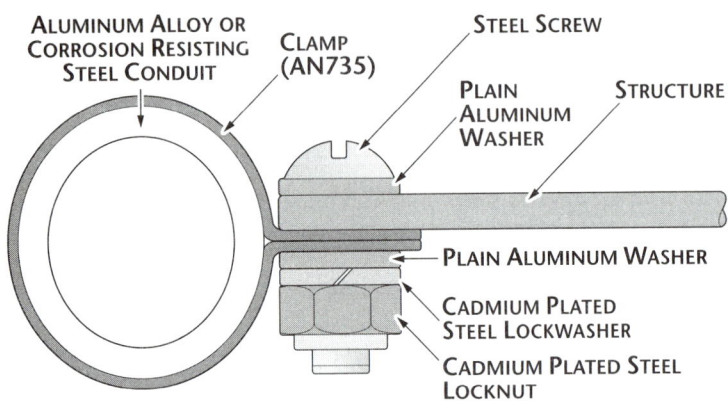

Figure 6-8. Bonding conduit to structure

aluminum jumper. Because of the ease with which aluminum is deformed, it is necessary to distribute screw and nut pressure by means of plain washers as shown. Figure 6-7 shows the arrangement of hardware for bonding with a copper jumper. No extra washers are used. If installation conditions require, use AN742 clamp (uncushioned) instead of AN735. Do not change any other hardware if this substitution is made.

Bonding Conduit to Structure

Bond aluminum alloy or corrosion-resisting steel conduit to structure as shown in Figure 6-8. If installation conditions require, AN742 clamp may be used instead of AN735, using same hardware.

Tightness of connections. Make sure that all connections are tight as evidenced by the split lock-washers being completely compressed.

> **CAUTION:** *When terminal is under head of screw or bolt (as shown in Figure 6-2), it is preferable not to install more than one terminal. A loose screw with two terminals may cause inadvertent operation of equipment.*

Bonding and grounding jumpers. To accomplish the purpose of bonding or grounding, it is necessary to provide a conductive path where direct electrical contact does not exist. Jumpers are used for this purpose in such applications as between moving parts, between shock-mounted equipment and structure, and between electrically conducting objects and structure. Keep jumpers as short as possible; if practical under three inches. Do not use two or more jumpers in series.

Fabricating bonding and grounding jumpers. Jumpers of tinned copper wire are fabricated in accordance with drawing MS25083. For smaller size wire, terminate with MS25036 insulated copper terminal lugs of appropriate size. Use standard tool MS25037 for crimping terminals to wire. For larger wire size, terminate with MS20659 uninsulated copper terminal lugs, crimped to the wire with AN3427 tool. Jumpers may also be fabricated of 1/16 inch ID #36 gage copper braid, using MS25036 terminal lugs. Instructions for use of crimping tools are given in Chapter 4.

Quick-disconnect jumpers. Where a quick-disconnect jumper is required, crimp an AN753 electrical disconnect splice into a copper wire jumper, fabricated as described in the previous paragraph, using the standard tool MS25037. Note that the disconnect splice is not centered in the jumper; but is installed so that the coupler remains on the short end when the jumper is disconnected.

mating surfaces of structure and tab are clean, and free of anodic film.

Connection to Cylindrical Surfaces

Make bonding or grounding connections to aluminum alloy, magnesium alloy or corrosion resisting steel tubular structure as shown in Figures 6-6 and 6-7. Figure 6-6 shows the arrangement of hardware for bonding with an

Section 5
Testing Bonds and Grounds

Resistance tests after connection. The resistance across a bonding or grounding jumper is required to be 0.1 ohms or less. Tests are made after the mechanical connection is completed, and consists of a multiohmmeter reading of the overall resistance between the cleaned areas of the object and the structure. See Figure 6-9.

Resistance Test Procedure

Measurements of the specified resistance value are made with a special calibrated low-range ohmmeter. Proceed as follows, observing the precautions emphasized in the meter instruction manual.

1. With the Function Control OFF, set the Range Control to the 0.1 ohms position.
2. Attach the instrument test-clips for intimate contact with the cleaned areas immediately adjacent to the jumper terminal lugs of object and structure.
3. Set the Function Control to CALIBRATE, then use the Calibration Adjustment Control to obtain a 0.1 ohm full scale deflection.
4. Set the Function Control to OHMS position, and note the bond (only) reading. It should be less than 0.1 ohm.
5. Return the spring loaded Function Control to OFF position, and remove the test clip leads and instrument.
6. If the specified resistance value has been obtained, the cleaned areas are now ready for refinishing.

Refinishing magnesium alloy surfaces. Within 24 hours after cleaning surface of magnesium alloys, apply a protective coat of zinc chromate brushing compound. Within one week after making and testing the connection, refinish cleaned area to match original surrounding finish.

Refinishing steel surfaces. Within one week after connection has been made, refinish cleaned area to match original surrounding finish.

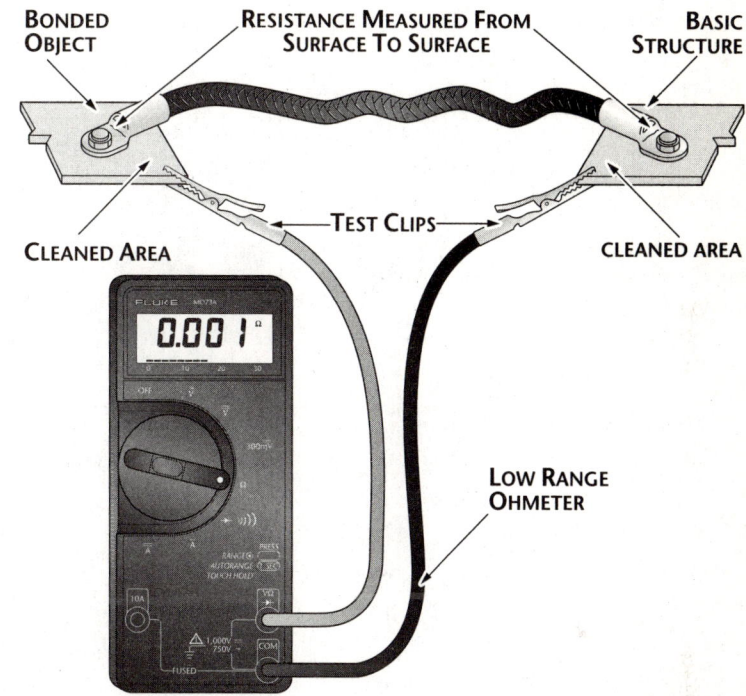

Figure 6-9. Special milliohmmeter and clip leads for testing bond resistance

Section 6
Refinishing

Refinishing aluminum alloy surfaces. Within one week after any area has been cleaned and connection made, refinish aluminum surfaces as follows:

1. Apply coat of zinc chromate brushing compound, and allow to dry.
2. When thoroughly dried, refinish cleaned area to match original surrounding finish.

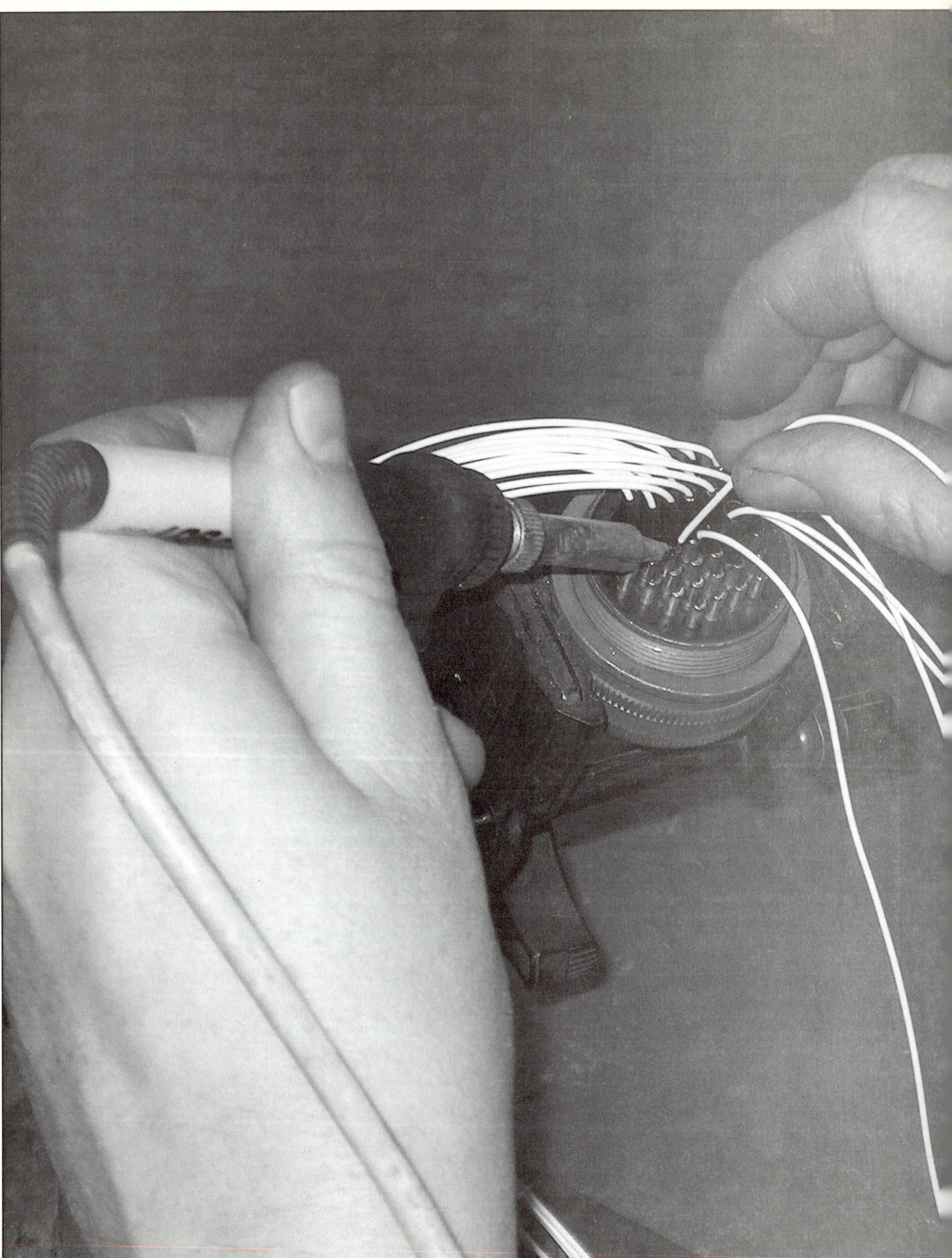

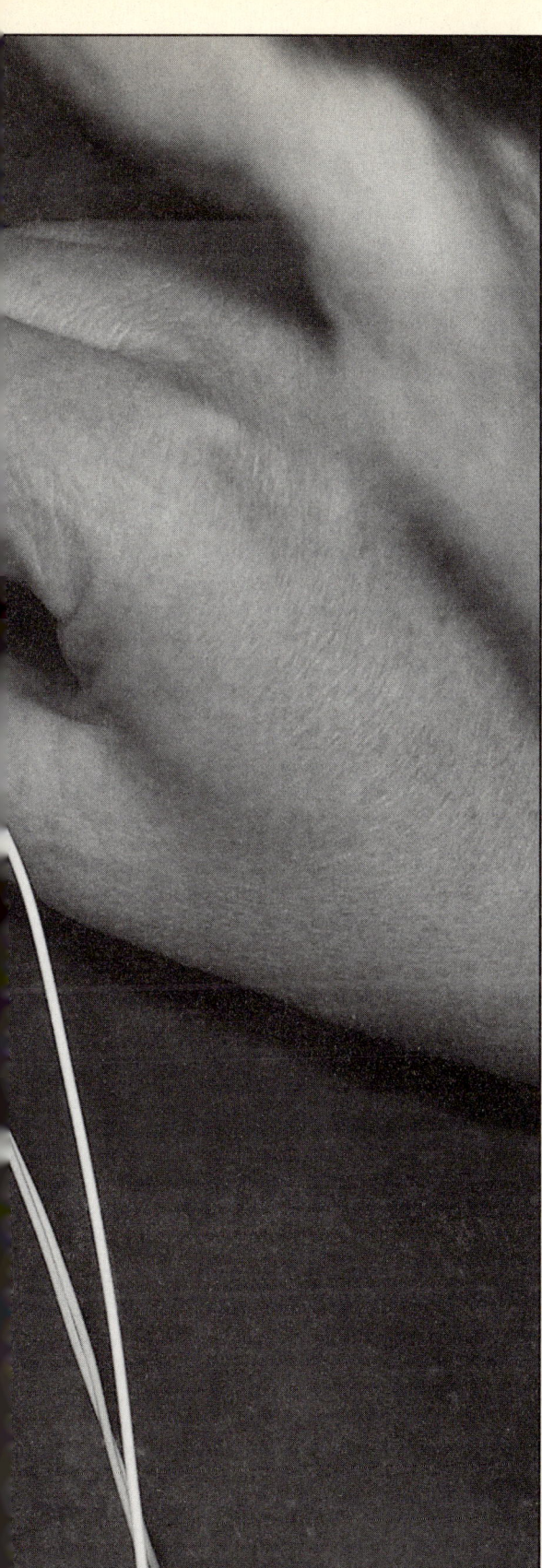

Chapter 7
SOLDERING *methods*

Section 1

Description and Identification

Soldered connections are used in aircraft electrical wiring to form a continuous and permanent metallic connection having a constant electrical value. The importance of establishing and maintaining a high standard of workmanship for soldering operations cannot be overemphasized.

This chapter describes the materials and equipment used in soldering aircraft interconnecting wiring. It also describes and illustrates preparation and care of equipment, procedures to be followed and the soldering techniques necessary to make a good soldered joint.

In addition, special materials, equipment and techniques used in soldering printed circuit assemblies are described where they differ from those used in general electrical soldering. In the repair of printed circuit assemblies, soldering is closely associated with repairs to the insulating base and conductor pattern, and with replacement of components. Therefore, typical procedures and techniques for making such repairs are included in this chapter.

Soldering. Soldering is the process of joining two (or more) metals together at a temperature lower than the melting points of the metals. In its molten state, solder chemically dissolves part of the metal surfaces to be joined. However, most metals exposed to the atmosphere acquire a thin film of tarnish or oxide; the longer the exposure the thicker the film will become. This film is present, even though it is not visible, and solder alone cannot dissolve it. A soldering flux with a melting point lower than the solder

> **Learning Objectives:**
> - Description and Identification
> - Heat Application Methods
> - Soldering Precautions and Procedures
> - Inspecting a Finished Solder Joint
> - Repair and Soldering of Printed Circuit Assemblies

Left. The soldering methods used for general aircraft wiring are essentially the same for both production soldering and for repair work.

must be used to "wet" the metal and allow the solder to penetrate it and remove the film. The flux melts first, removing the tarnish or metallic oxide, and also prevents further oxide from forming while the metal is being heated to soldering temperature. The solder then melts, floating the lighter flux and the impurities suspended in it to the outer surface and edges of the molten filler. The solder cools and forms an alloy with the metal. Most of the flux is burned away during the soldering process; any residue is removed by appropriate cleaning methods.

The soldering methods used for general aircraft wiring are essentially the same for both production soldering and for repair work. For printed circuit assemblies, production methods and repair methods are different. In production, a dip soldering method is used, where several connections are made at the same time. Soldering repairs, however, are made individually using techniques similar to those used for soldering general wiring, with special precautions to prevent thermal damage to the heat-sensitive, closely packed circuit elements.

Soft solder. Soft solder is an alloy consisting of various combinations of tin and lead with silver and other additives, which melts at temperatures below 700 degrees F. It may be in bar form to be melted for tinning, or in the form of rosin-cored solder for use with a soldering iron or other heating means.

Soft solder used in aircraft electrical wiring conforms to the requirements of Federal Specification QQ-S-571. For general applications at low temperatures (up to 248 degrees F max.), use composition Sn60 (60% tin, 40% lead) to solder tin-coated copper wire and coaxial cable. For silver-coated copper wire in high temperature applications (up to 375 degrees F max.) use a lead-silver mixture, composition Ag 2.5 or Ag 5.5. Do not confuse high temperature soft solder with the hard solder described in the following paragraph. For soldering printed circuit boards, use a eutectic solder (63% tin, 37% lead) with a silver additive of one to three percent. Rosin-cored solder, tubular type, 1/32 inch diameter is recommended for printed circuits.

Hard solder. Hard solder, often called brazing alloy, is a silver alloy, Federal Specification QQS-561, which melts at temperatures ranging from 700 to 1600 degrees F. Hard solder is used when greater mechanical strength or exposure to higher temperatures is required. Hard solder is commonly used in the aircraft electrical system for soldering thermocouple connections (refer to Chapter 5). Hard solder is not used on printed circuits.

Flux. Flux is a chemical reducer used for surface conditioning before and during the soldering process. With soft solder, use only water-white rosin, dissolved to a paste-like consistency in denatured alcohol, (Mil-Spec MIL-F-20329). With hard solder, use borax, or similar material, mixed to a paste with water, (Federal Specification O-F-499). A special solder sometimes used in thermocouple connections is described in Chapter 5 on page 5-8.

Typical soldering operations. Following are examples of typical soft-soldering operations used in aircraft electrical wiring:

> **NOTE:** *Hard-soldering procedures are described in Chapter 5.*

- **Tinning:** Wires or cables preparatory to joint soldering and to fuse ends; contact pins and inside surfaces of solder cups; shielded wire braid, after twisting, to fuse, terminate and connect.

- **Soldering:** Wires and cables, previously tinned, inserted into solder cups of terminals, or mechanically wrapped on shaped lugs and post or hooked terminals; twisted connections, or broken wire for emergency repair; printed circuit conductor pattern defects, or component leads and lugs to conductor pattern terminal areas.

- **De-soldering:** Soldered joints prior to re-making; printed circuit component connections to remove component for replacement.

Section 2
Heat Application Methods

Soldering iron. The most commonly used method of heat application for soldering joints in aircraft electrical wiring is by means of an electrically heated hand-held soldering iron. In addition to the conventional iron, a pencil iron or a soldering gun are frequently used. See Figure 7-1. Pencil irons, except for their smaller size, are identical to conventional irons, and are used for precision soldering of small units and miniature assemblies. Soldering guns, because they heat quickly, are excellent for intermittent use.

Resistance soldering. See Figure 7-2. Resistance soldering is frequently used in large volume production, where the operation is standardized. In this method, a low voltage transformer is used and the metal to be soldered is heated by the resistance to a flow of electric current. The work is gripped between two electrodes, completing the circuit and heating the metal for soldering. In another application, a carbon pencil is used as one electrode, and the metal to be soldered forms the other electrode; when contact is established through the carbon pencil intense heat is generated at the point of contact. Resistance soldering is well adapted to the soldering of small parts, or for congested assemblies where it is desired to restrict heat to a small part of the assembly.

Torch soldering. Torch soldering is used where a high heat is required, as in silver soldering. This process is also suitable for soft-soldering large work which is not part of an assembly, or when the part to be soldered can be removed for soldering. For example, wires may be torch-soldered to large contacts which have been removed from MS connectors. Torch soldering is not suitable for soldering small parts.

Dip soldering. Dip soldering is the process of immersing connections in molten solder; one or more connections can be made in a single operation. This process is used on printed circuits, where the conductor pattern is one side of the board, and the components on the opposite side. Joints are mechanically secured, dipped first into liquid flux and then into molten solder.

ver coating on the tip. (See Figure 7-4, next page).

CAUTION: *Do not allow the iron to come up to full temperature before starting the tinning operation.*

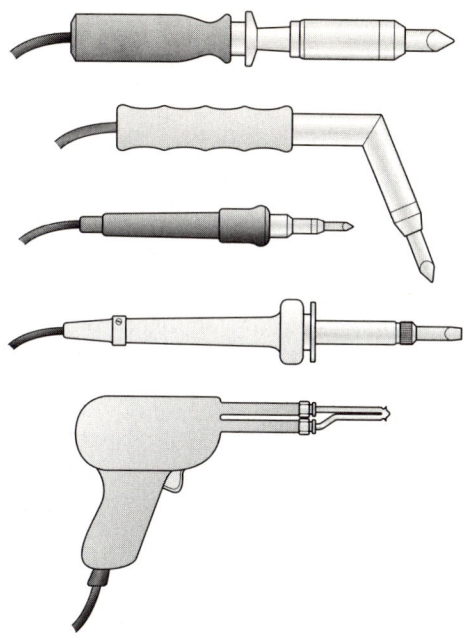

Figure 7-1. Types of hand soldering irons

Section 3
Soldering Precautions and Procedures

For successful, effective soldering, the soldering iron tip must be tinned to provide a completely metallic surface through which the heat may flow readily from the iron to the metal being soldered. If no tinning is present, the iron will oxidize and the heat cannot flow through. Copper has a very high rate of heat conductivity, but copper tips oxidize quickly, and must be frequently cleaned and re-tinned. If a tip has become badly burned and pitted because of overheating, replace it.

Preparing the soldering iron. Before using the soldering iron, prepare it as follows:

1. With the iron shut off, file each working surface of the soldering iron tip with a double-cut mill file until it is smooth and of a bright copper color. See Figure 7-3. Remove copper fuzz from dressed edges with a file card.

2. Plug in the iron and apply cored solder just as the bright dressed copper color is turning to a pigeon-blue, bronze, oxide color. This will allow the flux to "wet" and clean the working area when the solder melts to form an even bright sil-

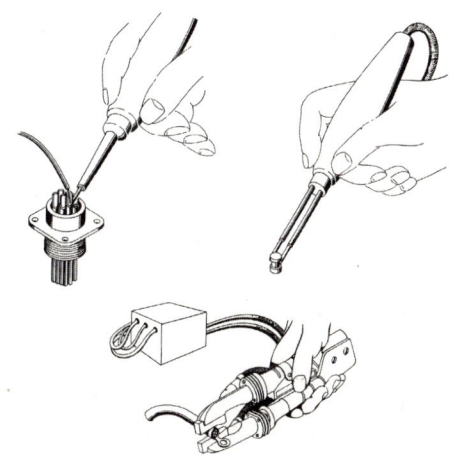

Figure 7-2. Resistance soldering

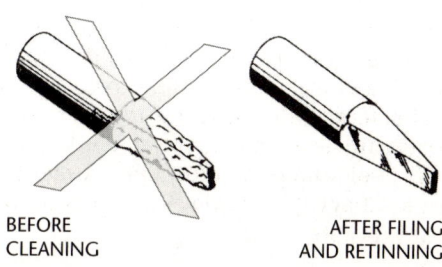

BEFORE CLEANING   AFTER FILING AND RETINNING

Figure 7-3. Soldering iron tip before and after cleaning

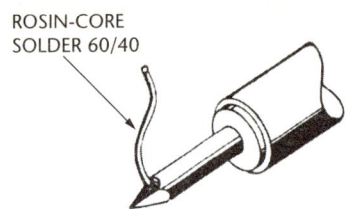

Figure 7-4. Tinning soldering iron tip

3. Wipe off excess solder with a damp sponge or cloth.

Some copper soldering iron tips used in production soldering are coated with pure iron to help prevent oxidation. Follow manufacturer's instructions for cleaning such irons. A clean damp cloth may be used to wipe the iron.

NOTE: *Do not file soldering iron tips coated with pure iron. Filing will ruin the protective coating. If the tip is pitted, replace it.*

Soldering iron maintenance. During use, just before each application, pass the soldering iron tip with a rotary motion through the folds of a damp cleaning sponge, or wipe on an asbestos wiping pad. This will remove the surface dross and excess solder from the working surface.

CAUTION: *Never shake or "whip" an iron to get rid of dross or excess solder droplets.*

Once a day remove the tip from the iron and clean out the black scale from the inside of the iron and from the tip with fine steel wool. When the iron or tip is new, coat the inside of the shank with dry flake graphite or anti-seize material to prevent freezing, and to insure maximum heat transfer. When replacing the tip, make sure to insert the tip to the full depth of the casing, seated firmly against the heating element.

Soldering Operation - General Precautions and Procedures

Regardless of the heating method used in the soldering process, a good connection will result only if the proper soldering techniques are followed, and certain precautions observed. The following instructions apply generally to soldering operations. Some special soldering techniques used in assembling or repairing printed circuit boards are listed on pages 7-7 through 7-9. Detailed procedures are given for soldering wires to MS connectors in Chapter 2, to coaxial connectors in Chapter 3 and for thermocouple connections in Chapter 5.

NOTE: *A quality soldered joint can be accomplished only on a mechanical connection of approved geometry, dress and dimensions.*

Cleanliness. Cleanliness is of the utmost importance in the soldering operation. If possible, soldering should be done in an area that is reasonably clean and free from excessive dust. Drafty areas should be avoided so that the soldering iron will not cool.

Parts contaminated with dirt, oil, grime, grease, etc. cannot be successfully soldered. Make sure that the parts are mechanically "bright-clean", before soldering. Clean the parts with a cloth or brush dipped in alcohol, carbon tetrachloride, trichlorethylene or other approved solvent. Badly corroded parts may be cleaned carefully by mechanical means such as fine abrasive paper, a wire brush or by careful scraping with a knife blade.

Pre-tinning. Wires to be attached to most electrical connectors must be pre-tinned. Follow the instructions given in Chapter 1 on pages 1-16 through 1-19.

Selection of flux and solder. Use only the solder and fluxes described on page 7-2.

CAUTION: *Do not use any corrosive flux in aircraft electrical wiring.*

Heating capacity. Use a soldering iron or other heating method of sufficient capacity to heat the metal being soldered to solder melting temperature.

Selection of soldering iron. The sole purpose of the soldering iron is to heat the joint to a temperature high enough to melt the solder. Select a soldering iron with a thermal capacity high enough so that the heat transfer is fast and effective. An iron with excessive heat capacity will burn or melt wire insulation; an iron with too little heat capacity will make a cold joint in which the solder does not alloy with the work. Soldering irons are available in wattage ranges from 20 to 500 watts. Irons with wattage ratings of 60, 100 and 200 watts are recommended for general use in aircraft electrical wiring. Pencil irons with a rating of 20 to 60 watts are recommended for soldering small parts. The soldering iron recommended for printed circuit soldering is a lightweight 55 watt iron with a 600°F Curie point tip control. This iron has a three-wire cord to eliminate leakage currents that could damage the printed circuits.

A soldering iron should also be suited to the production rate. Do not select a small pencil iron where a high steady heat flow is required. A soldering gun is useful for intermittent work, but is not suitable for high speed production

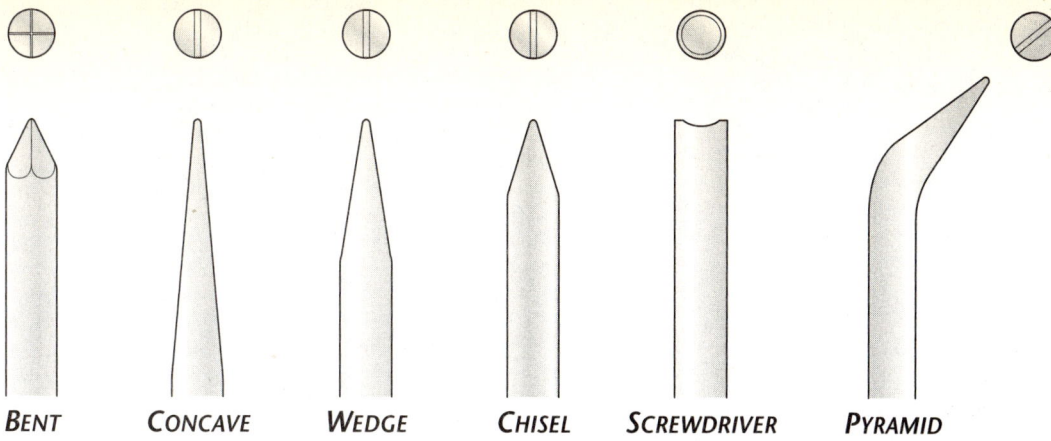

Figure 7-5. Soldering iron tip shapes

work because of the warm-up time required each time the trigger is depressed.

Choice of soldering tip. Select the tip best suited for the size and shape of the work being soldered. Some common tip shapes are shown in Figure 7-5. Soldering iron tips are available in sizes from 1/16 inch to 2 inches in diameter. For general use a tip of 1/4 inch to 3/8 inch diameter is recommended. For printed circuit soldering use a long shank tip of 1/16, 1/8, 3/32 or 3/16 inch diameter; Screwdriver, chisel and pyramid shapes are recommended.

Soldering iron. Before starting the soldering operation, make sure that the iron tip is clean, smooth and well-tinned. See pages 7-3 through 7-4 for instructions on preparation and maintenance of soldering iron. When resistance soldering equipment is to be used make sure that probes are clean.

Securing the joint. Whenever possible make sure that the joint is mechanically secure before soldering. When this is not possible, as with MS connector contacts, make sure that the joint is held rigid during the cooling period.

Application of heat and solder. Apply flux-core solder at the exact point between the metal and the soldering iron, as shown in Figure 7-6 and hold the iron directly against the assembly. Melt the solder on the joint, not on the iron. Place the soldering iron firmly against the junction; if heavy "rocking" pressure is necessary, either the iron does not have sufficient heat capacity for the job, or it has not been properly prepared, or both.

Heat application time. Do not apply heat to the work any longer than the time necessary to melt the solder on all parts of the joint.

Amount of solder. Do not use any more solder than necessary. Do not pile up solder around the joint; this is wasteful, and results in joints difficult to inspect. Care should be exercised with silver coated wire to prevent wicking during solder application.

Soldering iron holder. When the soldering iron is not in actual use during operations, keep it in a holder such as is shown in Figure 7-7. This will protect the operator against burns, and the iron against damage.

Protection against overheating. Do not allow the iron to overheat. Disconnect the iron when it is not in use between operations, or use a heat dissipating stand, which will keep the iron at a constant temperature.

Cooling the solder joint. When the solder joint is made, hold the work firmly in place until the joint has set. Disturbing the finished work will result in a joint mechanically weak, and with high electrical resistance. Allow solder joints to cool naturally. Do not use liquids or air blasts.

Cleaning. If the correct amount of solder is used and procedure instructions followed carefully, there should be little or no excess flux remaining on the finished joint. If cleaning is necessary, remove excessive flux by brushing the joint with a stiff brush dipped in methyl alcohol, methyl isobutyl ketone or a similar approved solvent. Use alcohol sparingly, and avoid contact between alcohol and wire insulation. For cleaning printed

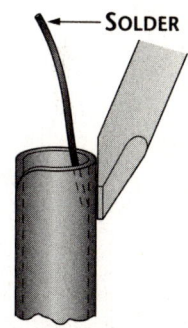

Figure 7-6. Position of soldering iron

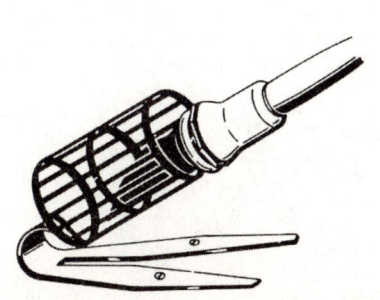

Figure 7-7. Soldering iron holder

circuit connections, use a cotton swab-stick for small areas, and a lint-free clean cloth for large areas and board edges.

Section 4

Inspecting a Finished Solder Joint

Acceptable solder joint. See Figure 7-8. A good soldered joint will have a bright silvery appearance, with smooth fillets and feathered, not sharp edges. The entire joint will be covered with a smooth even coat of solder, and the contour of the joint will be visible.

Unacceptable solder joint. Any of the following indicate a poor solder joint, and are cause for rejection:

- Dull gray, chalky or granular appearance, evidence of a cold joint.
- Hair, cracks, or irregular surface, evidence of a disturbed joint.
- Grayish, wrinkled appearance, evidence of excessive heat.
- Partially exposed joint, evidence of insufficient solder.
- Scorched wire insulation or burned connector inserts.
- Globules, drips or tails of solder.

If any of the above are present in a finished solder joint, the joint should be taken apart, parts cleaned, and the entire soldering operation repeated, using fresh solder and flux.

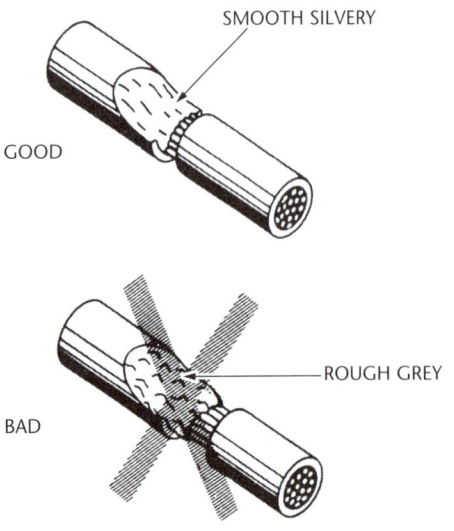

Figure 7-8. Good and bad soldered connections

Section 5

Repair and Soldering of Printed Circuit Assemblies

Printed circuit assemblies are used in the aircraft electrical system to save space and weight, to increase reliability and to facilitate replacement. Typical locations are in equipment cases at area termination points of the interconnecting wiring, and in junction boxes at break-out points in the electrical system.

A printed circuit board consists of a conductor pattern bonded onto an insulating base. Components are mounted on the opposite side from the conductor, and are attached to it by leads through drilled holes or eyelets in the base. A printed circuit board may have conductor patterns on both sides. Each printed circuit board assembly is identified by a number printed on the board, referencing the applicable circuit drawing or schematic.

Circuit Assembly Repair

Feasibility of repair. Many printed circuit assemblies are repaired and re-soldered to restore them to their original condition when they have been found to be defective or damaged. If the damage or defect is serious or extensive, it may not be practical to rework the board. It is not considered feasible to repair a printed circuit assembly if any of the following conditions could result:

- Repair is impractical because of the extent of the damage.
- The repair would decrease the life expectancy of any component.
- The repair could damage adjacent components or circuitry.
- The operation of the equipment would be changed by the repair.

In any of the above cases, replace the printed circuit assembly with a new identical assembly.

General Precautions and Procedures

Handling precautions. Printed circuit assemblies require very careful handling during repair and soldering operations, or when being installed as a replacement assembly. When handling assemblies withdrawn from cases and junction boxes or from replacement stores observe the following precautions:

- Keep the printed circuit assembly in a padded plastic or paper protective bag while awaiting use or repair. Keep in a cool dry place.

- Carry the assembly by its handle (if present) or by the edges.
- Do not handle or lift the assembly by any of its mounted components; this can result in broken leads.
- Do not stack one board on top of another. Always support the board on its long free edge.
- Never flex, bend or force the base during removal, repair or replacement.
- Avoid touching the face of the board with the hands, particularly the conductor or exposed contacts of plug-in assemblies and test points. Body acids produce corrosion and cause high resistivity that can affect the performance of the circuit.
- During repair, use an appropriate holding fixture to support the board by its edges, as shown in Figure 7-9. Provide adequate support underneath the board to offset the force of drilling, scraping and component removal or replacement.
- Make sure that all tools used for mechanical repairs are clean. Use only pliers with smooth jaws and radiused edges.
- Clean with short gentle puffs of low pressure, oil-free, dry air. Avoid violent bursts. Do not clean by ultrasonic methods this may affect the laminate.

De-soldering printed circuit components. When it is necessary to remove components from a printed circuit assembly, their leads or lugs must first be de-soldered, so that the leads or lugs can be readily withdrawn through the board to free the components. The procedures to be used and the precautions to be observed are as follows:

1. To melt the solder, apply the soldering iron tip to the fillet while slowly counting to four, then remove the iron.

   CAUTION: *Do not exceed this time.*

   Repeat as necessary, allowing the work to cool between applications. If it is desired to remove the component intact, as for testing, place a heat sink clamp on the lead, as close as possible to the component body before applying the iron.

2. Use a solder sucker to suck up the solder as it melts. See Figure 7-10a. Compress the bulb, place the tip directly on the liquid fillet, and release the bulb. Repeat as necessary.

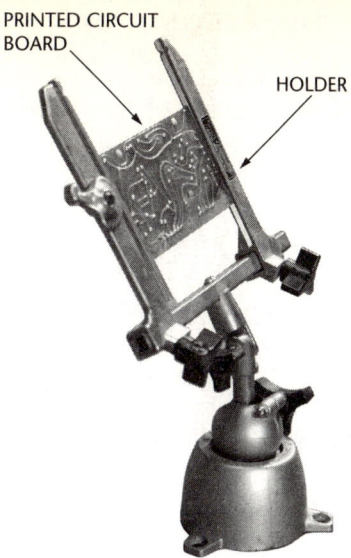

Figure 7-9. Holding fixture for printed circuit assembly

3. In open areas, or to remove large terminal areas of soldering, a piece of 1/8 inch tubular electro-tinned braid may be used to remove the liquid solder as shown in Figure 7-10b. Flatten 1/4 inch of the braid end, dip sparingly into liquid flux, and place this end against the fillet. When the iron is applied, the solder will be drawn up into the braid as it melts. Cut off the solder-saturated part of the braid and repeat if necessary.

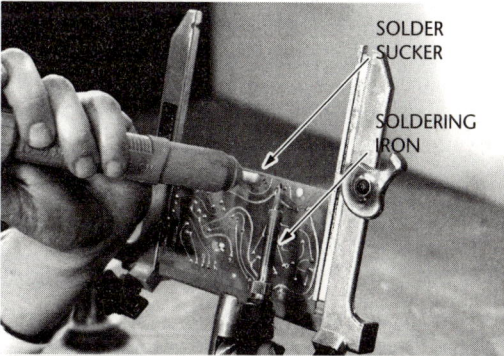

A. USE OF SOLDER SUCKER IN DESOLDERING OPERATION

B. REMOVING SOLDER WITH COPPER BRAID

Figure 7-10. Desoldering printed circuit components

Soldering Procedures

Working space on printed circuit board assemblies is limited because of tightly packed components. The following special procedures will be helpful in soldering operations under these conditions:

1. Fix the board in the holder in such a way that gravity will aid in forming the fillet.
2. Select a tip shape to suit the angle of approach to the work. Use a screwdriver tip for an approach perpendicular to the board; use a pyramid shape when the approach is at an angle. See Figure 7-11.
3. Rest the elbow, arm or wrist against the bench top or holding fixture to assist in directing the iron to the work.
4. When soldering, coil the fine diameter cored solder into a small helix for easier handling, and feed from the center of the coil.
5. Apply the flat side of the iron against the work with a deliberate touch, not to exceed a slow count of four, until experience dictates the practical limit for the condition.
6. If the solder does not "take", allow the connection to cool, and reapply the iron tip, after adding a small amount of liquid flux. If this does not produce a satisfactory joint, remake the joint.

Soldering Precautions

When soldering printed circuits, the completed solder fillet must mechanically bond the component lead or lug to the foil terminal area, and electrically provide a low resistance path between the two. The soldering operation must be carried out without damage to the laminate, foil, and adjacent components. Observe the following precautions in addition to the general precautions listed earlier on pages 7-3 through 7-5.

- Make sure that leads are properly dressed and fixed in position before soldering.
- Use the smallest tip size and shortest heat application time possible to avoid damage to board, foil and components.
- Use only the solder and flux described on page 7-2.
- Use heat sinks to dissipate excessive heat. Excessive temperature will cause the base/foil laminate to discolor, delaminate, blister or burn; board components can be damaged or suffer a change in value from overheating.
- Protect areas adjacent to the soldering operation with lightweight solder shield, cut and shaped around components.
- Avoid excessive solder or splatter.
- Be careful not to accidentally touch the iron to the cored solder. This may cause the solder to splatter, damaging the work and resulting in painful burns to the operator.

Detailed Procedures

The following paragraphs give instructions for typical feasible repair and soldering operations to defective or damaged circuit assemblies. Observe all precautions listed in the previous paragraphs.

Removal of protective sealing (conformal) coating. Before starting any repair, it is necessary to remove the protective coating in the defective area to expose and clean the defect. The procedure is as follows:

1. In open base areas with no adjacent heat sensitive components, remove the coating with a chisel-tip soldering iron, using a push motion as shown in Figure 7-12a. Keep the iron moving steadily to prevent heat damage to base or conductor pattern, and to avoid unwanted de-soldering.

   CAUTION: *Do not attempt removal in a single pass. Complete removal in one area at a time.*

2. Clean the area with a scrub brush and approved solvent to remove traces of coating or base material particles.
3. Wipe of excess solvent with cotton swabs, and dry with a lint-free cloth.
4. In congested base areas where heat damage to adjacent components must be avoided, use a hook scraper, with a pull motion as shown in Figure 7-12b.
5. Finish and clean out with an angle scraper or knife point. Smooth any remaining rough areas with file or knife blade.

   NOTE: *When repairs have been completed, the protective coating must be restored to the reworked area.*

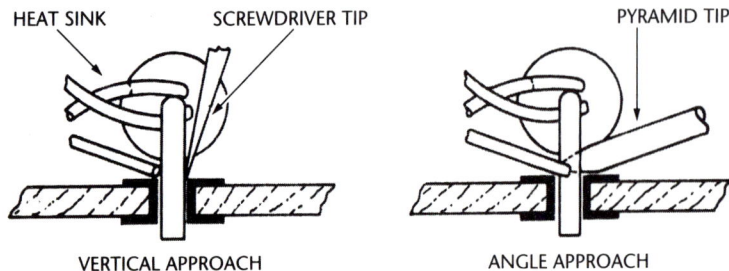

Figure 7-11. Soldering iron approach to work

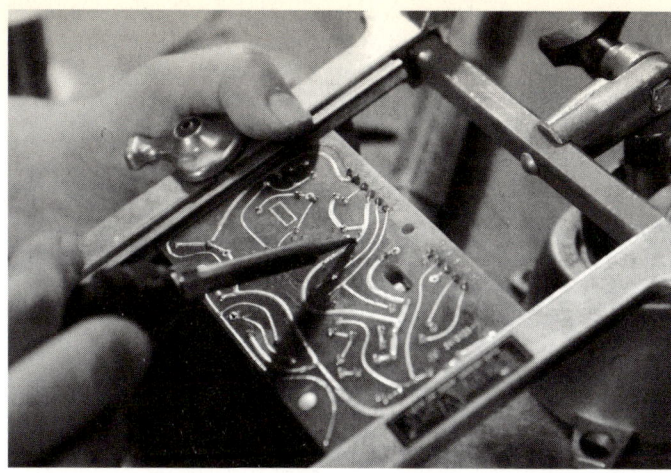
A. REMOVING COATING WITH SOLDERING IRON

B. REMOVING COATING WITH SCRAPER

Figure 7-12. Removing protective coating

Base laminate repairs. Before making repairs to conductor patterns, or replacing defective components, repair any physical defects in the base laminate such as chipped edges, edge cracks and gouges or other surface damage. It is not considered feasible to repair base laminate having the following defects:

- Ruptured conductor pattern, annular rings or terminal areas.
- Loss of support under printed contacts or connectors.
- Dangling components.
- Inadequate guide edge or mounting provisions.

Repair of conductor pattern. Feasible repairs to the conductor pattern are of two types: repair of minor defects such as fine cracks, pin holes, ragged edges, scratches, gouges and corrosion; and repair of major defects where the conductor foil, or its bonding is damaged to such an extent that part of it must be replaced with a wire segment. It is not considered feasible to repair annular rings or terminal areas that are incomplete or not bonded to the base. When making repairs to the conductor pattern, observe the following special precautions:

- Make sure the minimum separation distance between adjacent conductors is not reduced.
- Make sure the effective cross-sectional area of the conductor foil is maintained.
- Make sure that the original conductor dress and routing is adhered to.

Repair of minor conductor defects. The procedure for repairing the minor defects described in the previous section is as follows: (See Figure 7-13a).

1. Bright-clean the defective area; remove all foreign particles by brushing.

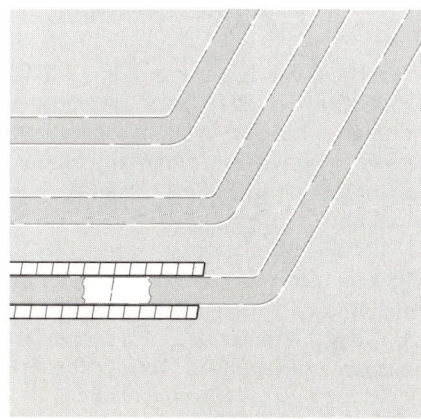
A. MINOR DEFECT PATCHED WITH SOLDER

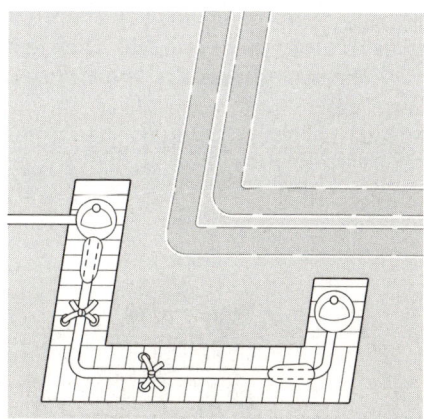
B. MAJOR DEFECT BRIDGED WITH WIRE

Figure 7-13. Repairing defective conductor pattern

2. Apply flux to the repair area; then apply solder to bridge or patch the defective area with a minimum 1/8 inch overlap.

3. Remove solder residue with a solvent-saturated swab, and dry the area.

Repair of major conductor defects. The procedure for repairing the major defects described is as follows: (See Figure 7-13b)

1. Measure width of damaged conductor, and select a wire of the correct diameter to replace it. This information will be found on the applicable printed circuit assembly drawing.

2. Remove all traces of the damaged part of the conductor from the base, using a sharp knife, scrapers and brushes.

3. Bright-clean the remaining ends of the conductor and remove foreign particles.

4. Cut the replacement wire to length, strip (if insulated) clean, trim and tin ends. Form the wire to the contour of the removed conductor.

5. Apply liquid flux to the wire and conductor ends, and solder.

6. Remove solder residue with a solvent-saturated swab, and dry the area.

7. If necessary, for long wire segments, tie the wire in place with 1/32 inch diameter lacing cord, through a pair of 1/16 inch diameter holes drilled in the base.

NOTE: *Short wire segments will be located and attached to the base by the final protective coating.*

Repair of conductor interfacial connections. Interfacial connections, such as swaged eyelets, and plated-through holes are used to connect the conductor patterns on opposite sides of the base. Surface cracks or imperfect soldering of these, or of component leads and "z" wires used in conjunction with them, may cause discontinuities in the electrical path. The repair procedure is as follows (See Figure 7-14):

1. Bright-clean the defective area on both sides of the board.

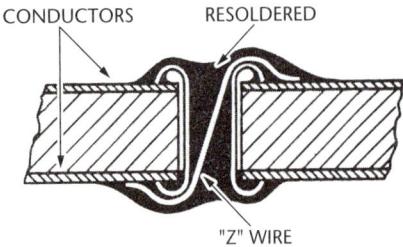

Figure 7-14. Repairing interfacial connection

2. De-solder the connection, using a soldering iron and solder sucker. Remove old fillets and contained solder from inside and around the hole, and from the component lead or "z" wire if present.

3. Inspect eyelet for tightness. If it is not tight, reset it.

4. If the through hole is an open hole and a "z" wire can be used, cut, form and insert it. Dress both ends on the foil centerlines.

5. Bright-clean both sides of the connection, and re-solder the connections to fill the opening, on both sides.

6. Remove solder residue with a solvent-saturated swab, and dry the area on both sides.

Replacement of defective circuit components. Defective components are removed from the printed circuit assembly, and replaced by a new component identical to the one removed. Components must be removed and replaced without damage to their attachments or to the surrounding area. Observe these special precautions:

- Make sure that position of replaced component on the board, lead dress and polarity are the same as the original. Position component so that its markings are visible, and that codes or legends on base surface are not hidden.

- Avoid pressure points where leads may contact other leads.

- Make sure minimum bend length dimensions for leads are observed.

- Provide stress relief for stiff leads, when used.

- Provide insulation where possibility of short circuits exists.

Replacement of lead mounted components. The procedures for replacing lead-mounted components such as resistors, capacitors, chokes, etc. are as follows (see Figure 7-15):

1. When the defective unit and its leads are to be entirely removed, cut the leads as shown and discard the component body. De-solder both leads, and extract stubs from component side. Hot ream both terminal holes from conductor side and clip any "icicles" on the opposite side. Remove solder residue and dry.

CAUTION: *When pulling leads through the board, pull against the side having the smallest foil area, to avoid loosening the foil.*

2. When defective unit body only is to be removed, cut the leads close to the body, and discard body. Erect the stubs and

bright clean in preparation for soldering.

3. When defective component is a carbon resistor, the body only is removed by destroying it, to leave the longest possible leads. Do this as follows:

   a. Position the holding fixture so that crushed body particles will fall free of other components.

   b. Use heavy diagonal pliers to crush the body at the center, steadying both leads with the other hand.

   c. Crush any body material adhering to the upper ends of the two leads.

   d. Erect the stubs and bright-clean in preparation for soldering.

   e. Remove all foreign particles and dust from the assembly.

   NOTE: *After each of the above procedures inspect the repair area on both sides of the base for overall integrity and next step readiness.*

4. When component leads have been completely removed as described in step 1, insert the replacement component leads in the mounting holes. Wedge and secure the component at the required base clearance and solder to the base.

5. When lead stubs have been left attached to the board as described in step 2, form loops at ends of replacement component leads, crimp around stubs and solder.

Replacement of lug mounted components.
The procedure for replacing lug-mounted components is as follows:

1. De-solder the lug from the base, and suck up liquefied solder with a solder-sucker.

2. If the lug is crimped over, re-heat the connection and straighten out the lug with a probe prong.

3. Apply the iron to the lug, then place the solder-sucker tip down around the lug to draw up any remaining solder.

4. With a sharp narrow knife blade, cut through any crystallized solder bridging the lug and foil terminal area. Make cuts next to and parallel with the lug, cutting into the lug base hole clearance.

5. Free the lug from the base.

   CAUTION: *Take care not to damage the conductor foil.*

6. Position the replacement component to pass the lugs through their base holes. Wedge and secure the component so the lugs cannot shift.

7. Trim and form the lugs as necessary.

8. Make sure lugs and terminal area are clean, then solder, using heat sink if necessary.

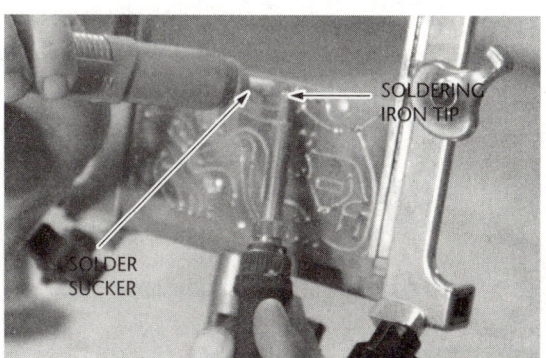

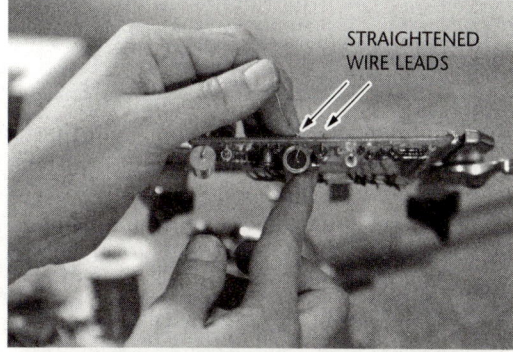

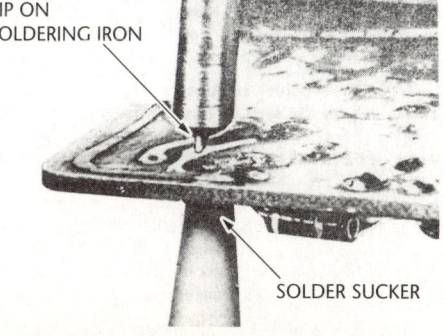

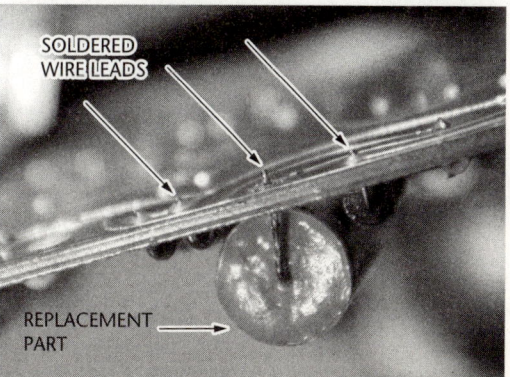

Figure 7-15. Replacing defective component

Chapter 8
ELECTRIC connector sealing

Section 1
Description and Identification

Sealing compound is used to moisture-proof the wiring connected to the backs of electric connectors, and to reinforce the backs of connectors against failure caused by vibration and lateral pressure that fatigue the wire at the solder cup. This process is commonly called potting. The sealing compound protects electric connectors from corrosion or contamination by excluding metallic particles and aircraft liquids, and also reduces the probability of arcover between pins on the back of the connector.

This chapter describes the potting compounds used on aircraft electric connectors, and gives instructions for preparing and storing the compound. Detailed instructions for potting MS electric connectors are given in Chapter 2.

Sealing compound in accordance with Mil-Spec MIL-S-8516 is a two-part thiokol rubber compound, consisting of a base and an accelerator (curing agent), packaged together. This compound is used to seal connectors located in areas where the ambient temperature does not exceed 200 degrees F.

> **NOTE:** *Sealing compound is not applied to environment proof type "E" and "R" connectors, coaxial connectors or connectors with crimped contacts.*

High temperature sealing compound. There are two commercial sealants currently used in aircraft electrical systems for high temperature applications; these are:

Learning Objectives:
- Description and Identification
- Preparation of Sealing Compound

Left. **Proper moisture proofing of connectors protects against harsh environmental conditions.**
Photo courtesy of Ka Ho, Ng - HKAEC

- PR#777 (Coast Pro-seal Company), a two-part polyurethane compound, designed to resist prolonged exposure at 350 degrees F and intermittent exposure at 400 degrees F.
- RTV-60 (General Electric); a two-part silicone rubber compound, which will resist temperatures up to 450 degrees F without deterioration.

General precautions. Potting compound is supplied in paired cans of base compound and accelerator. Use only the accelerator supplied with the base compound. Substitution may produce a sealant with substantial electrical properties. To avoid errors, store base and accelerator together in the carton.

Make sure that the entire amount of accelerator is mixed into the entire amount of base. Any change in proportion will affect electrical properties of the sealant, and may affect the work life, rate, of cure and hardness of the compound.

The sealant contains a small quantity of volatile flammable solvent. Observe adequate ventilation and fire precautions during mixing and storage.

Section 2
Preparation of Sealing Compound

Hand mixing procedure. Add the accelerator to the base compound as follows:

1. Remove lid from accelerator container and stir contents slowly into a smooth creamy paste with a clean spatula, wood tongue depressor or putty knife.

   WARNING: *Wear rubber gloves to avoid excessive skin contact, and clean hands thoroughly after handling.*

2. Cut top from base compound container and stir contents until material has a smooth texture. This is necessary to recombine material that may have settled out.

3. Combine accelerator and base material and thoroughly agitate or mix until no accelerator streaks or traces of unmixed material are visible. Mix slowly, do not beat or whip; fast mixing may cause excessive amounts of air to become trapped in the compound. Mixing normally requires five to eight minutes. Continued scraping of the sides and corners of the bottom of the container will insure complete mixing (See Figure 8-1).

4. Determine if mixing is complete by spreading a drop of the mixture very thinly on a piece of white paper with a knife blade or similar instrument. Close examination should not reveal any specks or streaks. Do not mix the sealant beyond the point where tests show the accelerator to be thoroughly mixed into the base compound.

5. When the mixing procedure has been completed, the sealant is ready for use and may be poured directly into the connector to be sealed. For details of connector sealing, see Chapter 2 page 2-32.

6. If the mixed compound is not to be used immediately, store it as directed on page 8-4.

Mechanical mixing procedure. Mechanical mixing should be done at 60 degrees F or lower to prolong the working life of the sealant. The procedure is as follows:

1. Hand mix the accelerator as described in step 1 of previous procedure. A paint shaker vibrating machine may be used if available. Shake for five to seven minutes.

2. If the base material is packaged in a metal call, cut off the top of the container using a mechanical can opener. This should leave a smooth wall without any burr at the top of the can.

3. Clamp base material container securely to

Figure 8-1. Hand mixing potting compound

drill press geared to 50 RPM minimum to 90 RPM maximum. Insert a mixing paddle fashioned from a drill rod and wire. See Figure 8-2.

4. Start drill press motor and slowly lower mixing paddle into the base compound to recombine any material that may have settled out.

5. Scrape all accelerator from its container and place it in the base material. Start drill press motor again and mix slowly for approximately two minutes. Stop machine, raise paddle and scrape container walls and paddle as clean as possible. Start the drill press, lower the mixing paddle again, and continue mixing for an additional three minutes.

6. Make thin spread of sealant on white paper as described in step 4 of the hand mixing procedure. If necessary, continue mixing in two-minute cycles followed by paper test until no traces of unmixed material are visible. The sealant is then ready for use.

7. If the mixed compound is not be used at once, store in accordance with the instructions on the next page for storing mixed compounds.

Machine mixing. See Figure 8-3. This machine will mix up to one quart of potting compound, and inserts the mixture under pressure into a cartridge for use with a sealant gun, if desired. The procedure is as follows:

1. Set the pressure regulator to the pressure recommended by the manufacturer of the compound being used.

2. Check that the dasher control valve lever is up, cartridge filler valve closed, (handle

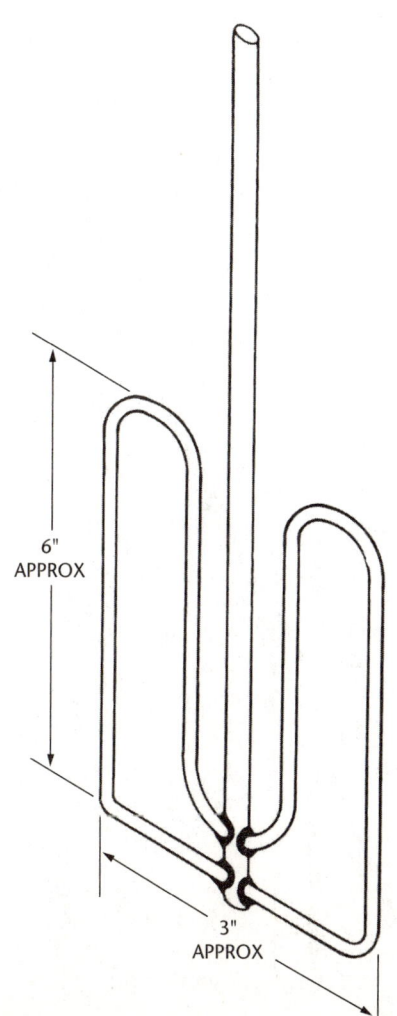

Figure 8-2. Mixing paddle for potting compound

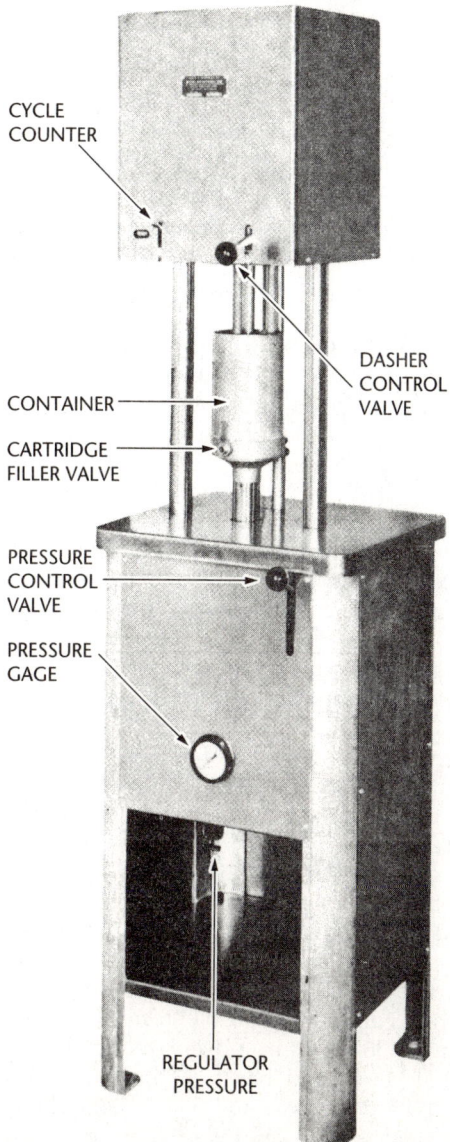

Figure 8-3. Machine for mixing potting compound

horizontal), and pressure control valve lever down.

3. Remove the container from the machine; place half the base material in the container, and add the accelerator in the center (not in contact with the container). Add the rest of the base material, and replace the container in the machine.

4. Raise the pressure control valve lever slightly so that the container rises slowly. When the container stops moving, push the lever all the way up, so that full pressure is acting against the material.

5. Set the cycle counter to zero, and move the dasher control valve lever down to start the mixing cycle.

6. After the counter indicates the completion of the required number of cycles, as recommended by the compound manufacturer, place a sealant gun cartridge, with plunger inserted, horizontally over the filler valve outlet, and open the filler valve while the machine is still mixing. Squeeze the cartridge slightly to force out air. When the dasher stops, the cartridge will be filled, and the container empty.

NOTE: *The machine should be thoroughly cleaned in accordance with manufacturer's instructions after each series of mixings.*

Preparation of high temperature sealing compound. Mix high temperature sealing compounds in the proportions as directed or supplied by the manufacturer, using one of the three methods just described.

Storage of unmixed sealing compound. Store base compound and accelerator in a cool place. Shelf life is approximately one year when stored below 75 degrees F.

CAUTION: *Do not store sealing compound at temperatures higher than 80 degrees F.*

Keep base compound and accelerator together in the carton as furnished. Note manufacturing date stamped on carton, and use oldest material first.

NOTE: *Do not use sealing compound that is over one year old.*

Storage of mixed sealing compound. Mixed potting compound can be stored for a maximum of 36 hours in a deep freezer at −20° F. Mixed compound has a 90 minute working life at 75 degrees F, and 50% relative humidity. This time is computed from the instant that the accelerator is added to the base compound and must include the time of mixing and use. If the mixed compound is to be stored, it must be cooled quickly and thawed quickly to avoid wasting the short working life. Thaw the mixed compound by blowing compressed air on the outside of the container. Never blow compressed air into the container or use heat to raise its temperature. The compound should be warm enough for use in 5 to 10 minutes.

Dispensers for sealing compound. Potting compound once mixed should be immediately poured into dispenser tubes made of polyethylene, TFE or aluminum. If necessary to store mixed compound in accordance with the previous paragraph, the compound should first be poured into the final dispenser tube.

Aluminum tubes (Figure 8-4), for use when handling small quantities of potting compound are similar to toothpaste tubes. These tubes are typically 2" in diameter and 10" long. Fold over the filling end and seal before storing. A simple key, as shown in Figure 8-4 can be used to windup the tube and force out the sealing compound.

Preparation of TFE-insulated wire for potting. Because of its chemical composition TFE insulation on high temperature electrical wire does not adhere well to potting compound. Pre-treated TFE wire is available from some wire manufacturers, or the TFE insulation may be treated with a sodium etch solution before potting. Follow the directions given by the manufacturer of the solution.

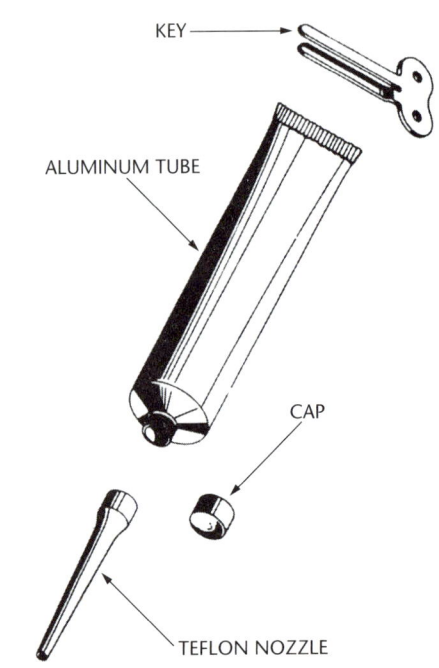

Figure 8-4. Tube dispenser for potting compound

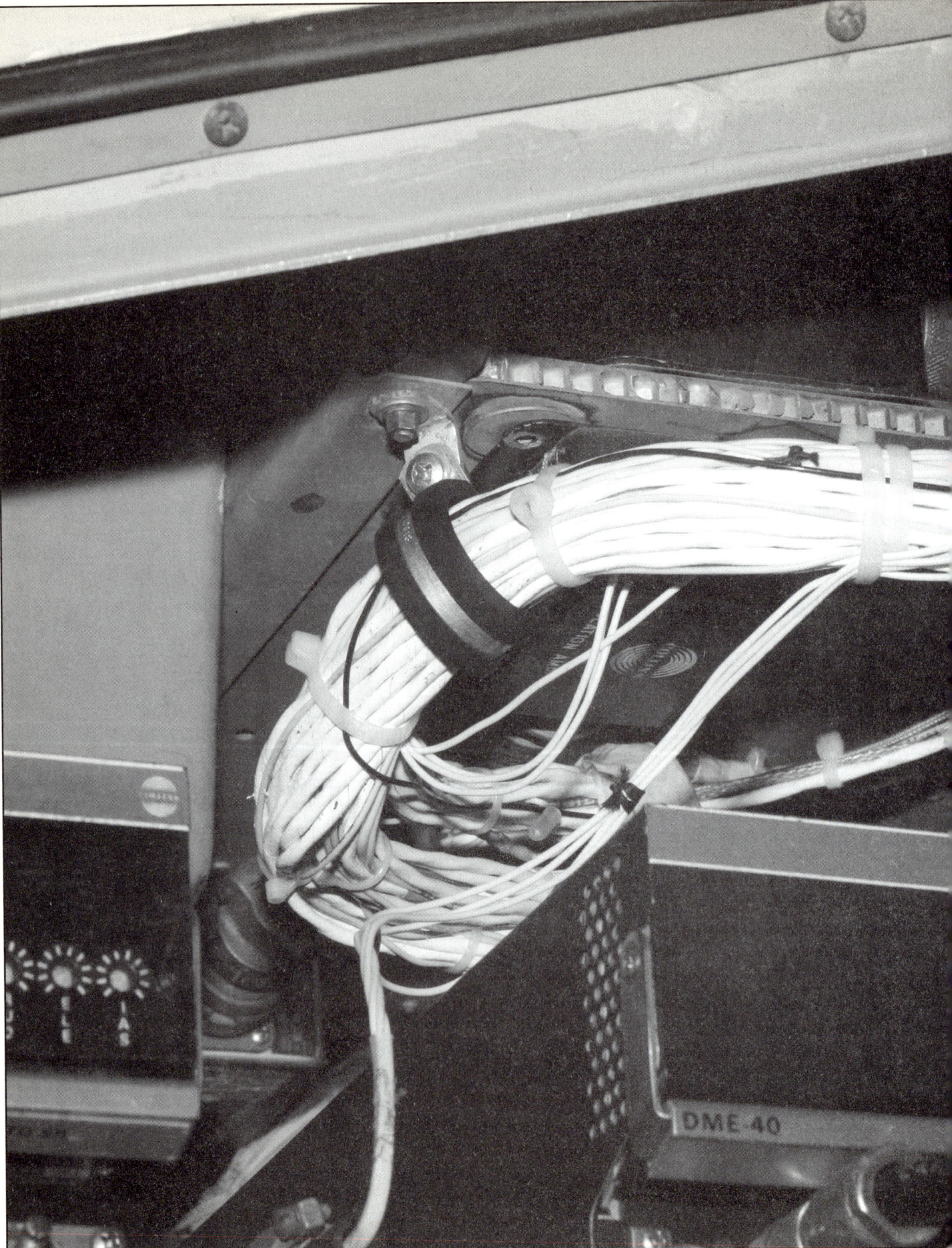

Chapter 9
INSTALLATION *of electrical hardware*

Procedures for installing equipment in aircraft are recommended in order to make installation easier, to standardize the methods used, and to provide the best possible protection for personnel and equipment.

This chapter describes the recommended procedures for installing bus bars (including preparation), conduit, junction boxes, protective devices and terminal boards in aircraft, It also describes methods of identification and protection and the correct use of hardware.

Learning Objectives:

- Preparation & Installation of Bus bars
- Installation of Conduit
- Installation of Protective Devices
- Installation of Terminal Boards

Section 1
Preparation and Installation of Bus bars

Bus bars are used in aircraft for power distribution. The most commonly used materials for bus bars are bare aluminum, plated aluminum or plated copper. Aluminum used for bus bars is EC (electrical) grade.

Preparation of bus bars. Bus bars for an aircraft electrical system must be clean and free from grease, dirt and oxide. Any of these at the electrical junction will cause the connection to heat up and fail. Bus bars are cleaned prior to installation in the aircraft, and are also treated to prevent or minimize oxidation after installation.

Preparation of unplated aluminum alloy bus bars. Clean unplated aluminum bus bars by immersing in Stoddard's solvent, or by wiping with a clean, soft cloth saturated with the solvent. Wipe dry with a clean, soft cloth.

Left. Clamps should be attached to the rigid surface of a structure so that the mounting screw is above the wire bundle.

Installation of Electrical Hardware

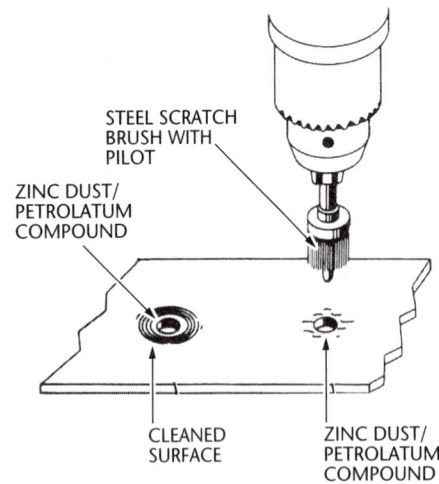

Figure 9-1. Scratch brushing unplated aluminum alloy bus bars

Nominal OD of Conduit	AN Clamp
3/16	AN742D3
1/4	AN742D4
3/8	AN742D6
1/2	AN742D8
5/8	AN742D10
3/4	AN742D12
1	AN742D16
1-1/4	AN742D20
1-1/2	AN742D24
1-3/4	AN742D28
2	AN742D32
2-1/2	AN742D40

Table 9-1 Support clamps for rigid or flexible bare aluminum conduit

After cleaning, treat all electrical contact surfaces as follows:

1. Cover contact surfaces completely with an even coating of petrolatum-zinc dust compound (50% petrolatum, 50% fine zinc dust, by weight)

2. Scratch brush the coated areas, using a rotary steel wire brush with a pilot as shown in Figure 9-1. Brush through the compound.

3. Remove most of the compound from bus bar by wiping lightly with a clean soft cloth

4. Examine bus bar to make sure that there are no steel brush bristles lodged in the aluminum

5. Apply a thin coating of clean petrolatum-zinc compound to contact surfaces. This compound is the same as that supplied in MS aluminum terminal lugs

NOTE: *Allow final coat of compound to remain on bus bar when installed. Excess will be squeezed out of connections and removed later.*

Preparation of plated aluminum and copper bus bars. Clean plated aluminum and copper bus bars thoroughly by immersing in Stoddard's solvent, or by wiping with a clean, soft cloth saturated with the solvent. Wipe dry with a clean soft cloth.

Repairing damaged plating. Examine contact surfaces of plated aluminum or copper bus bars for damage to plating. Reject damaged aluminum bus bars and return for rework. Repair slight damage to plated copper bus bars by tinning with a soldering iron, or by brush plating. Thoroughly wash and dry brush plated areas.

CAUTION: *Do not attempt to repair plating on aluminum.*

Mounting hardware. See Figure 9-2. When installing a copper bus bar, always place a cadmium plated steel plain washer between the bus bar and the lockwasher or self-locking nut. When installing an aluminum alloy bus bar, place an aluminum alloy plain washer between the bus bar and the lock washer or self-locking nut.

Insulation. Insulate the bus bar from structure, junction box, or support with a fiberglass, phenolic or other rigid insulating stand-off as shown in Figure 9-2. Do not use any moisture-absorbing material.

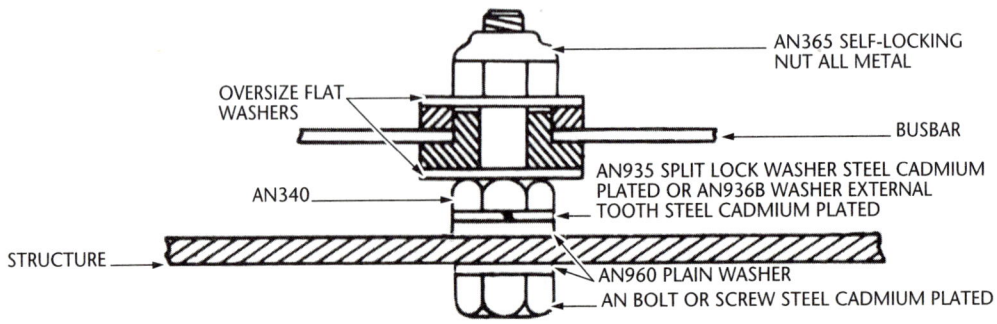

Figure 9-2. Mounting bus bars to structure

MS Clamp Number	Nominal ID of Clamp (Inches)
MS 21919D5	.313
MS 21919D6	.375
MS 21919D7	.438
MS 21919D8	.500
MS 21919D9	.563
MS 21919D10	.625
MS 21919D11	.688
MS 21919D12	.750
MS 21919D13	.813
MS 21919D14	.875
MS 21919D15	.938
MS 21919D16	1.000
MS 21919D19	1.888
MS 21919D21	1.313
MS 21919D23	1.438
MS 21919D25	1.563
MS 21919D27	1.688
MS 21919D29	1.812
MS 21919D31	1.938
MS 21919D33	2.062
MS 21919D35	2.188
MS 21919D37	2.312
MS 21919D43	2.688
MS 21919D45	2.812

Table 9-2 Support clamps for rubber covered flexible aluminum or brass conduit

Protection. Install bus bars inside panels, junction boxes or in protected areas when possible. If this cannot be done, protect the bus bar with vinyl tubing tied in place after connections have been made. See Chapter 10 page 10-12 for details.

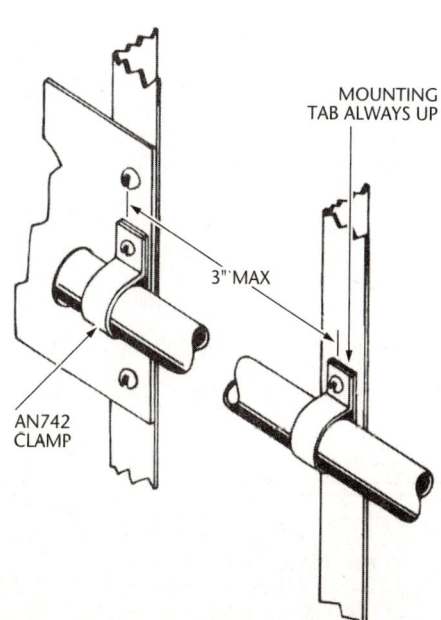

Figure 9-4. Spacing clamps for rigid metallic conduit

Section 2
Installation of Conduit

General. Metallic conduit, both rigid and flexible, used in aircraft to protect electrical wires and cables, is installed before the wiring is routed through it. Non-metallic conduit is installed at the same time the wiring is routed.

Supporting hardware. Attach conduit to aircraft structure with clamps. For rigid or flexible bare metallic conduit, select a plain aluminum clamp (AN742) of the proper size from Table 9-1. For rubber covered flexible metallic conduit, selected a cushioned clamp (MS21919) of the proper size from Table 9-2. Measure the OD of conduit to determine selection of size from Table 9-2.

Installation of supporting hardware. Attach clamps to a rigid surface of structure, so that there will be no relative motion between conduit and aircraft structure. Install clamps so that the mounting screw is above the conduit as shown in Figure 9-3. Install clamps so that conduit slants downward toward one end, when the aircraft is on the ground. If one end is attached to equipment, slant conduit toward the open end away from the equipment.

CAUTION: *Do not tighten clamps on conduit so that they damage or collapse the conduit. Reject and replace conduit that has been collapsed.*

Spacing of supports. See Figure 9-4. Support rigid metallic conduit with a clamp close to each end, and at maximum spacing of 3 feet along entire conduit run. Support flexible metallic conduit with clamps close to each end, and spaced 6 inches minimum to 24 inches maximum apart. Spacing of support clamps within the limits given is determined by conditions of structure.

Bending limitations. Rigid metallic conduit is bent as required before installation. Refer back to the section on rigid conduit fabrication. Form bends in flexible metallic conduit to radii given in Table 9-3.

CAUTION: *Make sure that conduit is not overstressed in installation, and that there is no strain on ferrules. Install conduit so that there will be no vibration flexing of the conduit at ferrules.*

Drainage. Do not drill drainage holes in metallic conduit. Drainage is provided for as described earlier. The procedure for making drainage holes in nonmetallic conduit is described in Chapter 10 on page 10-5.

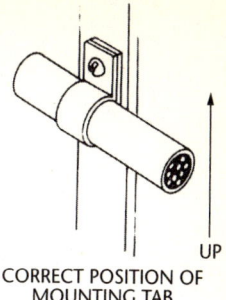

CORRECT POSITION OF MOUNTING TAB

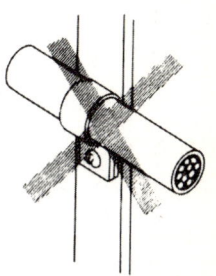

Figure 9-3. Installation of supporting clamps for conduit

Nominal ID of Conduit (Inches)	Minimum Bending Radius Inside (Inches)
3/16	2-1/4
1/4	2-3/4
3/8	3-3/4
1/2	3-3/4
5/8	3-3/4
3/4	4-1/4
1	5-3/4
1-1/4	8
1-1/2	8-1/4
1-3/4	9
2	9-3/4
2-1/2	10

Table 9-3. Minimum bending radii for flexible aluminum or brass conduit

9-4 | Installation of Electrical Hardware

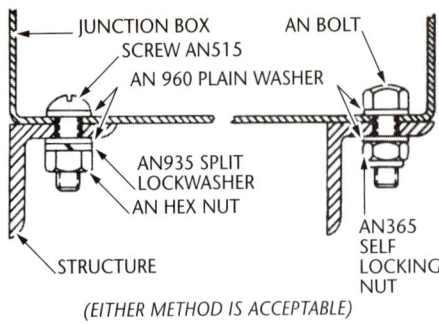

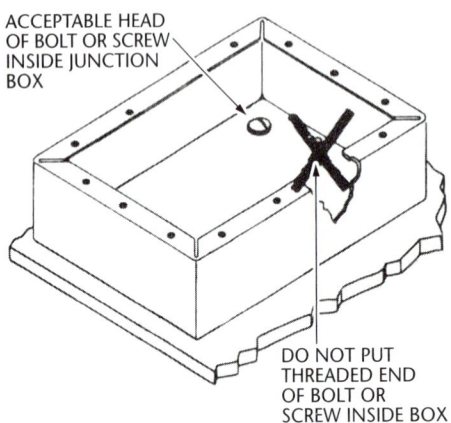

Figure 9-5. Attaching junction box to structure

Bonding or grounding conduit. Bond or ground metallic conduit to structure at each terminating or breaking point, by means of a plain metal clamp, or clamp and jumper. Test bond or ground as described in Chapter 6 on page 6-7.

Section 3
Installation of Junction Boxes

Junction boxes are containers, with hinged or removable covers, used in aircraft to provide a protected area for electrical power distribution equipment such as bus bars and terminal boards. The material of junction boxes is either metal or hard fiberglass.

Mounting hardware. See Figure 9-5. Use standard AN bolts or screws of the appropriate size to attach junction boxes to aircraft structure. Insert screws or bolts so that the head of the screw or bolt is inside the junction box.

> **CAUTION:** *Do not install attaching hardware so that threaded part of the screw or bolt protrudes inside the junction box, as the sharp thread edges will damage wire insulation.*

Insulation. The inside of metallic junction boxes is coated with white glyptal or similar material to insulate wiring from the metal, to improve visibility and to make inspection easier. Non-metallic junction boxes need not be so insulated. If this coating is damaged during the installation procedure, repair the damaged parts with the same material as used in the original installation. When a new metallic box is installed, make sure the insulating coating is present and undamaged.

Junction box covers. Junction box covers may be hinged, or attached by means of screws. Screw threads must not extend into the box. The sharp threads may cut wire insulation. If covers are not hinged, secure the cover to the box with an insulated bead chain, or No. 14 wire, as shown in Figure 9-6. Make this attachment outside the box so that when the box is closed the chain or wire will not interfere with the wiring.

> **NOTE:** *If covers are bent during installation or repair, straighten them before final attachment.*

Preparation of wire entry holes. Determine the outside diameter of the wire, or wire bundle, and make sure that the opening is at least 1/8 inch larger in diameter to allow for later enlargement of the bundle. Use a box connector and cable clamp to protect wiring if this is indicated on the engineering drawing. When a box connector is not used, protect the edges of the entry hole with plastic or fiber grommets. See Figure 9-7.

Drainage of junction boxes. Provide one or more drainage holes 3/16" diameter mini-

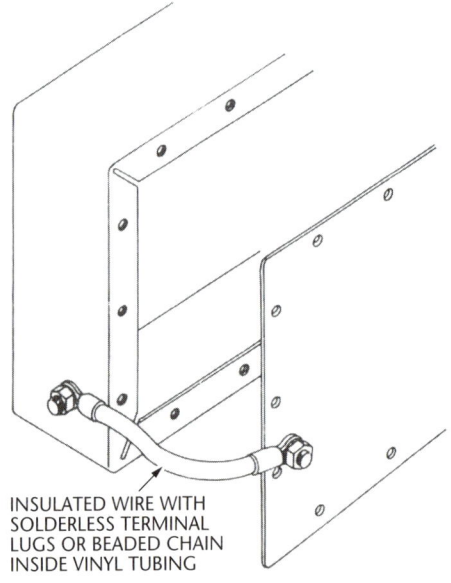

Figure 9-6. Attaching cover to junction box

Installation of Electrical Hardware | 9-5

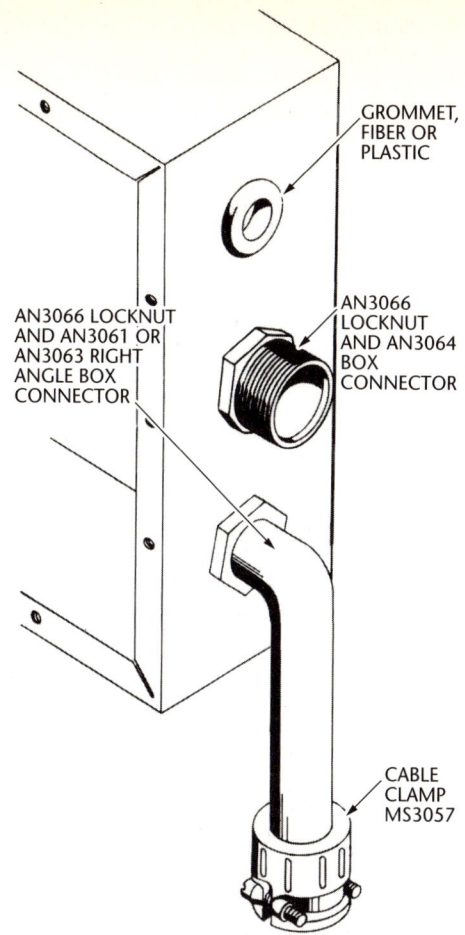

Figure 9-7. Wire entry holes in junction box

Section 4

Installation of Protective Devices

Protective devices are items of electrical equipment such as circuit breakers, fuses, etc., installed in aircraft to protect the electrical system against overloads caused by short circuits or other faults. See Chapter 11 for description of fuses and current limiters used in aircraft electrical systems.

Mounting hardware. If attaching hardware is furnished with the protective device, use it. If no attaching hardware is furnished, mount the protective device with standard AN cadmium plated steel screws or bolts, of the appropriate size. When replacing a protective device, use hardware exactly the same as in the original installation except that a longer screw may be used if necessary.

CAUTION: *Do not use self-tapping screws to mount protective devices.*

Mounting with through bolts or screws. When possible, attach protective devices to aircraft structure or other support with through bolts or screws. Install a plain washer under the head of the bolt or screw, and a plain

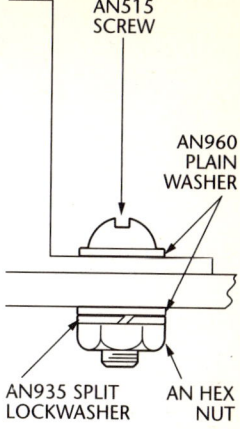

Figure 9-8. Mounting protective devices

mum, at the lowest point of the junction box when the aircraft is on the ground. After drilling drainage holes in metal junction boxes, deburr the edges of the hole with a deburring tool or a file.

CAUTION: *Do not drill holes in vapor-tight junction boxes.*

Vapor-tight boxes. Vapor-tight junction boxes in aircraft are identified as such on the covers. When doing work of any kind on vapor-tight boxes carefully follow the instructions given in the aircraft manufacturer's handbook of maintenance instructions for the specific airplane model.

Identification. If junction boxes as originally installed are not identified, it is not necessary to do so. If the junction box does have identification marking, make sure marking is replaced as in original.

Bonding or grounding junction boxes. Bond or ground junction boxes to structure by direct metal-to-metal contact or by means of a bonding jumper. Test bond or ground as described in Chapter 6 on page 6-7.

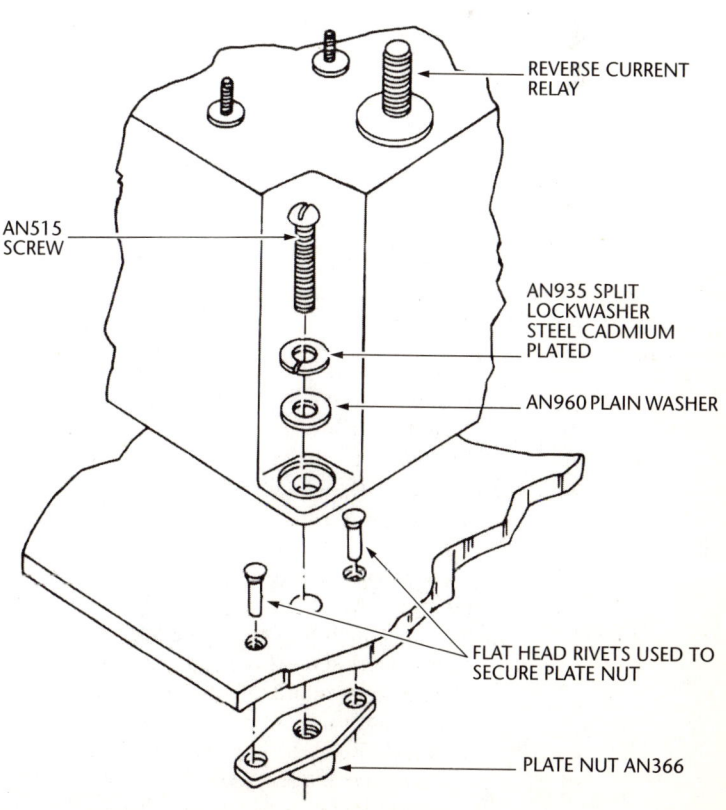

Figure 9-9. Typical mounting hardware for protective devices

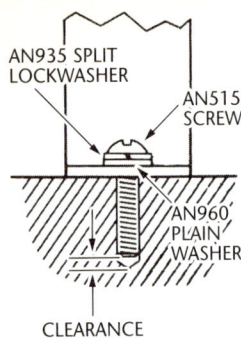

Figure 9-10. Determining screw length for mounting into blind holes

washer and a split lock-washer under the nut. See Figure 9-8.

Mounting into tapped hole or nut plate. When it is necessary to install a protective device with a screw into a tapped hole or a nut plate, install a plain washer and a split lock-washer under the screw or bolt head. See Figure 9-9 on the previous page.

Mounting into a blind hole. When a protective device must be attached with a screw into a blind tapped hole, make sure that the screw will give maximum thread engagement without bottoming in the hole. The length is determined as follows: See Figure 9-10.

1. Select a screw of approximately the correct length and install a plain washer and lock washer on it.

   NOTE: *Do not mount the protective device when determining screw length.*

2. Insert the screw into the blind tapped hole, and thread it in until it bottoms in the hole. Plain and lock in washers should remain free.

3. Back out the screw two turns, and measure the length of screw between the under surface of the head and the top washer.

4. Measure the thickness of the mounting part of the protective device, and subtract it from the measurement obtained in step 3. If the device to be mounted is thicker than the dimension obtained in step 3, repeat steps 1, 2, and 3, using a longer screw.

5. Subtract dimension obtained in step 4, from the overall length of the screw; this will give the maximum length that can be used without bottoming in the tapped hole. If the final length is not a standard

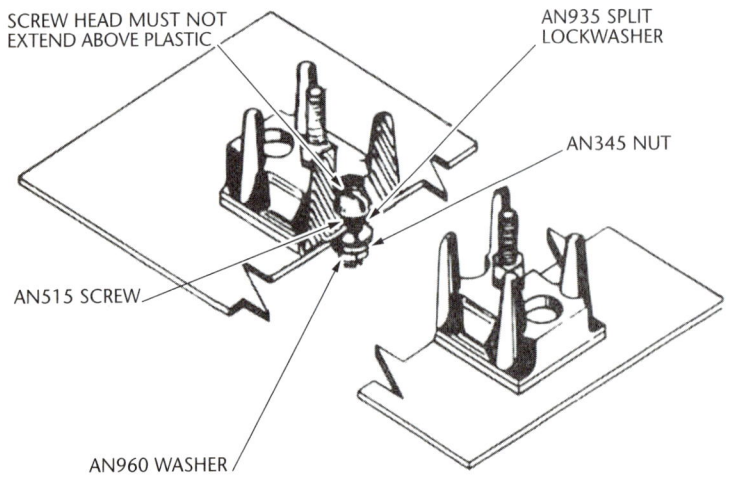

Figure 9-11. Mounting of terminal board

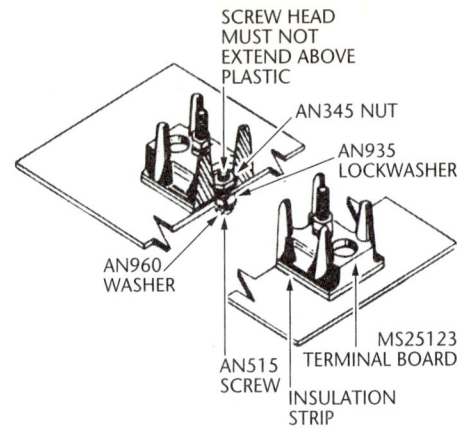

Figure 9-12. Alternate mounting of terminal board

length, use the next shorter standard screw.

Mounting circuit breakers. Mount switch circuit breakers so that when the switch breaker is in the off, or open position, the handle will be down or to the rear.

Mounting relays. Mount relays so that foreign particles cannot fall between the terminals, and so that liquid cannot accumulate inside the cover.

Special precautions for bonding or grounding connections. When a bond or ground connection is made through the mating surfaces of structure and mounting pad, prepare the contacting surfaces as described in Chapter 6 on page 6-4, before attaching the device to structure.

Protection. If possible mount protective devices in junction boxes or protected areas. If this is not possible, and the devices are to be installed in locations where they may be subject to damage or where the terminals may be dangerous to personnel, provide a cover to go over the protective device.

Protective coating for electrical connections. Electrical connections to protective devices may be protected against moisture, corrosion and damage from metal particles by the application of a strippable polyurethane coating. A typical commercial compound of this type is PR1532, made by the Products Research Corporation. The procedure for application is as follows:

1. Clean surfaces to be coated with a clean soft bristle brush.

2. Thoroughly mix the two parts of the compound together, in the proportions recommended by the manufacturer.

3. Using the injection nozzle of a pressure

Installation of Electrical Hardware | 9-7

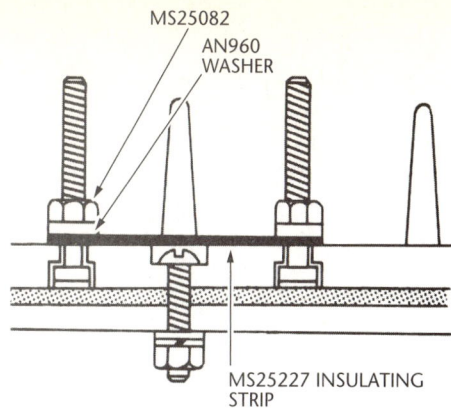

Figure 9-13. Insulation of terminal board

gun, apply a 1/16 inch minimum thick coating over the connection, and extending 1/8 inch beyond the connection all around.

4. Cure as recommended by the manufacturer. Protect the coated parts during the cure by covering with polyethylene sheeting.

Identification. Make sure that each protective device is identified by a plate or decal, permanently attached to adjacent aircraft structure. If the location of a protective device is changed make sure that the identification marking is also relocated, and completely visible. Make sure the new identification marking is exactly the same as the original.

Section 5

Installation of Terminal Boards

Terminal boards are used in aircraft to provide junction points of good electrical conductivity for circuits which are not frequently disconnected.

Mounting hardware. Use standard AN cadmium plated steel hardware of the appropriate size.

Method of attachment. Install mounting screws so that the screw protrudes through the bottom of the terminal board, as shown in Figure 9-11. The length of the screw should allow for some protrusion beyond the nut. Pass a steel scale or other flat piece of metal over the top of the nut. If it passes over freely, the screw is too short, and is to be replaced with the next longer length. Protrusion of screw should not exceed two threads.

Alternate method of attachment. If it is not possible to install the mounting screw from the top of the terminal board, install it from the back as shown in Figure 9-12. In this case, the end of the screw should project just beyond the top of the nut, but do not use a screw that will extend beyond the level of the terminal board mounting surface.

Insulation. See Figure 9-13. Place an insulating strip over each mounting screw, long enough so it will go over the two adjacent terminal studs under the washers and nut which secure each stud.

Attaching bus bar to terminal board. When a bus bar is to be attached to a terminal board mount the bus bar directly on top of the nut which holds the terminal stud in place. Terminal lugs are then installed on top of the bus bar as described in Chapter 10 on pages 10-10 through 10-11.

> **NOTE:** *If aluminum bus bars are removed, and are to be replaced, examine the bus bar for deformation before replacing it. If there is any deformation, discard it, and install a new bus bar.*

Protection of terminal boards. Where possible, mount terminal boards inside junction boxes or other enclosures. If this is not possible, and the terminal board is located where it may be damaged, or may be dangerous to personnel, provide a cover. Use terminal board cover MS17777 on the MS25123 terminal board. Use terminal board cover MS18029 on the MS27212 terminal board. Attach no more than two terminal lugs on the stud which is to be used for mounting the cover to the terminal board. If no cover is available, the terminal board may be protected by a wrapping of vinyl sheeting. Use a piece of vinyl sheet large enough to make a generous lap over the studs. Punch holes in the vinyl sheet, install it over grounded studs, and fasten with nuts and washers.

Identification. Each terminal board in the aircraft electrical system is identified by the letters TB followed by a number that is the number of the individual board. Each stud on the terminal board is identified by a number adjacent to it, with the lowest number in the series at the end nearest the terminal board identification number. (See Figure 9-14). The identification may be marked on the aircraft structure to which the terminal board is attached, or may be on an identification strip cemented to the structure, under the terminal board. When a terminal board is replaced, do not remove the identification marking unless it has been damaged. In that case, replace the identification marking exactly as in the original, in accordance with the applicable wiring diagram.

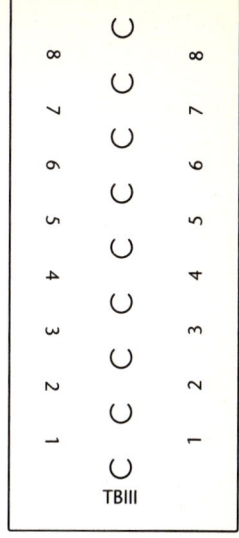

Figure 9-14. Identification of terminal board

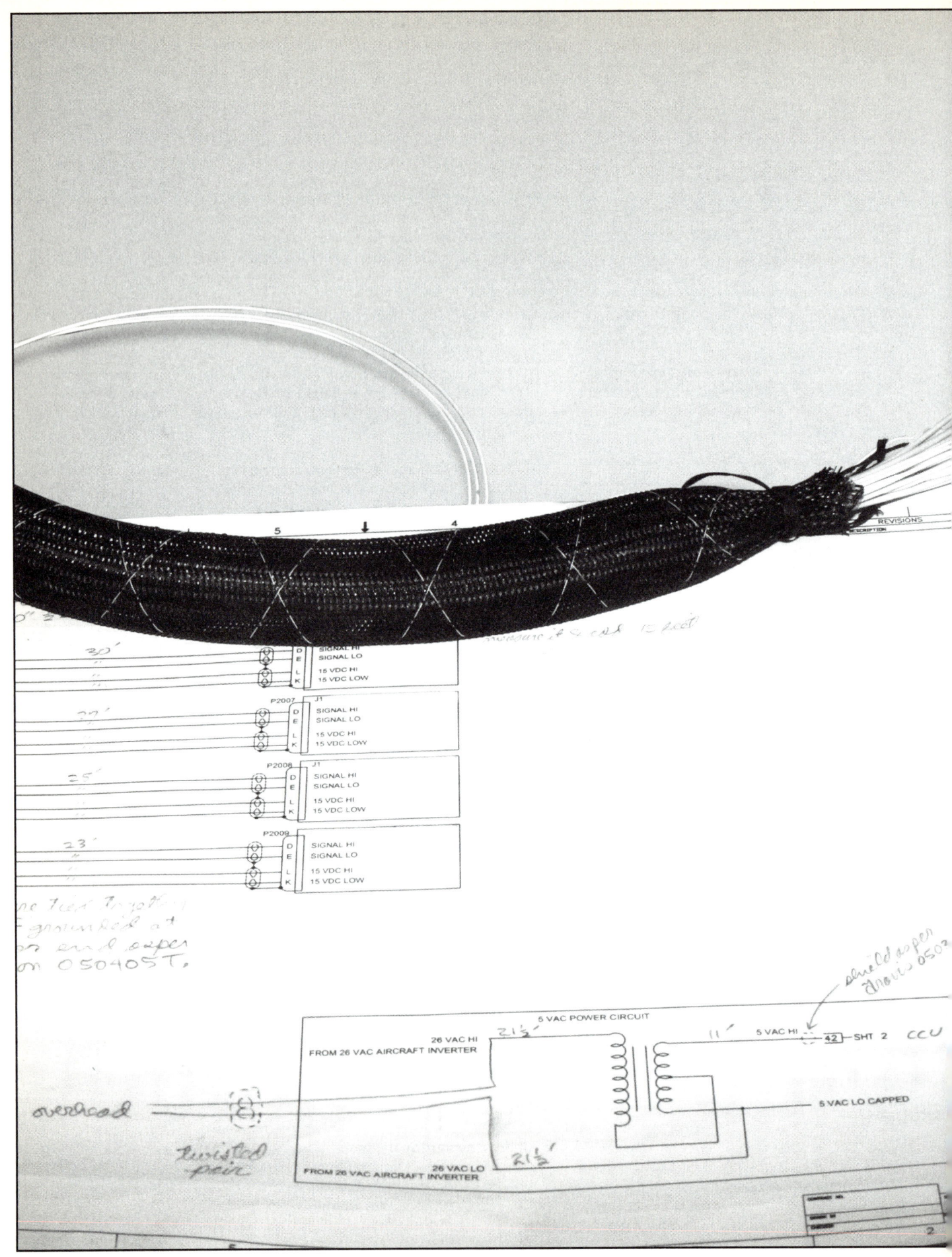

Chapter 10
ELECTRICAL wiring installation

Section 1

Description and Identification

This chapter describes recommended procedures for installing electrical wiring and related accessories in aircraft.

- **Open wiring.** Any wire, wire group or wire bundle not enclosed in conduit.

- **Wire group.** Two or more wires going to the same location, tied together to retain identity of the group.

- **Wire bundle.** Two or more wire groups, tied together because they are going in the same direction at the point where the tie is located.

- **Wire harness.** A wire group or bundle tied together as a compact unit (open harness), or contained in an outer jacket (enclosed harness). Wire harnesses are usually pre-fabricated and installed into the aircraft as single assembly.

- **Electrically protected wiring.** Those wires, which are included in the circuit protection against overloading such as fuses, circuit breakers or other limiting devices.

- **Electrically unprotected wiring.** Those wires generally from generators to main bus distribution points) which do not have protection such as fuses, circuit breakers or other current limiting devices.

Wire types. The wires most commonly used in aircraft electrical systems are in accordance with the specifications listed in Table 10-1. See Chapter 1, Table 1-1 for details of conductor, insulation, voltage and temperature.

Learning Objectives:

- *Wire Groups and Bundles*
- *Connections to Terminal Boards and Bus bars*
- *Installation of Wires in Conduit*
- *Installation of Connectors*
- *Installation of Wire in Junction Boxes*
- *Lacing*
- *Tying*
- *Procedures for Lock, Shear and Seal Wiring*

Left: Connecting and routing wires and bundles in accordance with the manufacturer's instructions is important.

1. 600 Volts or Under			
General Purpose	Aluminum	High Temperature	Fire Resistant
MIL-W-5086 Types I, II and III	MIL-W-7072	MIL-W-7139 MIL-W-8777 MIL-W-18678, Type E *MIL-W-22759 MIL-W-27300	MIL-W-25038
—	2. Over 600 Volts		
	MIL-W-5086, Type IV MIL-W-16878, Type EE *MIL-W-22759		
	3. Cabled Shielded and Jacketed		
	MIL-C-5767 MIL-C-7078 MIL-C-27500		

*MIL-W-22759 covers wire rated at 600 volts-and 1000 volts.

Table 10-1. Wire types

Section 2

Wire Groups and Bundles

Wire separation. Mil-Spec MIL-W-5088 restricts the grouping or bundling of certain wires such as electrically unprotected power wiring, and wiring to duplicate vital equipment. Do not add such wires to existing bundles unless specifically authorized.

Size of wire bundle. Good practice and most approved installations generally limit the size of a wire bundle to 75 wires, or two inches diameter, whichever is smaller.

Identity of groups within bundles. When several wires are grouped at junction boxes, terminal boards, panels, etc., retain the identity of the group within a bundle by spot ties, as shown in Figure 10-1.

Combing wires. Comb out all wires, except those listed on this page, so that wires will be parallel to each other in group or bundle. A useful tool for combing out wires is shown in Figure 10-2. Make this tool from a piece of 1/8 inch nylon or other smooth insulating material. Be sure all sharp edges are rounded to protect wire insulation.

Twisting wires. When specified on applicable engineering, drawings, twist together the following wires:

- Wiring in vicinity of magnetic compass or flux valve
- Three-phase distribution wiring
- Other wires (usually sensitive circuit avionic wiring) as specified on engineering drawings

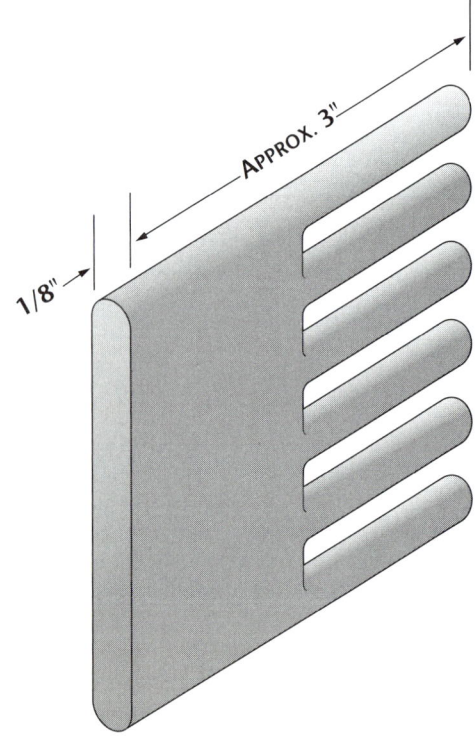

Figure 10-2. Comb for straightening wires in bundles

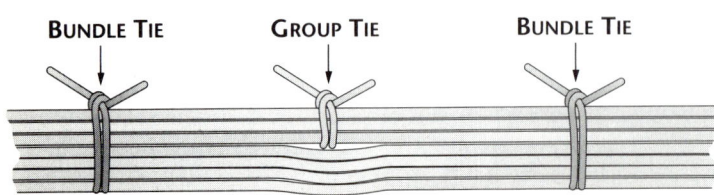

Figure 10-1. Group and bundle ties

Wire Size	#22	#20	#18	#16	#14	#12	#10	#8	#6	#4
2 Wires	10	10	9	8	7-1/2	7	6-1/2	6	5	4
3 Wires	10	10	8-1/2	7	6-1/2	6	5-1/2	5	4	3

Table 10-2. Twists per foot

Twist wires so they lie snugly against each other, making approximately the number of twists given in Table 10-2. Check wire insulation for damage after twisting. If insulation is torn or frayed, replace the wire.

Spliced connection in bundles. Locate spliced connections in wire groups or bundles so that they can be inspected. Stagger splices as shown in Figure 10-3, so that the bundle does not become excessively enlarged. Make sure that all non-insulated splices are covered by shrinkable tubing or by plastic sleeves securely tied at both ends.

Slack. Do not install single wires or wire bundles with excessive slack. Slack between support points such as cable clamps should normally not exceed 1/2 inch. See Figure 10-4. This is the maximum that it should be possible to deflect the wire with moderate hand force. This may be exceeded if the wire bundle is thin and the clamps are far apart but the slack must never be so great that the wire bundle can touch any surface against which it may abrade. Allow a sufficient amount of slack near each end for any or all of the following:

- To permit ease of maintenance
- To allow replacement of terminals at least twice
- To prevent mechanical strain on the wires, cables, junctions and supports
- To permit free movement of shock-and-vibration mounted equipment
- To permit shifting of equipment for purposes of maintenance, while installed in the aircraft

Bend radii. Bend individual wires to a minimum radius of ten times the outside diameter of the wire, except at terminal boards where the wire is suitably supported at each end of bend, a minimum radius of three times the outside diameter of the wire is acceptable. Bend wires in wire bundles to a minimum radius of ten times the outside diameter of the largest wire in the bundle.

CAUTION: *Never bend coaxial cable to a smaller radius than six times the outside diameter.*

When it is not possible to hold the bending radius of single wires to the above limits, enclose wire in tight plastic tubing for at least two inches each side of the bend.

Routing and Installation

General instructions. Install wiring so that it is mechanically and electrically sound, and neat in appearance. Wherever practical, route wires and bundles parallel with, or at right angles to the stringers or ribs of the area involved, as shown in Figure 10-5, next page.

> **NOTE:** *Route coaxial cable as directly as possible. Avoid unnecessary bends in coaxial cable. Locate attachments at each frame rib on runs along the length of the fuselage, or at each stiffener on runs through the wings.*

General Precautions

When installing electrical wiring in aircraft, observe the following precautions:

- Do not permit wire or wire bundles to have moving, frictional contact with any other object.
- Do not permit wire or wire bundles to contact sharp edges of structure, holes, etc. (Refer to page 10-5).
- Do not use any installing tools other than those specifically authorized.

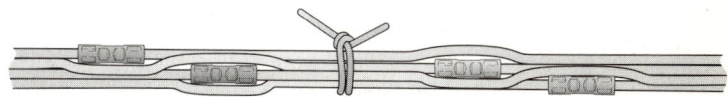

Figure 10-3. Staggered splices in wire bundle

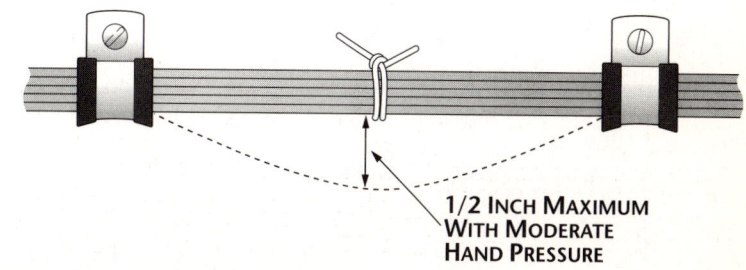

Figure 10-4. Slack between supports

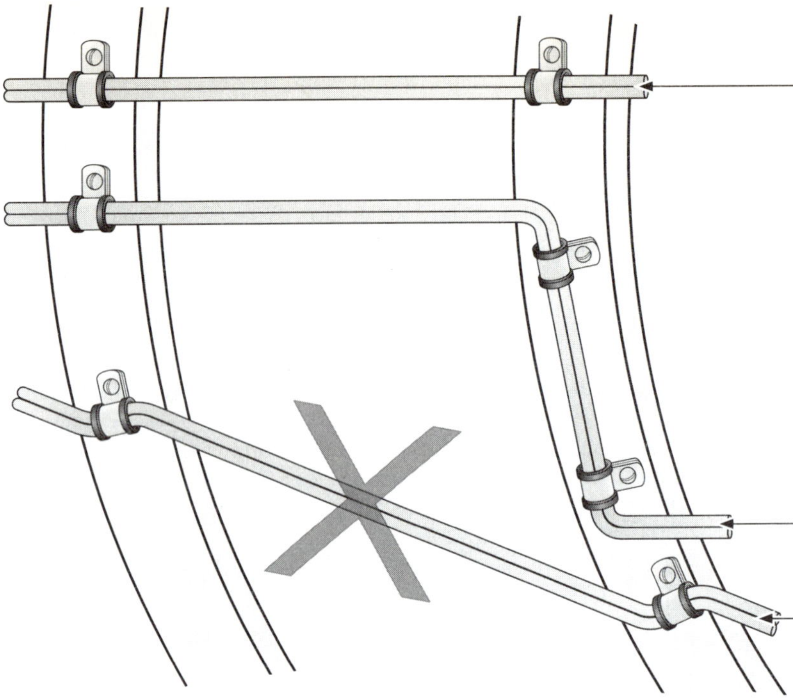

CORRECT: BUNDLE IS AT RIGHT ANGLE TO RIB STRUCTURE

CORRECT: BUNDLE IS PARALLEL TO RIB STRUCTURE

INCORRECT: BUNDLE ANGLES ACROSS RIB STRUCTURE

Figure 10-5. Routing bundles

- Do not damage threads of attaching hardware by overtightening or cross-threading.
- Do not subject wire bundles to sharp bends during installation (refer to page 10-3).
- Do not allow dirt, chips, loose hardware, lacing tape scraps, etc. to accumulate in enclosures or wire bundles.
- Do not hang tools or personal belongings on wire bundles.
- Do not use installed wire bundles or equipment as footrests, steps, or handholds.
- Do not compensate for wires that are too long by folding wire back on itself and hiding such folds within bundles.
- Do not twist or pull wire bundles during assembly or installation so pins are pulled from connectors, or connectors or wires otherwise damaged.
- Do not stretch wires to mate connectors; allow sufficient slack to permit easy mating.

Support. Bind and support wire and wire bundles to meet the following requirements:

- Prevent chafing of cables.
- Secure wires and wire bundles routed through bulkheads and structural members.
- Fasten wires in junction boxes, panels and bundles for proper routing and grouping.
- Prevent mechanical strain that would tend to break the conductors and connections.
- Prevent arcing or overheated wires from causing damage to mechanical control cables.
- Facilitate re-assembly to equipment and terminal boards.
- Prevent interference between wires and other equipment.
- Permit replacement or repair of individual wires without removing the entire bundle.
- Prevent excessive movement in areas of high vibration.

Protection. Install and route wires and wire bundles to protect them from the following:

- Chafing or abrasion
- High temperature
- Use of wire bundles as handholds, footrests or steps, or as support for personal belongings and equipment
- Damage by personnel moving within the aircraft
- Damage from cargo stowage or shifting

- Damage from battery acid fumes, spray or spillage
- Damage from solvents and fluids
- Abrasion in wheel wells where exposed to rocks, ice, mud, etc.

Protection against chafing. Install wires and wire groups so they are protected against chafing or abrasion in locations where contact with sharp surfaces or other wires would damage the insulation. Damage to the insulation may result in short circuits, malfunction or inadvertent operation of equipment. Use MS cable clamps to support wire bundles at each hole through a bulkhead. See Figure 10-6. If wires come closer than 1/4 inch to edge of hole install a suitable grommet in hole as shown in Figure 10-7.

> **CAUTION:** *Do not depend on vinyl sleeving as protection against abrasion or chafing, or as a substitute for good routing practice.*

Protection against high temperature. To prevent wire insulation deterioration, keep wires separate from high temperature equipment such as resistors, exhaust stacks, heating ducts, etc. The amount of separation is specified by engineering drawings. If wires must be run through hot areas, insulate the wires with high temperature material such as fiberglass or TFE. Additional protection in the form of conduit may be specified by engineering.

> **CAUTION:** *Never use a low temperature insulated wire to replace a high temperature insulated wire.*

Many coaxial cables have soft plastic insulation such as polyethylene. These are especially subject to deformation and deterioration at elevated temperatures. Avoid all high temperature areas with these cables.

Protection against personnel and cargo. Install wire bundles so they are protected by the structure. Use structure or conduit to prevent pinching against the air frame by cargo. Locate wire bundles so that personnel are not tempted to use sections of the wire runs as handholds or ladder rungs.

Protection against battery acids. Never route any wires below a battery. Inspect wires in battery areas frequently. Replace any wires that are discolored by battery fumes.

Protection against solvents and fluids. Avoid areas where wires will be subjected to damage from fluids. Wires and cables installed in aircraft bilges shall be installed at least six inches from the aircraft centerline.

If there is a possibility that wire without a protective outer jacket may be soaked in any location, use plastic tubing to protect it. This tubing should extend past the wet area in both directions and be tied at each end if the wire has a low point between the tubing ends. The lowest point of the tubing should have a 1/8 inch drainage hole as shown in Figure 10-8. Punch the hole in the tubing after the installation is complete and the low point definitely established. Use a hole punch to cut a half circle. Be careful not to damage any wires inside the tubing when using the punch.

Protection in wheel wells. Wires located in wheel wells are subject to many additional hazards such as exposure to fluids, pinching, and severe flexing in service. Make sure that all wire bundles are protected by sleeves of flexible

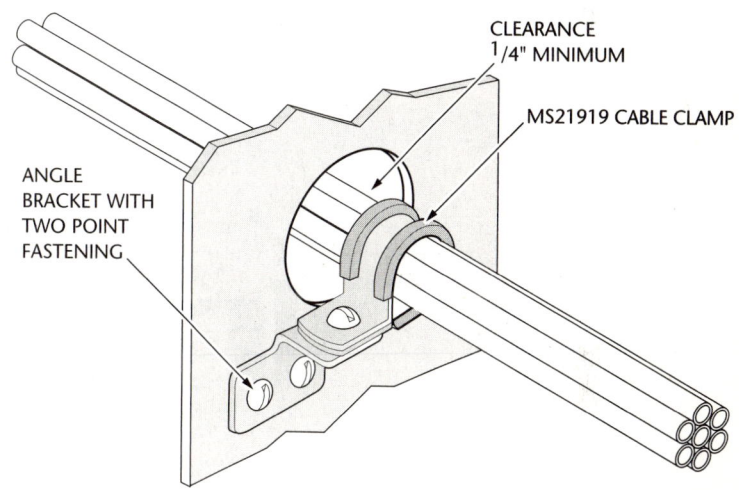

Figure 10-6. Cable clamp at bulkhead hole

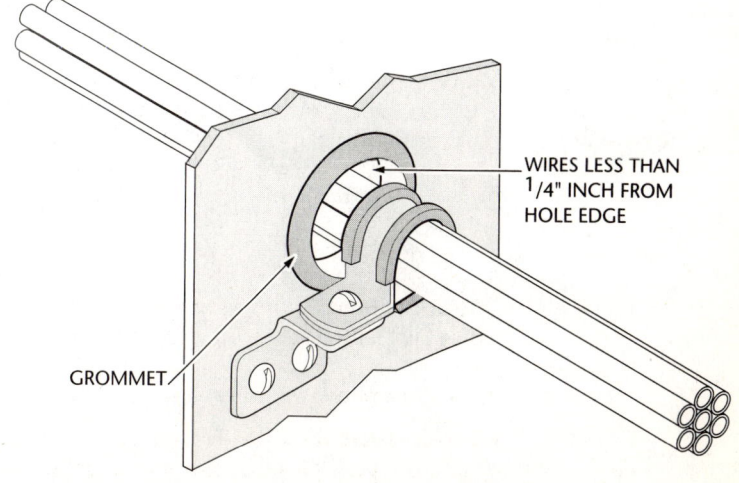

Figure 10-7. Cable clamp and grommet of bulkhead hole

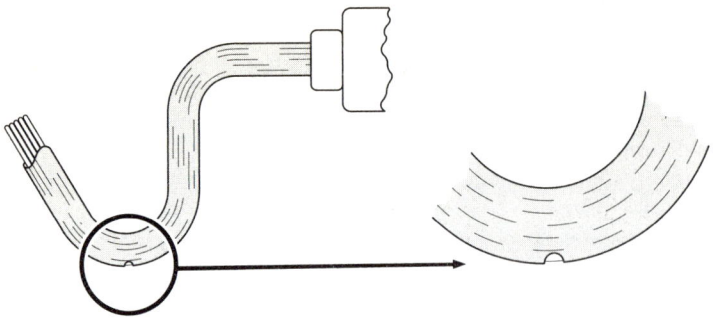

A drainage hole 1/8 inch diameter at lowest point in tubing.

Make the hole after installation is complete and lowest point is firmly established.

Figure 10-8. Drainage hole in low point of tubing

tubing securely held at each end. There should be no relative movement at point where flexible tubing is secured. Inspect these wires and the insulating tubing carefully at very frequent intervals. Replace wires and/or tubing at the first sign of wear. There should be no strain on attachments when parts are fully extended but slack should not be excessive.

Separation from plumbing lines. When wiring must be routed parallel to combustible fluid or oxygen lines for short distances, maintain as much fixed separation as possible; six inches or more. Route the wires on a level with, or above, the plumbing lines. Space clamps so that if a wire is broken at a clamp it will not contact the line. Where a six-inch separation is not possible, clamp both the wire bundle and the plumbing line to the same structure to prevent any relative motion. If the separation is less than two inches but more than 1/2 inch use a nylon sleeve over the wire bundle to give further protection. Use two cable clamps back to back as shown in Figure 10-9 to maintain a rigid separation only (not for bundle support).

> **CAUTION:** *Do not route any wire so that it can possibly come closer than 1/2 inch to a plumbing line.*
>
> **WARNING:** *Never support any wire or wire bundle from a plumbing line carrying flammable fluids or oxygen. Clamps may be used only to insure separation.*

Route wiring to maintain a minimum clearance of three inches from control cables. If this is not possible, install mechanical guards to prevent contact of wiring with control cables

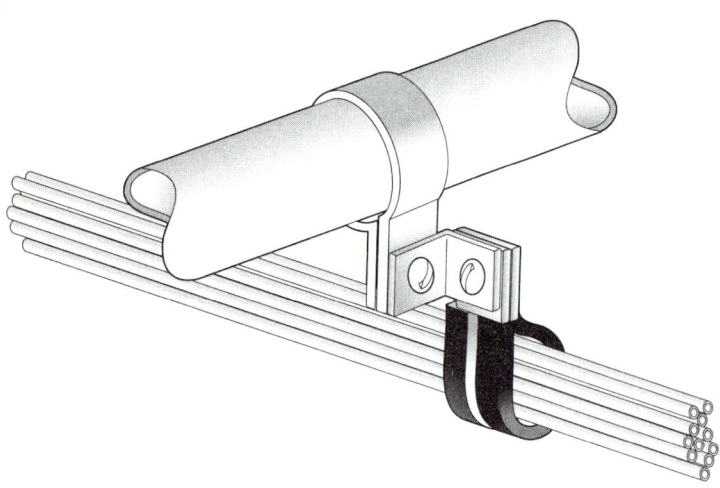

Figure 10-9. Separation of wires from plumbing lines

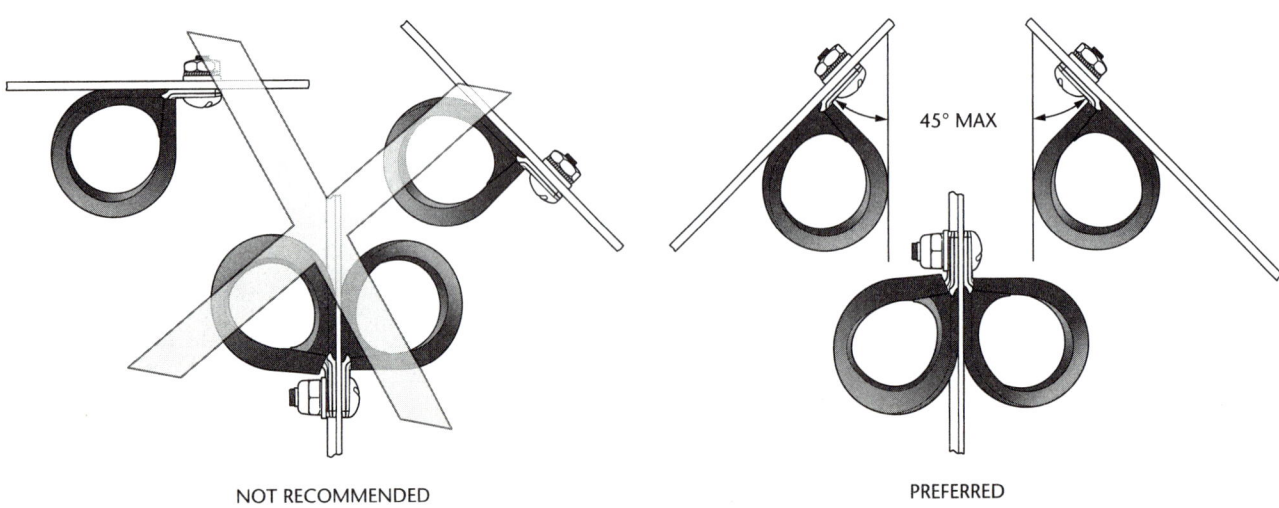

NOT RECOMMENDED

PREFERRED

Figure 10-10. Preferred angle for cable clamps

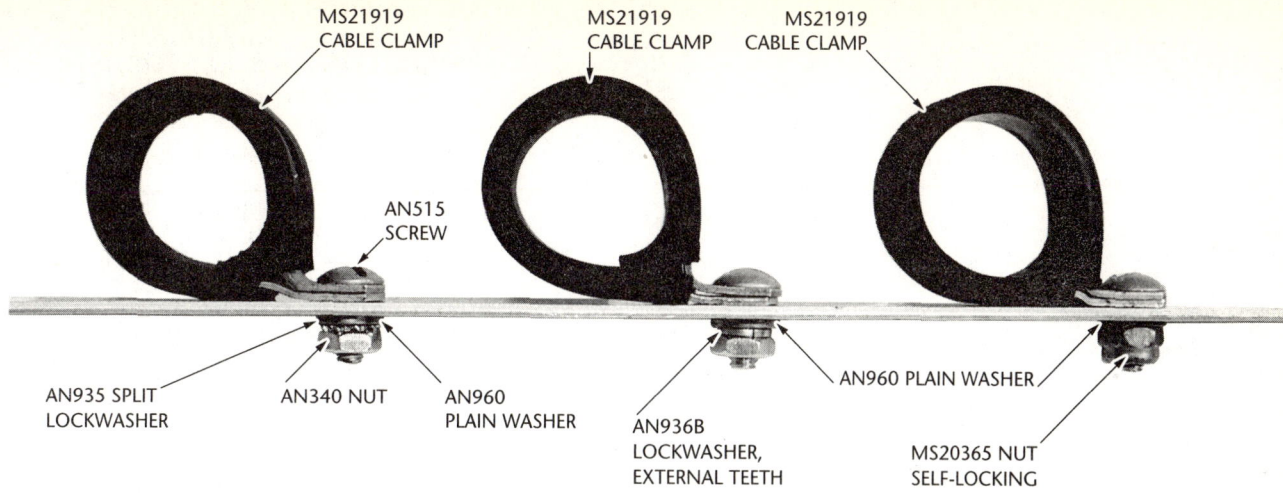

Figure 10-11. Typical mounting hardware for MS21919 cable clamps

Installation of Cable Clamps

Install MS21919 cable clamps as shown in Figure 10-10. The mounting screw should be above the wire bundle, if possible. It is also desirable that the back of the cable clamp rest against a structural member. Use hardware as shown in Figure 10-11 to mount cable clamps to structure. Be careful not to pinch wires in cable clamp. If the wire bundle is smaller than the nearest clamp size or if a clamp of the proper size is not available, wrap the wire bundle with the necessary number of turns of vinyl tape so that the bundle will be held securely in the clamp.

NOTE: *MS21919 cable clamps are cushioned with insulating material to prevent abrasion of wires. Never use metal clamps without cushions to hold wires.*

Nylon cable clamps, MS25281, may be used to support wire bundles up to two inches in diameter in open wiring, or inside junction boxes and on the back of instrument panels. When installing nylon cable clamps, use a large diameter metal washer under the screw head or nut adjacent to the clamp.

CAUTION: *Do not use nylon cable clamps where the ambient temperature may exceed 235°F.*

Mount cable clamps directly to "Z" members of structure. Use angle bracket with two mounting screws if structural member is angle as shown in Figure 10-12.

A tool to facilitate the installation of cable clamps is shown in Figure 10-13. Similar to conventional multiple slip joint pliers, the tool compresses and holds the clamp with the securing bolt in place while a nut is being installed on the bolt. The tool is particularly useful for installing clamps in restricted areas, and for installing groups of two or three clamps.

Installing cable clamps to tubular structure. Use AN735 clamps without cushions

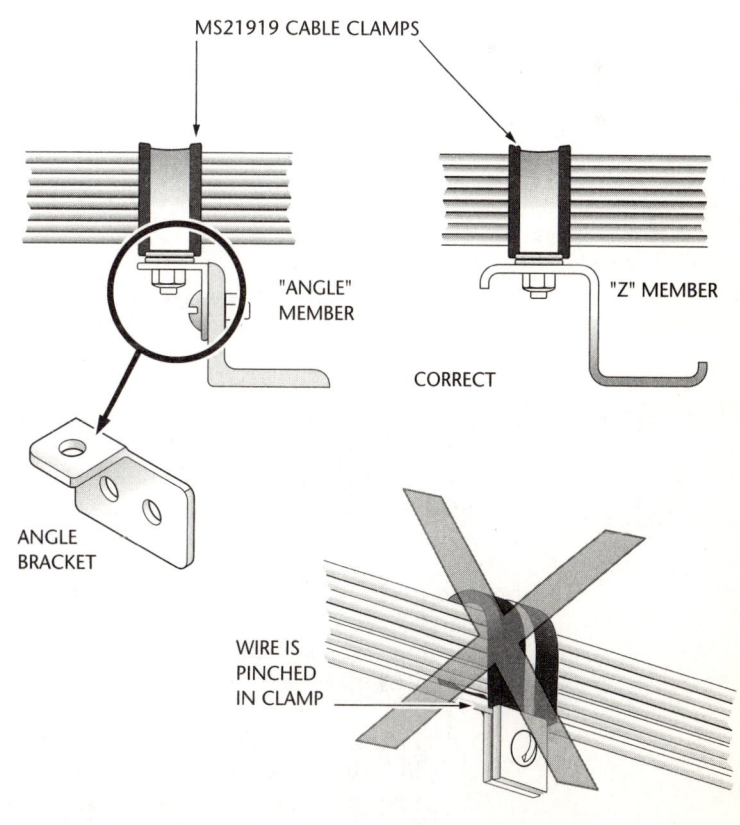

Figure 10-12. Attaching cable clamp to structure

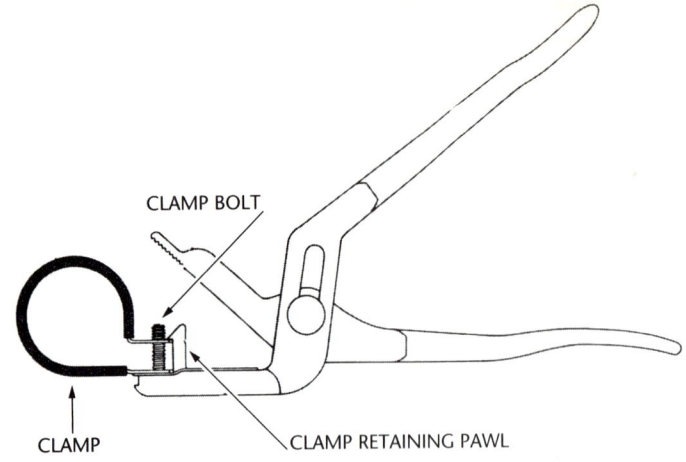

Figure 10-13. Tool for installing cable clamps

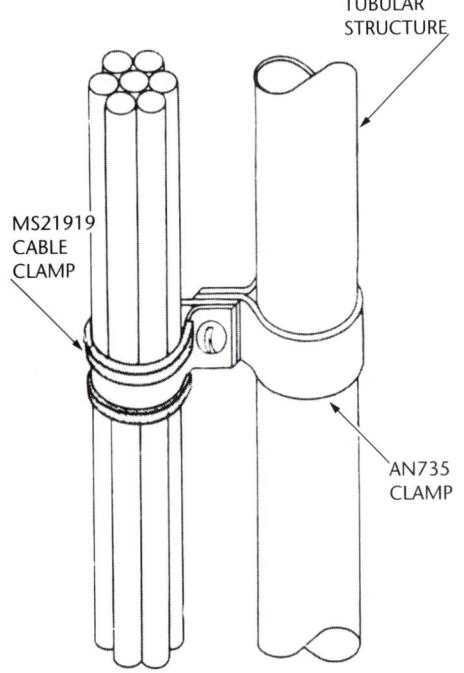

Figure 10-14. Installing cable clamps to tubular structure

Standard	Material	Upper Temperature Limit
MS35489	Rubber, Hot Oil and Coolant Resistant	250°F
MS35490	Rubber, General Purpose	250°F
MS21265 and MS21266	Nylon	235°F
MS21265 and MS21266	TFE	500°F

Table 10-3. Grommets-temperature limitations of material

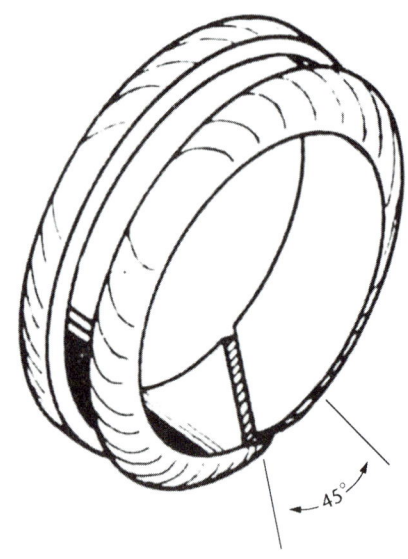

Figure 10-15. Split grommet

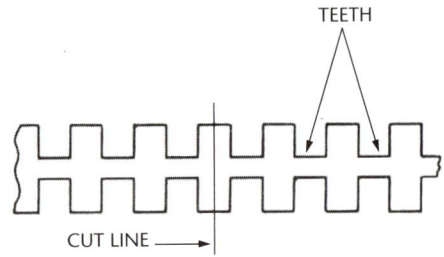

Figure 10-16. Cutting caterpillar grommet

for clamping to tubular structure. The clamps must fit tightly but should not deform when locked in place. Attach wire bundle in MS21919 cable clamp to the AN735 clamp with AN hardware as shown in Figure 10-14.

Installing grommets. MS grommets are available in rubber, nylon and TFE. Select grommet suitable for the environmental conditions from Table 10-3.

If it is necessary to cut a nylon grommet in order to install it, make the cut at an angle of 45 degrees as shown in Figure 10-15. Cement the grommet in place with general purpose cement, with the cut at the top of the hole. When installing caterpillar grommets, cut the grommet to the required length, making sure to cut square between the teeth as shown in Figure 10-16. Cement the grommet in place with general purpose cement, with the cut at the top of the hole.

Section 3

Connections to Terminal Boards and Bus bars

Connecting terminal lugs to terminal boards. See Figure 10-17. Install terminal lugs on terminal boards MS25123 or MS27212 in such a way that they are locked against movement in the direction of loosening. See Table 10-4 for MS27212 terminal board specifications.

Hardware for Wiring Terminal Boards

MS25123 terminal boards have studs secured in place with an AN960 flat washer and an MS20341 steel nut. Place copper terminal lugs directly on top of the MS20341 nut. Follow with an AM960 flat washer, an AN935 split steel lockwasher, and either an MS20341 steel nut or an MS21044 self-locking all-metal nut. See Figure 10-18 for details of this assembly.

> **NOTE:** Do not eliminate the AN935 split lockwasher even when using an MS21044 self-locking nut.

MS27212 terminal boards have studs molded in, so do not require hardware for attaching studs to the terminal board. Use same hardware for installing terminal lugs as for MS25123 terminal boards (see Figure 10-18).

Place aluminum terminal lugs over an MS25440 plated flat washer of the correct size for terminal and stud from Table 10-5. Follow the terminal lugs with another MS25440 flat washer; an AN935 split steel lockwasher and either an MS20341 nut or an MS21044 self-locking all-metal nut. See Figure 10-19 for details of this assembly.

> **CAUTION:** Do not place any washer in the current path between two aluminum terminal lugs or between two copper terminal lugs.

Terminal Board MS Part Number	Stud Thread	Number of Studs	Cover Part Number
MS27212-1-20	6-32UNC-2A	20	MS18029-1-20
MS27212-2-16	10-32UNF-2A	16	MS18029-2-16
MS27212-3-8	1/4-28UNF-2A	8	MS18029-3-8
MS27212-4-8	5/16-24UNF-2A	8	MS18029-4-8
MS27212-5-8	3/8-24UNF-2A	8	MS18029-5-8

Note: Terminal boards and covers are procured in full lengths with number of studs indicated. Cut to suit needs at installation.

Table 10-4. MS27212 terminal boards and covers

MS Number	Terminal Size	Stud Size
MS25440-3	8, 6, 4	No. 10
-4	8, 6, 4, 2, 1, 1/0	1/4
-5	8, 6, 4, 2, 1, 1/0, 2/0	5/16
-6	8, 6, 4, 2, 1, 1/0, 2/0	3/8
-6A	3/0, 4/0	3/8
-8	2, 1, 1/0, 2/0, 3/0, 4/0	1/2

Table 10-5. Washers for use with aluminum terminal lugs

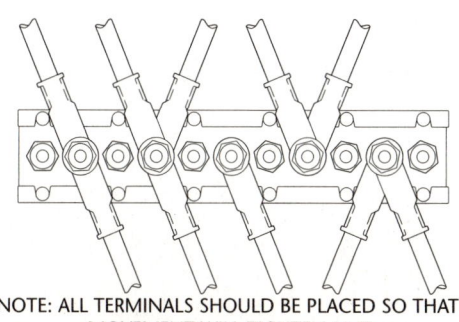

Figure 10-17. Connecting terminals to terminal board

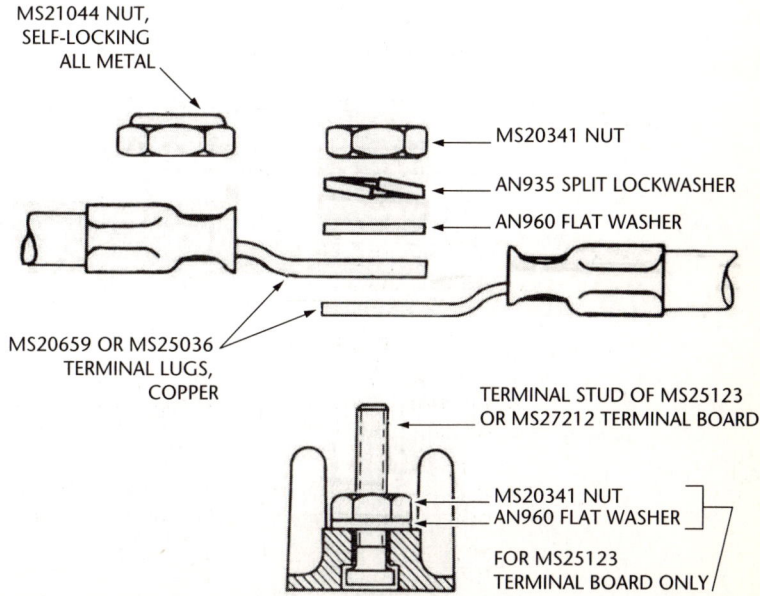

Figure 10-18. Hardware for wiring terminal boards with copper terminals

10-10 | Electrical Wiring Installation

CAUTION: *Never place a lockwasher directly against the tongue or pad of an aluminum terminal or bus bar.*

To join a copper terminal lug to an aluminum terminal lug, place an MS25440 flat washer over the nut which holds the stud in place; follow with the aluminum terminal lug, another MS25440 washer, the copper terminal lug, AN960 plain washer, AN935 split steel lockwasher and MS20341 plain nut or MS21044 self-locking nut. See Figure 10-20 for details.

Installation torques for large copper terminals. Use a torque wrench to tighten nuts on 3/8" and larger diameter studs to insure sufficient contact pressure. The tightening torques for steel studs are as listed in Table 10-6.

Installation torques for aluminum terminals. Use a torque wrench to tighten nuts over any stack-up containing an aluminum terminal lug. The tightening torques for steel studs are listed in Table 10-7.

Connecting terminal lugs to bus bars. In order to obtain maximum efficiency in the transfer of power, the terminal lug and the bus bar should be in direct contact with each other, so that the current does not have to go through any of the attaching parts, even if these are good current-carrying materials. As illustrated in Figures 10-21 through 10-24 the above applies whether the terminal lugs and the bus bar are of the same or of different materials.

Cleaning bus bars when making connections. Clean all bus bar areas before making new connections or replacing old connections. See Chapter 9 for procedure to be followed in cleaning bus bars. As noted on page 10-2, the cleaned surface of an aluminum bus bar is coated with a petrolatum-zinc dust compound, which is left on the surface while the connection is made.

Hardware for connection to bus bars. Cadmium plated steel hardware (except as noted below) is used to secure terminals to bus bars. Use split lockwashers under hex nuts and under self-locking nuts. Use plated

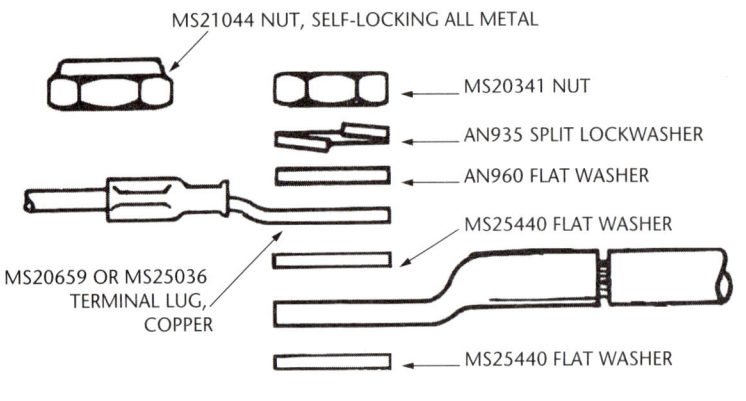

Figure 10-19. Hardware for wiring terminal boards with aluminum terminals

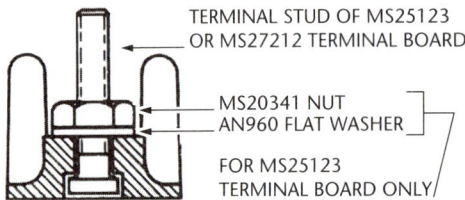

Figure 10-20. Hardware for wiring terminal boards with combination of terminals

	Inch-Pounds of Torque	
Stud Size	Plain Nuts	Self-locking Nuts
3/8-24	110-120	115-125
1/2-20	135-150	150-170

Table 10-6. Installation torques for copper terminal

Stud Size	Inch-Pounds of Torque
#10	32
1/4	100
5/16	150
3/8	250
1/2	480

Table 10-7. Installation torques for aluminum terminals

steel plain washers between lockwashers and copper terminals. Use plated brass flat washers (MS25440) between lockwashers and aluminum terminals. As shown in Figures 10-21 through 10-24, the head of the screw or bolt can be located on the terminal side or the bus bar side as required to simplify the installation.

Use a cadmium plated steel split lockwasher (AN935) under the head of every bolt or screw and also under the nut, as shown.

Use plated brass flat washers (MS25440) in contact with aluminum. The washer diameter must be at least equal to the tongue diameter of the terminal. See Table 10-5. Do not select a washer so large that it will ride on the barrel of the terminal. After tightening connection use soft cloth to wipe off excess petrolatum-zinc compound left in place in accordance with information on previous page.

General Precautions

Precautions when replacing existing connections. Observe the following precautions when replacing existing terminal lug connections to bus bars.

- Check all flat washers. Replace bent washers.

Replace washers that have scratched plating or paint on faying surface.

- Clean bus bar connection areas by approved methods (See Chapter 9).
- Check plated copper terminal lugs before connecting to an aluminum bus bar. If plating is scratched, replace terminal lug.

Connecting two terminals to same point on bus bar. Terminal lugs must always be in direct contact with bus bar. As shown in Figure 10-25 (next page), connect one terminal lug to top of bus bar and the other to bottom.

NOTE: *Terminal lug offset is positioned so that barrel cannot contact bus bars. This allows proper seating of tongue on bus bar.*

Protection of bus bars against accidental shorting. Bus bars are usually enclosed in panels or junction boxes to protect them against accidental shorting. If the bus bars are not enclosed it is desirable to use some protective coating. A good protective coating, which is easily applied, is MIL-S-8516 sealing compound. This is applied thickly with a spatula or short bristled brush to the cleaned bus bar prior to assembly of connections. Mask all areas where connections will be made. Use pressure sensitive tape for masking. See detailed instructions for applying and curing sealing

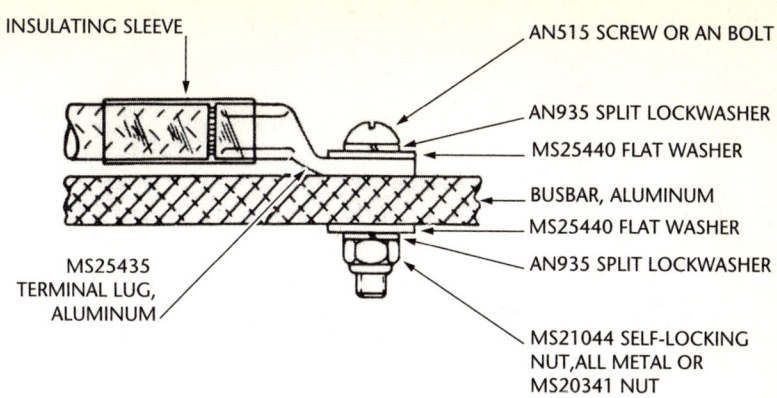

Figure 10-21. Connecting aluminum terminal to aluminum bus bar

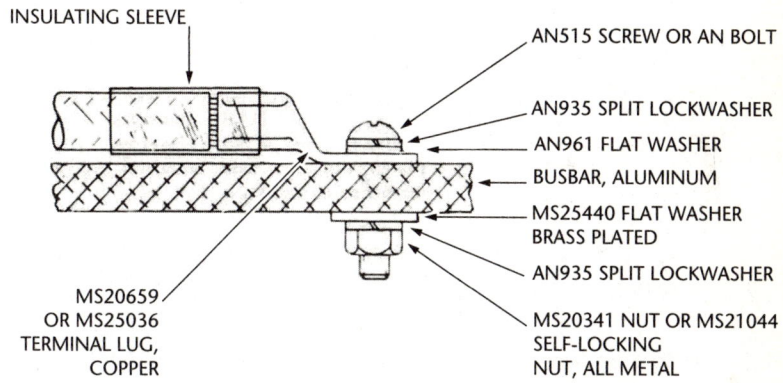

Figure 10-22. Connecting copper terminal to aluminum bus bar

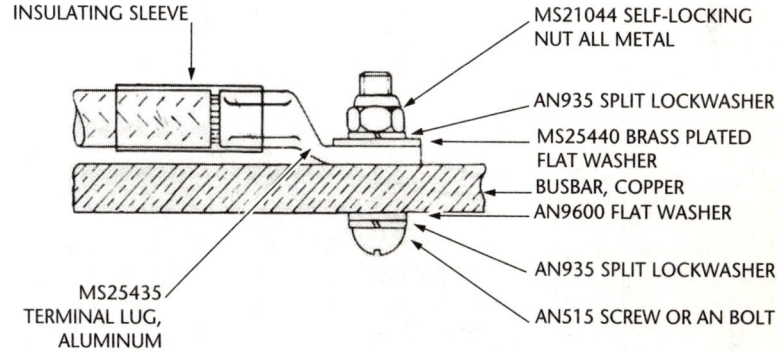

Figure 10-23. Connecting aluminum terminal to copper bus bar

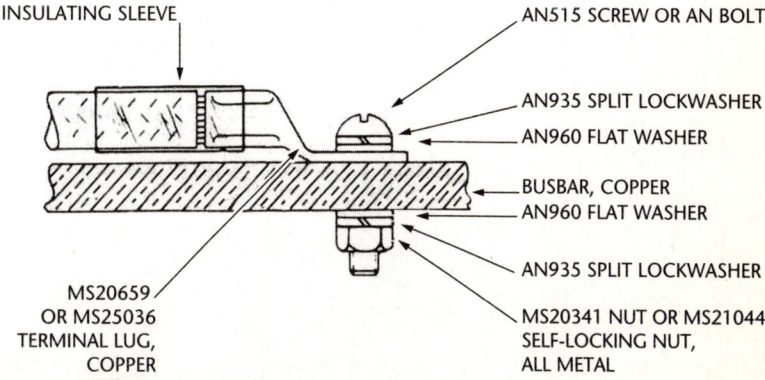

Figure 10-24. Connecting copper terminal to copper bus bar

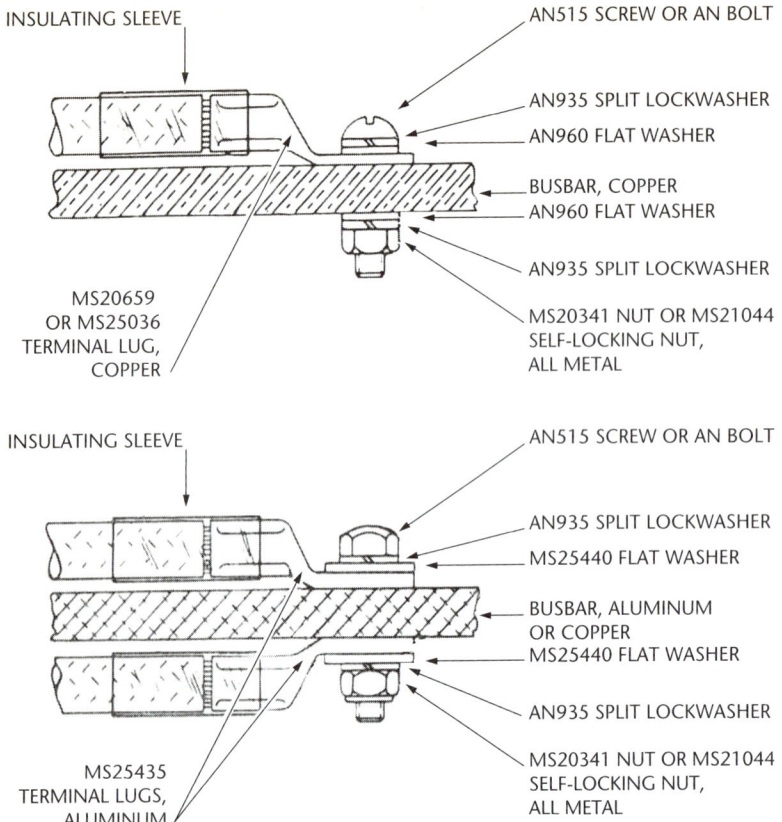

Figure 10-25. Connecting two terminals to same point on bus bar note

compound in Chapter 8 and in Chapter 2, on page 2-32. Remove masking tape after sealing compound is cured by cutting into compound next to tape with a razor blade and peeling tape from the masked area.

Bus bars can also be protected by slitting a piece of vinyl tubing and wrapping it around the bus bar after all connections are made. Select vinyl tubing, which has large enough diameter to permit a generous overlap when tying it in place. See Figure 10-26 for cutting and tying details.

Connecting terminal lugs to equipment. When connecting wired terminal lugs to terminals on switches, relays and other equipment, the terminal lugs may be bent at the barrel-tongue junction if necessary to permit installation. When bending is required, keep the bend radius as large as possible, while keeping the bend angle as small as possible.

CAUTION: *Do not bend terminal lugs to an angle greater than 90 degrees. Do not subject terminal lugs to more than one bending operation.*

Section 4

Installation of Wires in Conduit

Conduit capacity. Measure the bundle wires before installing in conduit. In accordance with MIL-W-5088 the bundle diameter must not exceed 80% of the internal diameter of the conduit. See Figure 10-27.

CAUTION: *No ties or splices are permitted inside a conduit.*

Feeding wires into conduit. Feed wires through a short length of conduit by taping the end of the bundle together and pushing it gently through. Longer runs of conduit or conduit with complex bends will require a leader. Make a leader out of a flannel or other soft cloth patch attached to a string long enough to pass completely through the conduit. The patch should fit loosely in conduit. See Figure 10-28. Use compressed air at no more than 35 psi to blow patch and attached string through the conduit. Tie wire bundle securely to string and tape over junction to cover all wire ends. Pull string through conduit while carefully feeding wires into other end. After wire is installed remove tape and detach string.

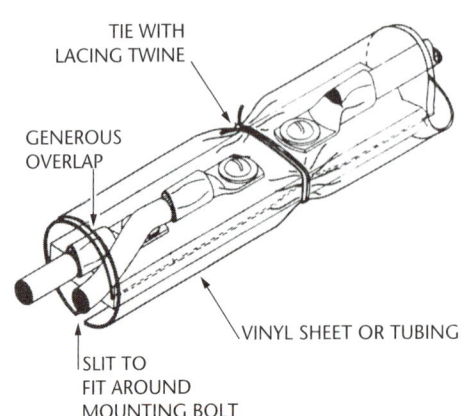

Figure 10-26. Vinyl tubing around bus bar

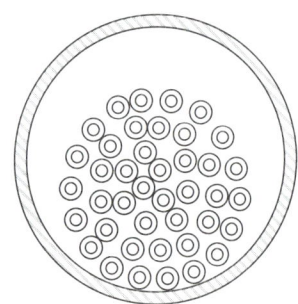

Figure 10-27. Conduit capacity

Supporting wires at end of rigid conduit.
Use an MS21919 cable clamp to support wires at each end of conduit. Place the cable clamp in a direct line with the conduit end to prevent chafing of wires at edge of conduit. Place cable clamp as close to end of conduit as practical, but never more than 10 inches away. See Figure 10-29.

NOTE: *Do not leave wire slack inside conduit. Wires should be free but not taut inside conduit.*

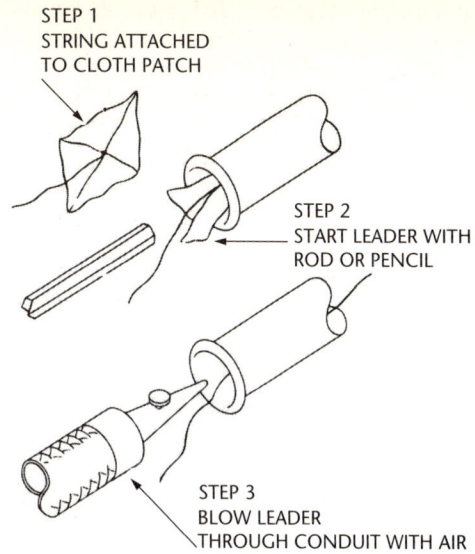

Figure 10-28. Leader for conduit

Section 5

Installation of Connectors

Assembly of connectors to receptacles.
Assemble connectors to receptacles as follows:

   CAUTION: *Do not use force to mate connectors to receptacles.*

1. Locate the proper position of the plug in relation to the receptacle by aligning the key of one part with the groove or keyway of the other part.

   CAUTION: *Do not twist wire bundle excessively to achieve proper matching of plug and receptacle.*

2. Start the plug into the receptacle with a light forward pressure and engage the threads of coupling ring and receptacle.

3. Alternately push in the plug and tighten the coupling ring until the plug is completely seated.

   CAUTION: *Do not hammer a plug into its receptacle. Never use a torque wrench, or pliers to lock coupling rings.*

4. Use a strap wrench to tighten coupling rings 1/16 to 1/8 turn beyond finger tight if space around connector is too small to obtain a good finger grip.

Disassembly of connectors from receptacles. Disassemble connectors as follows:

1. Use strap wrench to loosen coupling rings which are too tight to be loosened by hand.

2. Alternately pull on the plug body and unscrew coupling ring until connector is separated.

   CAUTION: *Do not pull on attached wires.*

3. Protect disconnected plugs and receptacles with caps to keep debris from entering and causing faults

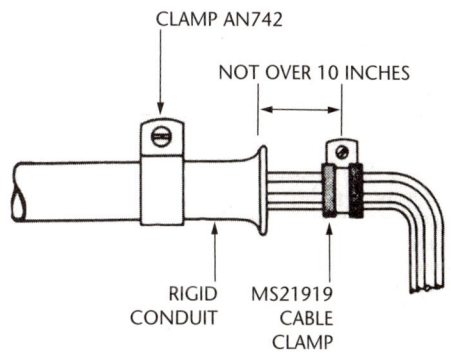

Figure 10-29. Support for wire at conduit end

Coding of connectors. As a design objective, receptacles whose plugs are interchangeable are not located in close proximity to each other. However, when installation requirements are such that these receptacles are in adjacent locations use clamps on the plug wires, or assemble plugs and receptacles to use one of the alternate insert positions, to make it physically impossible to connect a plug into the wrong receptacle. Also, color-code the connector plug body and the flange or mounting area of the receptacle.

- Use one bright color, such as red, green or yellow for each matching pair.

- Paint only the shell of plugs-not the coupling rings.

- Paint only the mounting flange of the receptacle.

NOTE: *Avoid painting the threaded surfaces or insulators of plugs or receptacles.*

Special precautions for connectors with resilient inserts. When assembling or installing miniature MS connectors with resilient inserts, observe the following special precautions:

- Before mating connectors, inspect to see that contacts are not splayed or bent. When mating connectors make sure that plug is inserted straight into the receptacle before tightening coupling ring.
- Avoid, where possible, locating connectors of the same shell size adjacent to each other, whether they have different insert arrangements or not.
- Locate receptacles where they are clearly visible and accessible to aid in keying and inserting plug. This will help to avoid bending receptacle pins while seeking proper polarization.

CAUTION: *Do not misconnect plugs and receptacles by forcing pins into the resilient insert, either by misalignment of properly mating connectors, or by joining connectors with identical shells but differently keyed insert arrangements.*

- When mating connectors with bayonet lock coupling, make sure that all locking rivets of the coupling are engaged.

Mounting connectors. Before mounting receptacles to the back of a panel or bulkhead, make sure there is sufficient clearance to couple the plug to the receptacle. Make sure that mounting hardware does not interfere with the installation of the locking ring.

Installing conduit on connectors. When installing a stepped-down conduit on the back shell of a connector having large wires (size No. 8 or larger), add an additional back shell to the connector before installing the conduit. This will allow the wire bundle to decrease in diameter gradually, and prevent sharp bends in the wires. (See Figure 10-30.)

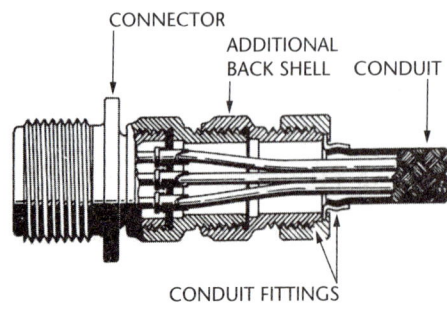

Figure 10-30. Installing conduit on connector back shell

Section 6

Installation of Wire in Junction Boxes

Lacing or tying in junction boxes. Wire bundles can be either laced or tied with spot ties. Lacing and tying procedures are described on page 15.

Support inside junction boxes. Use MS21919 cable clamps to support wires across hinged doors to that wires will be twisted and not bent when the door is opened. See Figure 10-31 for correct and incorrect methods of support.

Attach wire bundles to walls of junction box to prevent chafing or abrasion against terminal studs or other items in box. Tie up slack (required for terminal rework) to prevent snagging.

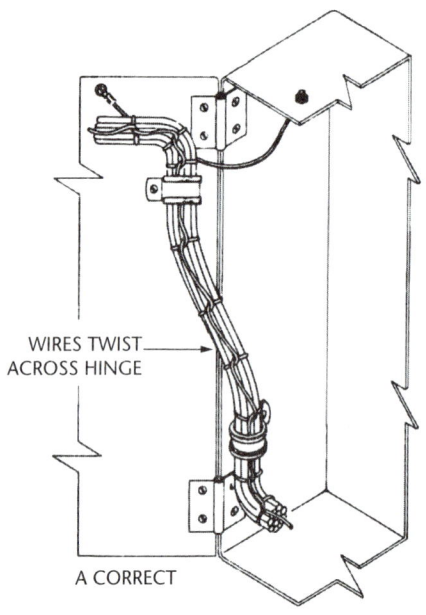

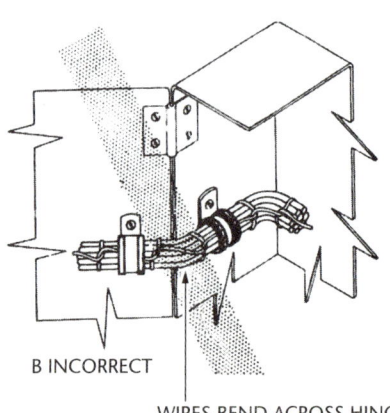

Figure 10-31. Support inside junction box

Section 7
Lacing and Tying

Description and Identification

Wire groups and bundles are laced or tied to provide ease of installation, maintenance and inspection.

This chapter describes and illustrates the recommended procedures for lacing and tying wire groups or bundles, using knots which will hold tightly under all conditions; and for installing self-clinching plastic cable straps.

Definitions

Tying. The securing together of a group or bundle of wires with individual ties at regular intervals around the group or bundle.

Lacing. The securing together of a group or bundle of wires, installed inside enclosures, by means of a continuous cord forming loops at regular intervals around the group or bundle.

Wire group. Two or more wires tied or laced together to give identity to an individual system.

Wire bundle. Two or more wires or groups tied or laced together to facilitate maintenance.

Materials. Use narrow flat tape wherever possible for lacing and tying. Round cord may also be used, but its use is not preferred because cord has a tendency to cut into wire insulation. Use cotton, linen, nylon or glass fiber cord or tape, according to temperature requirements. Cotton or linen cord or tape must be prewaxed to make it moisture and fungus resistant. Nylon cord or tape may be waxed or unwaxed; glass fiber cord or tape is usually not waxed.

> **CAUTION:** *Use only flat braided waxed nylon tape to lace or tie coaxial cables, or bundles containing coaxial cables.*

Use either vinyl or glass fiber pressure-sensitive tape, according to temperature requirement. Use pressure-sensitive tape only when its use is specifically permitted.

Molded nylon self-clinching cable straps may be used where the temperature is not expected to exceed 350°F.

Section 8
General Precautions

When lacing or tying wire groups or bundles, observe the following precautions:

a. Lace or tie bundles tightly enough to prevent slipping, but not so tightly that the cord or tape cuts into or deforms the insulation. Be especially careful when lacing or tying coaxial cable, which has a soft dielectric insulation between the inner and outer conductors.

> **CAUTION:** *Do not use round cord for lacing or tying bundles that contain coaxial cable.*

b. Do not use ties on that part of a wire group or bundle located inside a conduit.

c. When tying wire bundles behind connectors, start ties far enough back from the connector to avoid splaying of contacts.

Section 9
Lacing

Continuous lacing may be used only on those wire groups or bundles that are to be installed in panels or junction boxes. Use double cord lacing on groups or bundles larger than one inch in diameter. Use either single or double cord lacing on groups or bundles one inch or less in diameter. Instructions for lacing groups that branch off a main bundle are given later in this chapter.

Single Cord Lacing

Lace a wire group or bundle with a single cord as follows (see Figure 10-32, next page):

1. Start the lacing at the thick end of the wire group or bundle with a knot consisting of a clove hitch with an extra loop.

2. At regular intervals along the wire group or bundle, and at each point where a wire or wire group branches off, continue the lacing with half hitches.

> **NOTE:** *Space half hitches so that the group or bundle is neat and securely held.*

3. End the lacing with a knot consisting of a clove hitch with an extra loop.

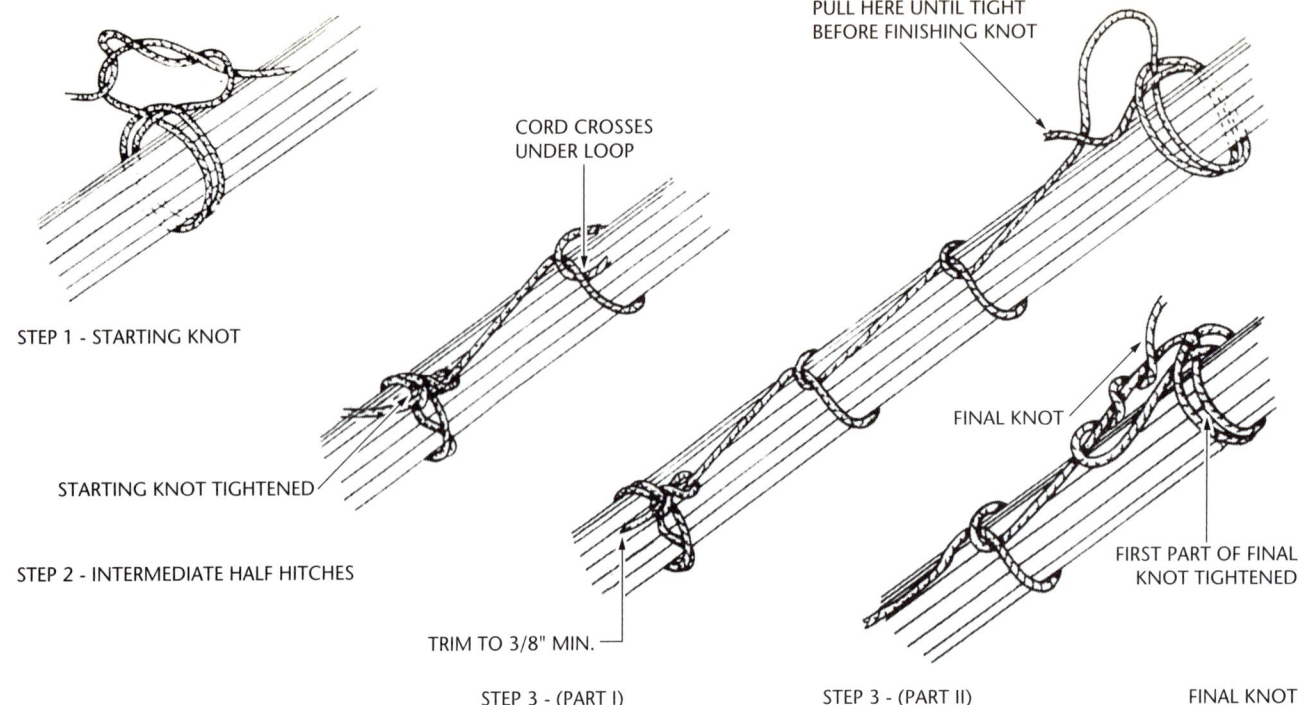

Figure 10-32. Single cord lacing

4. Trim the free ends of the lacing cord to 3/8 inch minimum.

Double Cord Lacing

Lace a wire group or bundle with a double cord as follows: See Figure 10-33.

1. Start the lacing at the thick end of the wire group or bundle with a bowline on a bight.

2. At regular intervals along the wire group or bundle, and at each point where a wire group branches off, continue the lacing with half hitches, holding both cords together.

NOTE: *Space half hitches so that the group or bundle is neat and securely held.*

3. End the lacing with a knot consisting of a half hitch, using one cord clockwise and the other counter-clockwise, and then tying the cord ends with a square knot.

4. Trim the free ends of the lacing cord to 3/8 inch minimum.

Lacing Branch-offs

Lace a wire group that branches off the main wire bundle as follows (see Figure 10-34):

1. Start the branch-off lacing with a starting knot located on the main bundle just past the branch-off point. When single cord lacing is used, make this starting knot as described in in the single cord instructions, step 1; and when double cord lacing is used, make it as described in the double cord instructions, step 1.

2. Continue the lacing along the branched-off wire group, using regularly spaced half hitches. Where a double cord is used, both cords are held together.

NOTE: *Space half hitches so that the group or bundle is neat and securely held.*

3. End the lacing with the regular knot used in single and double cord lacing, as described in the respective sections, step 3.

4. Trim the free ends of the lacing cord to 3/8 inch minimum.

Section 10
Tying

Tie all wire groups or bundles where supports are more than 12 inches apart. Space ties 12 inches or less apart.

Making ties. Make tie as follows:

1. Wrap cord around wire group or bundle, as shown in Figure 10-35.

2. Make a clove hitch, followed by a square

knot with an extra loop.

3. Trim free ends of cord to 3/8 inch minimum.

Temporary ties. Temporary ties are used to aid in making up and installing wire groups or bundles. Use colored cord to make temporary ties, and remove these ties when the installation is complete.

CAUTION: *Cut temporary ties with scissors or diagonal pliers only. Do not use a knife or other sharp edged instrument that may damage the insulation.*

Tying wire groups into wire bundles. Tie wire groups into bundles as described in the previous procedure, treating the wire groups as though they were individual wires.

Tying sleeves to wire groups or wire bundles. Secure sleeves to wire groups or bundles by tying as described previous procedure.

Securing with pressure-sensitive tape. See Figure 10-36 (next page). When use of pressure-sensitive tape is permitted, install the tape as follows:

1. Wrap tape around wire group or bundle three times, with a two-thirds overlap for each turn.

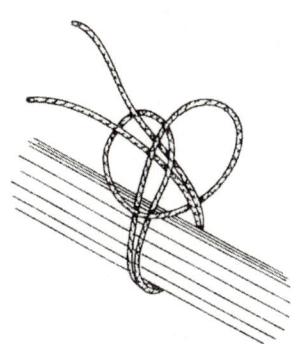

STEP 1   STARTING KNOT - BOWLINE ON A BIGHT

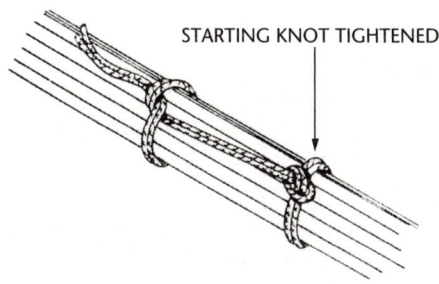

STEP 2   INTERMEDIATE HALF HITCHES

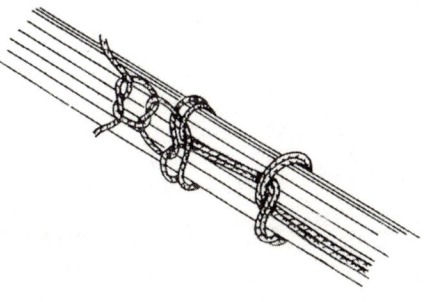

STEP 3   FINAL KNOT

Figure 10-33. Double cord lacing

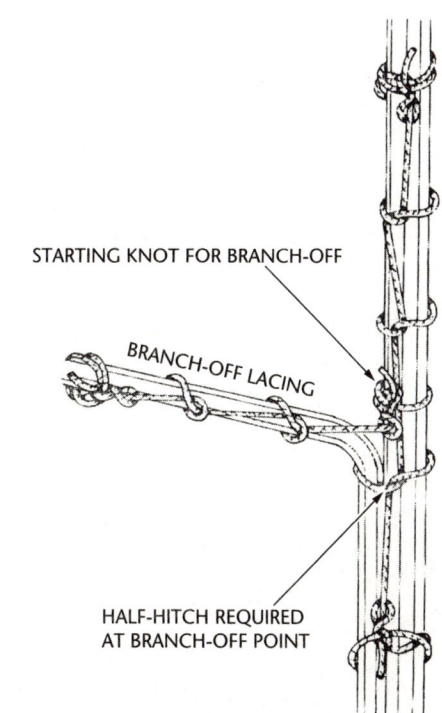

Figure 10-34. Lacing a branch-off

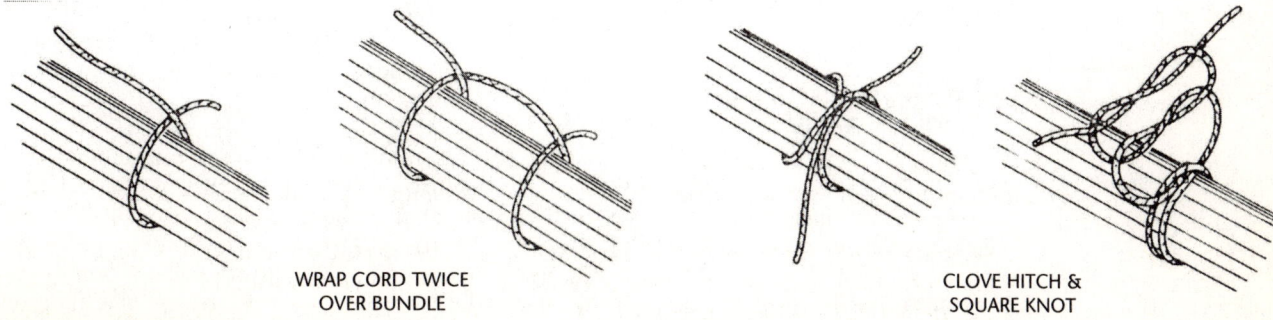

WRAP CORD TWICE OVER BUNDLE

CLOVE HITCH & SQUARE KNOT

Figure 10-35. Making ties

Electrical Wiring Installation

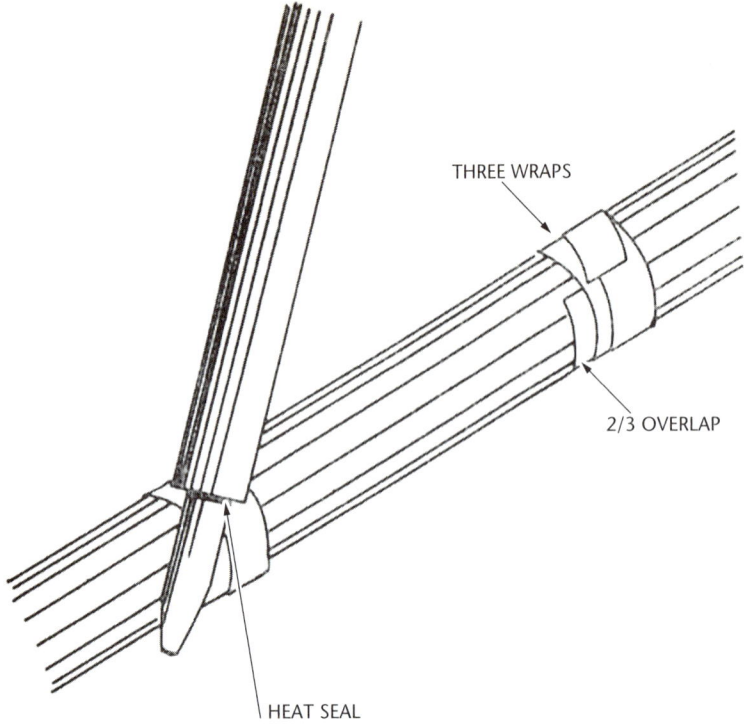

Figure 10-36. Securing with tape

Color Dash No.	Color
-0	Black
-1	Brown
-2	Red
-3	Orange
-4	Yellow
-5	Green
-6	Blue
-7	Purple
-8	Gray
-9	Natural

Table 10-8. Cable strap colors

MS Part No.	Type	For Bundle Dia (Inches)	MS Tool No.
MS17821-1	Thin	1/16 - 1 3/4	MS17823
MS17821-2	Thin	1/16 - 4	MS17823
MS17821-3	Thick	3/16 - 3 3/4	MS17824
MS17822-1	Thin	3/8 - 1 3/4	MS17823

Table 10-9. Self-clinching plastic cable and installing tools

2. Heat-seal the loose tape end with the side of a soldering iron heating element.

CAUTION: *Do not use tape for securing wire groups or bundles which may require frequent maintenance.*

Self-clinching cable straps. These are adjustable lightweight flat nylon straps, with molded ribs or serrations on the inside surface to grip the wire. They may be used instead of individual cord ties for fast securing of wire groups or bundles. The straps are of two types: MS17821, a plain cable strap, and MS17822, similar to MS17821, but with the addition of a flat surface for identification marking. Refer to Chapter 1 for use of cable straps to identify wire bundles. Both types of cable straps are available in ten colors as listed in Table 10-8.

Installing self-clinching cable straps. Use the hand tools listed in Table 10-9 to install plastic cable straps on wire bundles. The procedure is as follows (see Figure 10-37):

1. Select a strap of the desired color for the application, and of a size suitable for the wire bundle from Table 10-9.
2. Slip the plastic strap around the bundle, with the ribbed side in.
3. Thread the tip of the strap through the eye in the strap boss, and pull the strap tight around the bundle.
4. Select the proper tool for the strap used from Table 10-9.
5. Pass the free end of the strap through the slot in the top of the tool, and push the tool snug against the strap boss.
6. Hold the tool firmly against the boss, and pump the tool handles gently until the strap feels tight in the tool and snug abound the bundle.
7. Press the cut-off release clip, and close the tool handles all the way.

Section 11

Wiring: Lock, Shear and Seal

Description and Identification

Electric connectors, emergency, devices and other pieces of electric equipment in aircraft are secured with wire when specified on engineering drawings in order to prevent accidental loosening.

This chapter outlines the recommended procedures for wiring MS electric connectors, and emergency devices such as switches, switch guards and handles which operate emergency releases, fire extinguishers etc. general practices for safety wiring are specified in drawing MS33540.

Electrical Wiring Installation | 10-19

Lock wire. Lock wire is a heavy twisted double-strand wire used to secure parts against inadvertent opening in all areas of high vibration such as the engine compartment. Electric connectors are lock-wired in high-vibration areas, which are normally inaccessible for periodic maintenance and inspection.

Shear wire. Shear wire is a lighter, single strand wire used to secure parts which may be subject to periodic disconnection, maintenance and inspection or for parts, which must be quickly removed.

Seal wire. Seal wire is a thin, easily breakable wire used as a seal on fire extinguishers, oxygen regulators and other emergency devices which must be quickly released for use, and to indicate whether these devices have been tampered with or used.

Section 12

Procedures for Lock, Shear and Seal Wiring

CAUTION: *Use only new wire; when replacing wired electrical connectors or emergency devices, do not re-use the old wire.*

Length. Use wire of the shortest length that will allow accomplishment of the procedures outlined in following paragraphs.

Double twist lock wiring. Use the double twist method of lock wiring as illustrated in Figure 10-38 for all equipment in areas of high vibration, and for electrical connectors in such areas that are inaccessible.

Single wire method. Use the single wire method shown in Figure 10-39 in all conditions specified for shear and seal wire as described on previous page. In addition, the single wire method may be used in areas hard to reach, and for small screws in a closely spaced pattern.

Twisting with pliers. When wire is twisted by hand, use pliers for the final twists to apply tension, and to secure ends of wire. Cut off part of wire gripped by pliers to remove rough edges.

CAUTION: *Make sure wire does not become kinked or nicked during twisting operation. If wire is damaged, replace with new wire.*

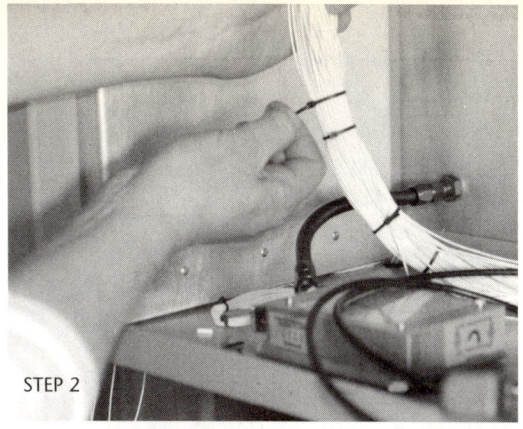

STEP 2

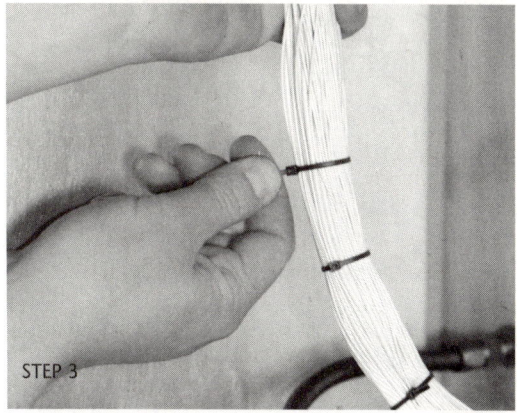

STEP 3

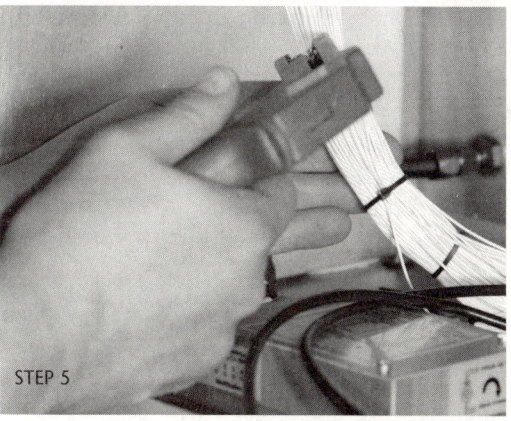

STEP 5

Figure 10-37. Installing self-clinching cable strap

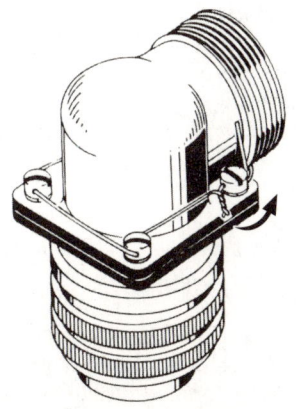

BEND PIGTAIL AROUND SCREW TO PROTECT PERSONNEL

Figure 10-39. Single wire method

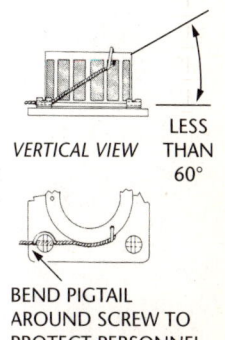

VERTICAL VIEW — LESS THAN 60°

BEND PIGTAIL AROUND SCREW TO PROTECT PERSONNEL

HORIZONTAL VIEW

Figure 10-38. Double twist lock wiring

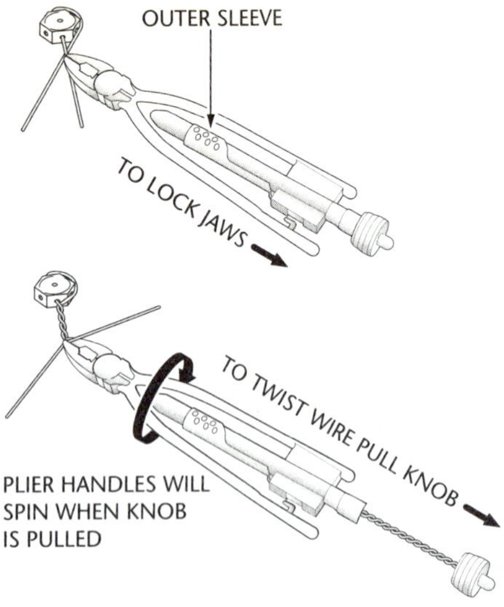

Figure 10-40. Use of wire twister

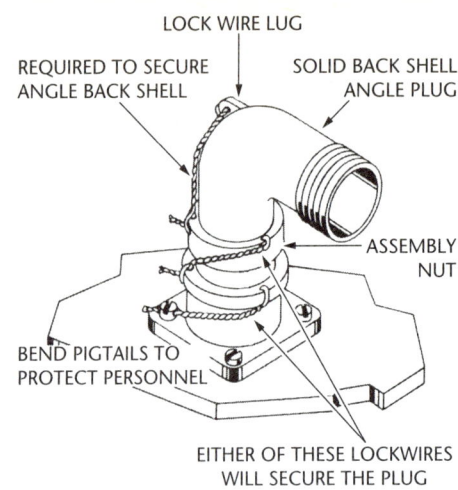

Figure 10-41. Wiring AN type connector

Twisting with special tools. Twist wire with a wire twister as follows (see Figure 10-40):

1. Grip wire in jaws of wire twister and slide outer sleeve down with thumb to lock handles.
2. Pull knob; spiral rod spins pliers and twists the wire.
3. Squeeze bundles together to release wire.

Tightness of wire. Install wire so that the wire will be in tension if the part loosens. Twist wire together so that it is tight, but do not overstress wire as it may break under load or vibration.

Specific Procedures for Lock, Shear and Seal Wiring

Lock wiring electrical connectors. Secure electrical connectors with lock wire only when specified on engineering drawings. Electric connectors are usually lock-wired in engine nacelles, areas of high vibration and in locations not readily accessible for periodic maintenance inspection. Connectors in these locations are identified by a painted/affixed red dot 1/2 inch in diameter on adjacent aircraft structure.

CAUTION: *Do not wire electric or RF connectors that have a mechanical lock, as lock wiring will act against the locking feature.*

Lock wiring AN-MS connectors. When specified on engineering drawings, lock wire AN-MS connectors as follows (see Figure 10-41):

1. Thread lock wire through wire hole in coupling ring.

NOTE: *If connector plug to be lock wired does not have a wire hole, remove coupling nut and drill a #56 (.046) diameter hole diagonally through the edge of nut, as shown in Figure 10-42.*

2. Twist wire, under slight tension, approximately 6 to 8 turns per inch, by hand, or by special tool, as described above. Twist wire right handed, so it will have a tightening effect.
3. Pull one end of twisted wire through hole in drilled fillister head screw on mounting flange of connector. Use a fillister head screw so located as to allow a 60° or smaller angle of the wire, as shown in Figure 10-38.

CAUTION: *Do not back off or over-torque mounting fillister head screws, in order to align holes for lock wiring.*

4. Form pigtail 1/4 to 1/2 inch (3 to 6 twists) with pliers.
5. Bend pigtail back toward body of connector, to prevent it from injuring personnel.

CAUTION: *Lock-wire all connectors individually. Do not lock wire one connector to another.*

Lock wiring connector to structure. If no screw is available for attaching lock wire, secure wire to drilled hole in structure not more than 6 inches from connector, as shown in Figure 10-43. Use same procedure as described on page 10-19.

Shear wiring split shell assemblies. Split shell connectors made by Amphenol are held together by two fillister head screws. Secure these screws as follow (see Figure 10-44):

1. Draw wire through hole in one screw.

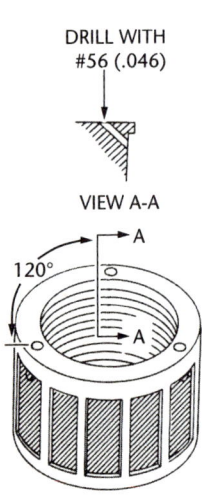

Figure 10-42. Drilling hole in coupling nut

Electrical Wiring Installation | 10-21

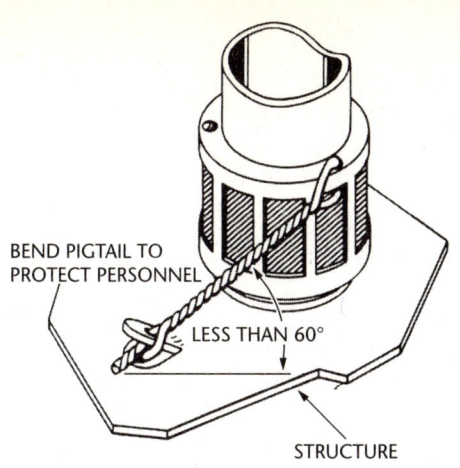

Figure 10-43. Lock wiring connector to structure

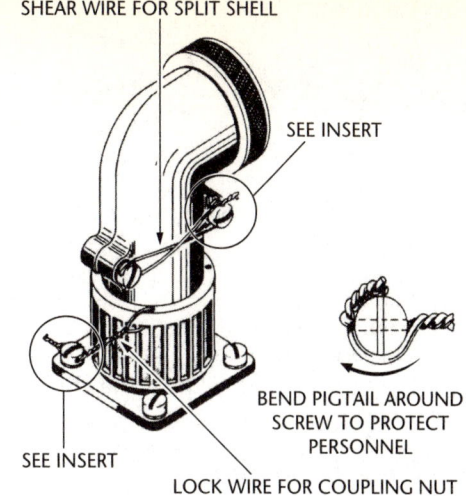

Figure 10-44. Wiring split shell assembly screws

2. Cross wire from left to right between screws, and draw through second screw.

3. Twist wires together with pliers, and bend back.

Wiring solid shell angle plugs. Angle plugs with solid back shells as made by Amphenol are in two parts, held together by four screws through mating flanges. Wire these screws with a single shear wire as shown in Figure 10-39. Solid shell angle plugs made by Bendix and Cannon have back shells held in place by assembly nuts. Install a double twisted lock wire between hole in assembly nut and lug on back shell as shown in Figure 10-41. If necessary to lock wire the plug itself, install a second double twisted wire between the assembly nut and the coupling nut or between the coupling nut and one of the receptacle mounting screws, as shown in Figure 10-41.

Seal wiring emergency devices. See Figure 10-45. Use single wire method to secure emergency devices. Make sure that wire is installed so that it can easily be broken, when required, in an emergency.

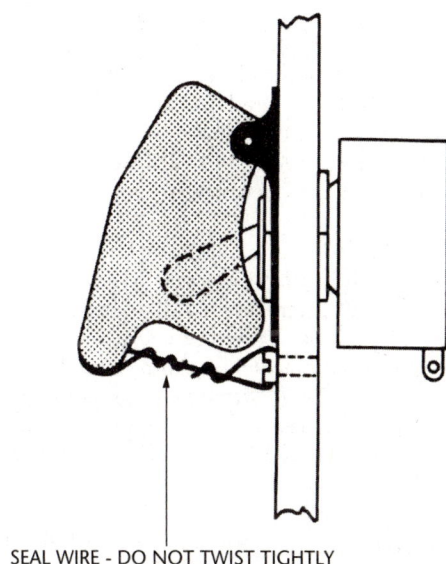

Figure 10-45. Seal wiring switch guard

tions. These two standards cover how data is transferred between components in an avionics system.

Data Bus Function

The avionics systems in aircraft use digital electronics to process the data and information. This equipment can be as simple as a radio transmitter/receiver combination or as complex as an auto-flight system that is integrated with the navigation and engine systems. The one thing that they have in common is the need for standard communications through a data bus.

It is important to understand that a data bus refers to the wiring that transmits the information between two or more line replaceable units (LRU), not the internal wiring of the LRUs. The

Section 13
ARINC Digital Bus

Over the years, as the avionics industry evolved a need for common communication standards developed. These standards would allow various systems to communicate with a common digital language.

The standards for data bus communications are the ARINC 429 and ARINC 629 specifica-

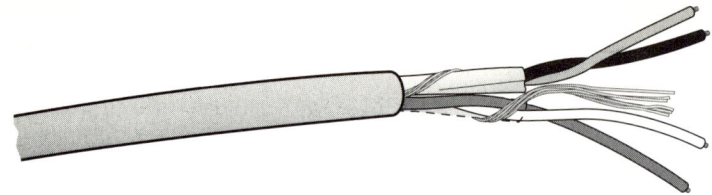

Figure 10-46. An example of ARINC 429 data bus wiring.

two standards, ARINC 429 and 629, may be found in the same aircraft, however they will never share the same physical data bus, rather they will have separate, unique, data buses.

Data bus description

A typical 429 digital data bus is a pair of 24 gauge copper wires. Each wire is plastic insulated and twisted together to minimize inductive problems. The wires are wrapped in a foil sheath that is wraped with a drain wire that is grounded at each end of the cable. The completed bus is insulated in plastic and can be seen in Figure 10-46. Ordinary aircraft wire cannot be used as a data bus, since there is no foil shielding, nor are the wire pairs twisted.

The ARINC 429 data bus is unidirectional, in other words, information only flows from the transmitter to the receiver. Also there can only be one transmitter, however there can be up to twenty receivers, in the system. Data bus systems are polarity sensitive and it is very important that each conductor be separate from the other and never switched. Wire A must be connected to terminal A and wire B to terminal B.

The difference between the 429 and the 629 data bus is that the 629 is bidirectional and can have up to 120 transmitter/receivers. The 629 bus can be either twisted pairs or fiber optic cable up to 100 meters (328 ft.) in length.

Bus Installation

Data bus wiring is installed using the same guidelines and precautions as used for single wires and bundles in any aircraft. The bus wiring is installed using clamps that secure the wiring without pinching, observe bend radii for the size of the wire and secure the wiring above any fluid lines. For general installation guidelines refer to the information in the previous sections of this chapter on wiring installation. In all installations, the manufacturer may have specific requirements that will take precedence over any textbook or general guidelines.

Terminations

ARINC data bus wiring can be terminated with any of the common multiple pin connectors that are used for avionics equipment. The more common types are the Sub D, nine and fifteen pin connectors that meet Mil-C-24308 specifications, shown in Figure 10-47.

Standard crimping and soldering methods for pins and sockets are applicable to all terminations used on data bus wiring.

NOTE: the use of inline splices and repairs should be avoided. Improper repairs can lead to attenuation and signal loss problems.

Troubleshooting and Repair

Keeping in mind that the data bus is nothing more than a twisted pair of wires, there are relatively few problems, in fact, many aircraft have operated for years without experiencing any data bus problems.

Most of the problems occur at the connectors. Just about every connector will wear out through repetitive use, vibration or corrosion. Since the data signals are very low power, a slight increase in connector impedance will create a faulty bus connection.

Inspection and testing

The first step in troubleshooting a data bus is a careful visual inspection for physical defects of the bus itself. If the bus or the wires are broken, pinched or exposed then the most obvious problems are solved. Next, visually inspect the pins and sockets of the connectors as part of the visual inspection. Using a

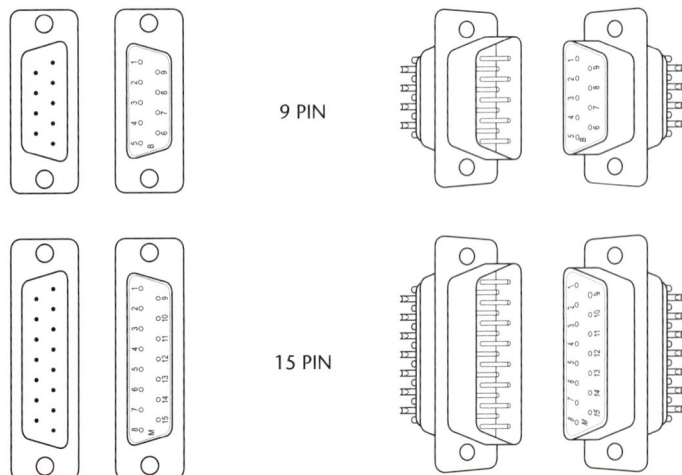

Figure 10-47. Sub-D connectors are commonly used for data bus terminations.

magnifying glass, look for bent or worn pins and deformed sockets, as well as evidence of corrosion.

When cleaning the connectors, use only the approved solvents and exercise care not to damage the sockets or bend the pins by forcing swabs down into them. Never use any type of abrasive materials or tools, since that will damage the platting, leading to further problems.

An ohmmeter can be used to detect an open in the bus wiring or a broken grounding shield. However, an ohmmeter test may show that the bus is intact, yet the bus may fail to properly transmit data due to high impedance.

The most common diagnostic tool is a data bus analyzer which can be hooked into the system to graphically represent the data between the transmitter and the receivers. An example is shown in Figure 10-48. Some analyzers can also transmit a data string to the receivers to verify their operation.

Signal faults

There are three other problems that can be found in a data bus that will affect the quality of the signal. These are:

- Interference
- Attenuation
- Signal dropouts

Interference. Interference in a data buss happens when an unwanted electrical signal is imposed into the wiring. Improper shielding of the data bus and exposure to a changing magnetic field is often the cause of this type of problem. Another reason may be an excessively strong electromagnetic field near a termination point or connector. In both of these cases the independent component is the cause and not the bus or the LRUs.

The solution is to isolate the bus from any high-tension wiring and maintain adequate clearance between the bus and any components that emit electromagnetic fields. Maintaining the integrity of connectors and the shielding at both the receiver and the transmitter will also help to eliminate problems.

Attenuation. Attenuation refers to a loss of signal strength, or a drop in the transmitting

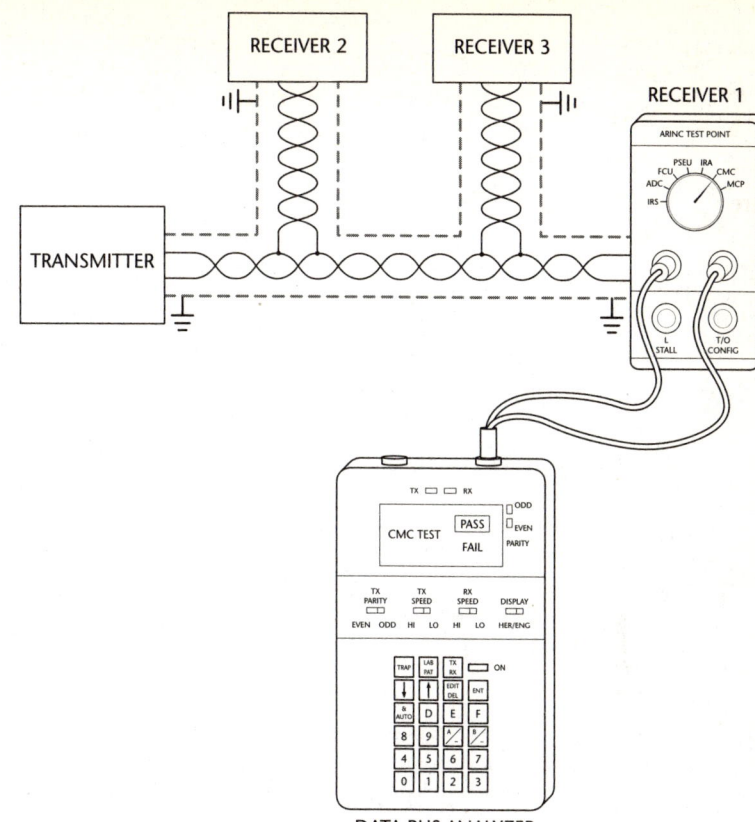

Figure 10-48. A typical data bus analyzer.

voltage, in the data bus. While there is some attenuation in every data bus due to the resistance of the wiring, when it becomes excessive it becomes a problem. Once the attenuation becomes high, the data receiver will not recognize the signal.

Something as simple as dirty, corroded pins and sockets or poorly soldered and crimped connectors can cause a significant increase in total attenuation. A pinched data bus or even the presence of moisture in the data bus can also be the reason for the attenuation.

Signal dropouts. When portions of a data signal are completely lost, a signal dropout has occurred. Dropouts are typically caused by bad and intermittent connections at either the transmitter or the receivers.

Intermittent connections often occur when vibration levels are high, or during turbulence or unusual flight attitudes. Often the problem can be located by shaking the wiring or wiggling the LRU.

Chapter 11

AIRCRAFT electrical system lamps

Section 1

Lamps and Fuses

Learning Objectives:
- Lamp Types
- Operating Precautions
- Aircraft Electrical Fuses

Lamps are used in aircraft interior general lighting, in instrument panels and as indicating lights. Lamps used in aircraft exterior lights include those for position, fuselage, wing inspection, landing and taxiing; the last three are of the sealed beam type.

This chapter describes and illustrates lamps used in both the interior and exterior circuits of aircraft.

Lamp types. The lamps most commonly used in aircraft exterior circuits are listed in Table 11-1 (next page), with their electrical and physical characteristics. The MS drawing dash numbers, listed in the second column, are usually the same as the commercial designation (trade number) of the lamp. Lamps commonly used in interior circuits are listed in Table 11-2 (next page). Lamp types are illustrated in Figures 11-1 and 11-2 (next page).

Abbreviations. Abbreviations to describe bulb shape, size and finish, and type of base are as follows:

- Bulb Shape:

	BULB SHAPE
G	Globular
GG	Grimes Globular
S	Straight
T	Tubular
PAR	Parabolic Alum. Reflector
R	Reflector

Left. Aircraft exterior lights include those for position, fuselage, wing inspection, landing and taxiing.

11-2 | Aircraft Electrical System Lamps

Part No.	Dash No.	Bulb	Base	Volts	Amps	Use (Typical)
MS15570	-303	G-6	S.C. Bay.	28	.30	Fuselage Lights
MS25231	-313	T-3-1/4	Min. Bay.	28	.17	Exterior Lights
MS25235	-311	S-11	S.C. Bay.	28	1.29	Fuselage Lights
M525238	-301	G-5	S.C. Bay.	28	.17	Anchor Light
MS25241	-4553	PAR-46	Scr. Term.	28	250W	Landing Lights
	-4581	PAR-46	Scr. Term.	28	450W	Landing Lights
	-4582	PAR-46	Scr. Term.	28	450W	Helicopter Land/Hover Lights
M525242	-4559	PAR-64	Scr. Term.	28	600W	Landing Lights
MS25243	-4502	PAR-36	Scr. Term.	28	50W	Taxiing; Wing Insp.
	-4505	PAR-36	Scr. Term.	28	50W	Position Light
M525309	-600	GG-10	S.C. Bay.	6.2	26W	Wing Lights
	-1506	GG-10	S.C. Bay.	6.0	21W	Wing Lights
M535478	-307	S-8	S.C. Bay.	28	.67	Tail Position Light
	-1680	S-8	S.C. Bay.	6	4.10	Tail Position Light
AN3120	-1047	RP-11	S.C. Bay.	26	2.7	Signalling Light
—	-1959	T-4 (Quartz)	Tab	28	150W	Position Light

*Amps unless otherwise noted: W-Watts

Table 11-1. Lamps used in aircraft lighting-exterior

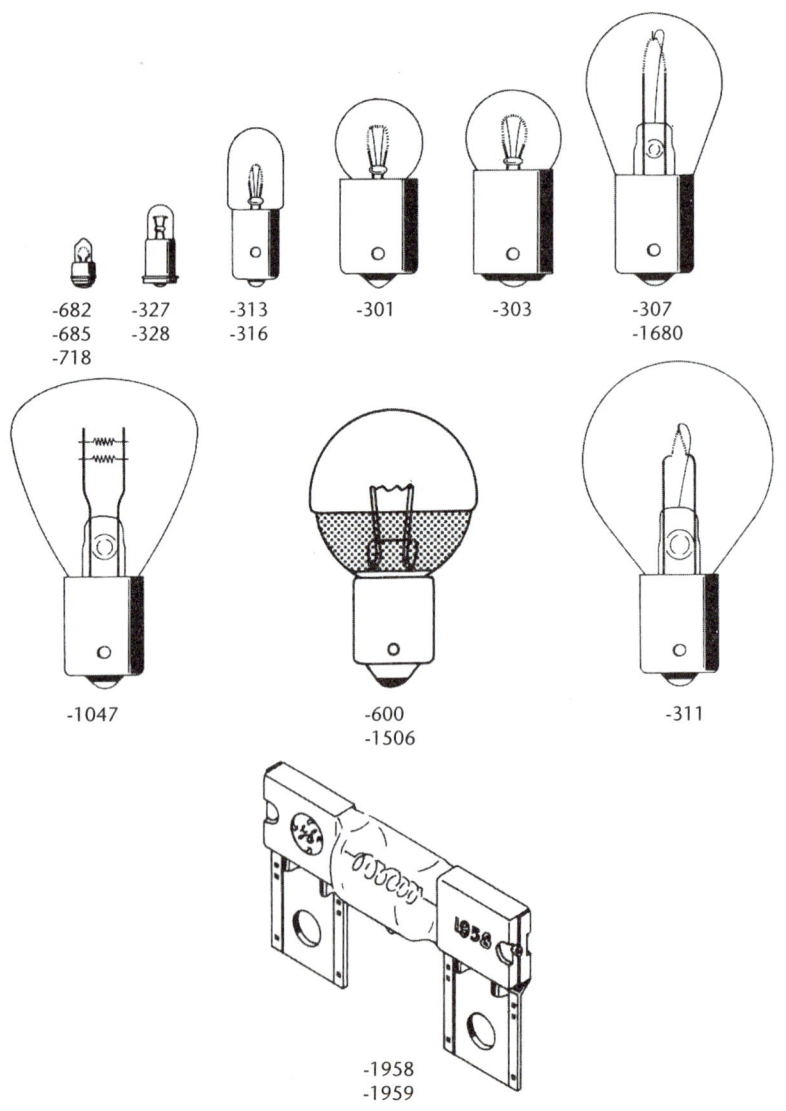

Figure 11-1. Lamps used in aircraft electrical systems

- Bulb size is indicated in one-eighth inches of greatest diameter. For example, T-8 is a tubular lamp, with a 1 inch diameter.
- Bulb finish is indicated by letters preceding the MS dash number as follows:

BULB FINISH	
R	Red
SB	Silvered Bowl

If no letter is present, the lamp is clear glass.

BASE TYPE	
Min. Bay.	Miniature Bayonet
S.C. Bay.	Single Contact Bayonet Candelabra
S.C. Mm.	Flg. Single Contact Miniature Flanged
S.C. Mid Flg.	Single Contact Midget Flanged
Scr. Term.	Screw Terminal
Sub. Mid. Flg.	Sub Midget Flanged

General precautions. When installing lamps in the aircraft electrical circuit, observe the following:

- Replace a burnt-out or damaged lamp with a lamp of the same MS number, or the approved alternate.

  NOTE: *The trade number, which is usually the same as the MS dash number, is stamped on the base of each lamp.*

- Make sure the glass bulb of the lamp is clean and free from grease and dirt. To help keep bulbs clean, avoid touching the glass bulb with bare hands if possible.
- Do not force a lamp into its socket. Check

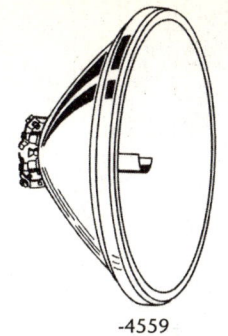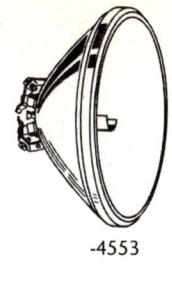

-4559   -4553   -4501 / -4502 / -4505

Figure 11-2. Aircraft lamps-sealed beam type

to see if the lamp will fit properly in only one position.

Section 2
Aircraft Electrical Fuses

Description and Identification

Fuses are current-limiting devices used in the aircraft electrical system to afford electrical protection against short circuits and system faults.

This section describes and illustrates types of fuses commonly used in aircraft electrical systems. Procedures for installing fuses and other protective devices are given in Chapter 9.

Fuses. See Figure 11-3 (next page). Fuses used in aircraft are of two types: the enclosed link type (current limiter) installed in a block type

Part No.	Dash No.	Bulb	Base	Volts	Amps	Use (Typical)
MS24515	-682	T-1	Sub. Mid. Flg.	5	.06	Instrument Panel
	-685	T-1	Sub. Mid. Flg.	5	.06	Instrument Panel
	-718	T-1	Sub. Mid. Flg.	5	.115	Instrument Panel
MS25231	-313	T-3-1/4	Min. Bay.	28	.17	Indicating Light
	-R313	T-3-1/4	Min. Bay.	28	.17	Indicating Light
	-316	T-3-1/4	Min. Bay.	6	.70	Instrument Panel
MS25235	-311	S-11	S.C. Bay.	28	1.29	Work Table Light
	-R311	S-11	S.C. Bay.	28	1.29	Interior Lighting
	-SB311	S-11	S.C. Bay.	28	1.29	Cabin Dome Light
MS25237	-327	T-1-3/4	S.C. Mid. Flg.	28	.04)	Indicating Light
	-328	T-1-3/4	S.C. Mid. Flg.	6	.02)	Instrument Panel
MS35478	-307	S-8	S.C. Bay.	28	.67	Cabin Dome Light
	-R307	S-8	S.C. Bay.	28	.67	Interior Lighting
	-SB307	S-8	S.C. Bay.	28	.67	Interior Lighting
MS25239	-4501	PAR-36	Scr. Term.	26	5.3	Flashing Signal

Table 11-2. Lamps used in aircraft lighting-interior

fuseholder, and the cartridge type, installed in the electrical system in an extractor post style fuseholder or in fuse clips. Fuses commonly used in aircraft electrical systems are listed in Table 11-3 by detailed Mil-Specs and MS drawings.

Fuseholders. See Figure 11-4. Extractor post fuseholders in accordance with drawings MS18091, MS18092, MS25474 and MS26572 are used in conjunction with cartridge type fuses. Block type fuseholders in accordance with drawings MS24000 and MS24001 are used with enclosed link fuses.

Identification. Enclosed link fuses are identified with the MS part number and the amperage rating. Cartridge type fuses are marked with the current and voltage ratings, style designation and characteristic letter as listed in Table 11-3.

General Precautions

When replacing fuses in the aircraft electrical system, observe the following precautions:

- Do not use tools to remove or insert fuses.
- Make sure that the new fuse has the same electrical features as the fuse being replaced.

  CAUTION: *Cartridge fuses marked F02 and F03 are 1-1/4 inches long and 1/4 inch diameter; fuses marked F05 and F06 are 1-1/4 inches long and 9/32 inch diameter. Do not interchange the two sizes.*

- Make sure that the plating on all metal parts is clean and intact.
- Make sure that the wire inside the replacement fuse exhibits continuity.
- Make sure that the replacement fuse has no cracks or breaks.
- Do not force a fuse into a holder that does not readily accept it; check that a fuse of the correct size is being used.

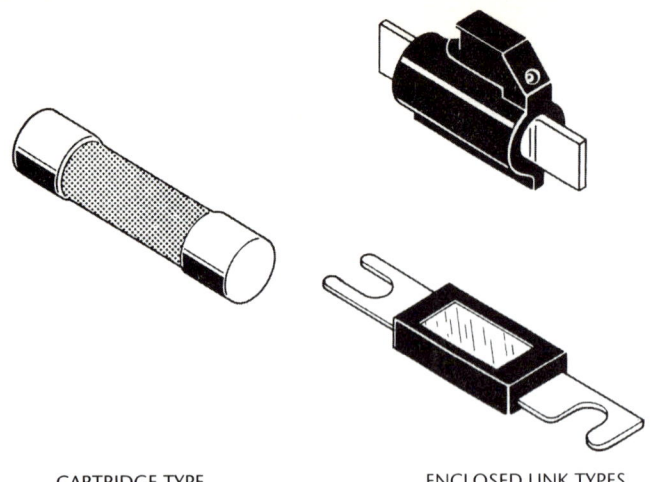

Figure 11-3. Typical aircraft fuses

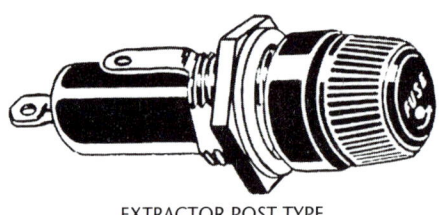

EXTRACTOR POST TYPE

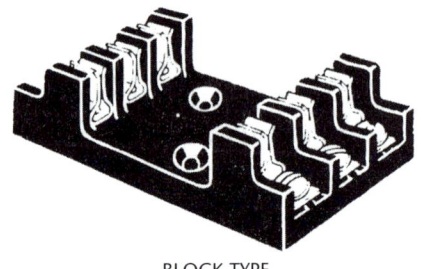

BLOCK TYPE

Figure 11-4. Typical fuse holders

Enclosed Link Type			
MS Part Number	Voltage Rating	Current Rating	Type
MS24124-5	115/200 vac	5 amps	A
-10		10 amps	A
-20		20 amps	A
-30		30 amps	A
-40		40 amps	A
-50		50 amps	A
-60		60 amps	A
MS24125-5	115/200 vac	5 amps	B
-10		10 amps	B
-20		20 amps	B
-30		30 amps	B
-40		40 amps	B
-50		50 amps	B
-60		60 amps	B

Table 11-3. Fuses used in aircraft electrical system

Number	Style	*Characteristic	Cartridge Type Max.Volts	Amperes	Replaces MS Number: Char. A	Char. B
MIL-F-15160/02	F02	A,B	250	1/100	90078-1-1	90078-16-1
		A,B	250	1/32	-2-1	-17-1
		A,B	250	1/16	-3-1	-18-1
		A,B	250	1/8	-4-1	-19-1
		A,B	250	1/4	-5-1	-20-1
		A,B	250	3/8	-6-1	-21-1
		A,B	250	1/2	-7-1	-22-1
		A,B	250	3/4	-8-1	-23-1
		A,B	250	1	-9-1	-24-1
		A	250	1-1/2	-10-1	—
		A	250	2	-11-1	—
		A	250	3	-12-1	—
		A	250	4	-13-1	—
		A	250	5	-14-1	—
		A	250	6	-15-1	—
		B	125	1-1/2	—	90078-25-1
		B	125	2	—	-26-1
		B	125	3	—	-27-1
		A,B	32	10	—	—
		A,B	32	15	—	—
		A,B	32	20	—	—
		B	32	5	—	—
		B	32	8	—	—
MIL-F-15160/03	F03	A,B	250	1	90079-1-1	90079-20-1
		A	250	3	-2-1	—
		A	250	5	-3-1	—
		A	250	8	-4-1	—
		A	250	10	-5-1	—
		A	250	12	-6-1	—
		A	250	15	-7-1	—
		B	250	1/100	—	90079-10-1
		B	250	1/32	—	-11-1
		B	250	1/16	—	-12-1
		B	250	1/8	—	-13-1
		B	250	15/100	—	-14-1
		B	250	3/16	—	-15-1
		B	250	1/4	—	-16-1
		B	250	3/8	—	-17-1
		B	250	1/2	—	-18-1
		B	250	3/4	—	-19-1
		A	125	20	90079-8-1	
		A	125	30	-9-1	
		B	125	3		
		B	32	5		
		B	32	8		
		B	32	10		
		B	32	12		
		B	32	15		
		B	32	20		
		B	32	30		
MIL-F-15160/05	F05	A,B	32	10	90081-1	90081-8
		A,B	32	15	-2	-9
		A,B	32	20	-3	-10
		A,B	32	25	-4	-11
		A,B	32	30	-5	-12
MIL-F-15160/06	F06	A	250	1	90082-1	
		A	250	2	-2	
		A	250	3	-3	
		A	250	5	-4	
		A	250	10	-5	
		A	250	15	-6	
MIL-F-15160/07	F07	A	250	1		
		A	250	2		
		A	250	3		
		B	125	1		
		B	125	2		
		B	125	3		
		A,B	32	5	90083-1	90083-10
		A,B	32	10	-2	-11
		A,B	32	15	-3	-12
		A,B	32	20	-4	-13
		A,B	32	30	-5	-14

*A Normal (normal interrupting capacity); for general circuit protection
B Time Lag; for circuits containing motors, and circuits where provision must be made for momentary surges.

Table 11-3 (cont'd). Fuses used in aircraft electrical system

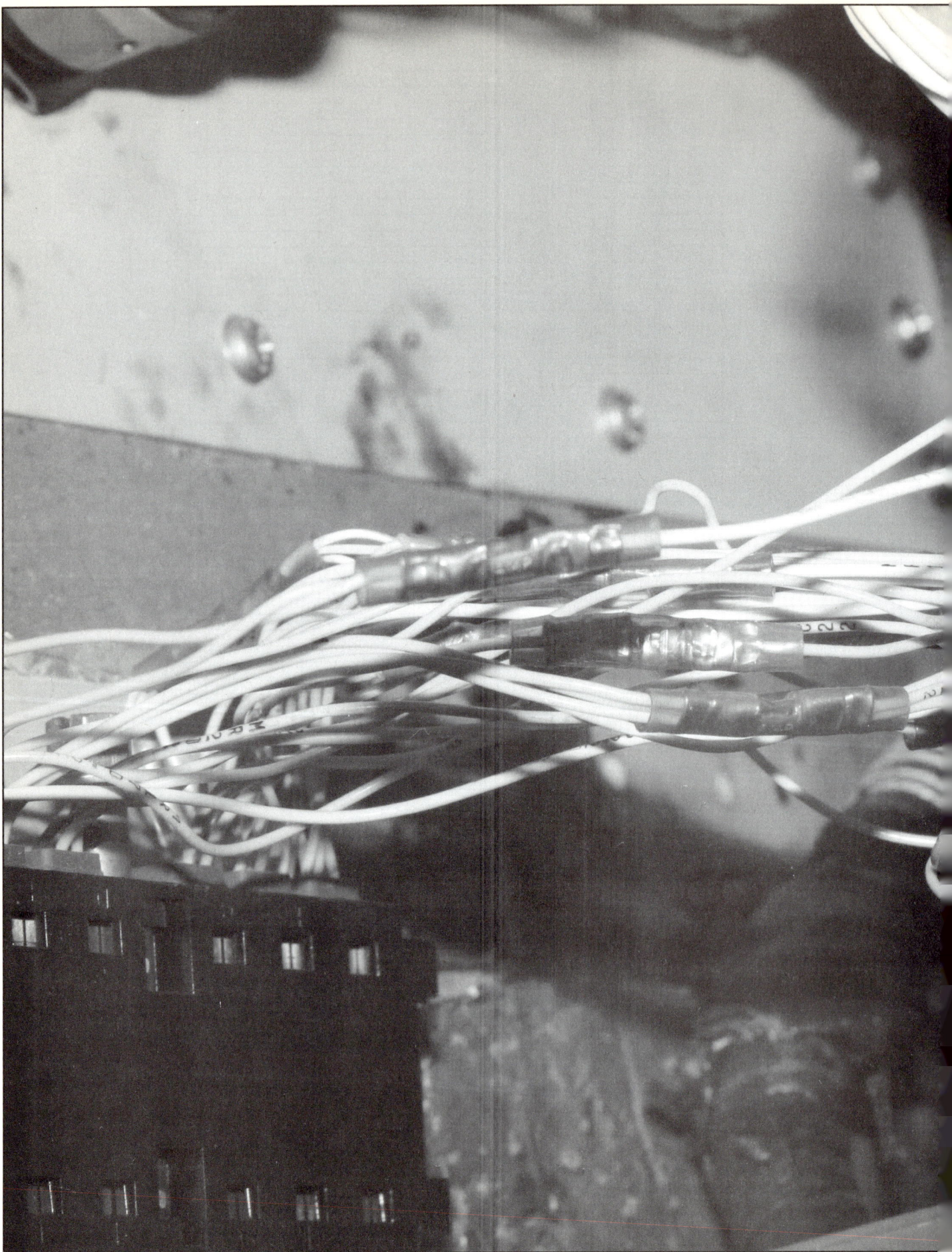

Chapter 12

EMERGENCY *repairs*

It is sometimes necessary to make emergency repairs to the aircraft electrical wiring system at advanced fields, where a minimum of tools and equipment are available.

This chapter describes and illustrates some recommended procedures for emergency repairs to broken or damaged copper wires, shielded and coaxial cable, electric connectors; and for replacing terminal board covers.

> **CAUTION:** *All repairs described in this chapter are temporary, for emergency use only. Replace all temporary repairs as soon as possible with permanent repairs.*

Learning Objectives:

- Repairing Broken or Damaged Wires
- Repairing Damaged MS Connectors

Section 1

Repairing Broken or Damaged Wires

Methods of Repairing Wire

Repair of broken wires is accomplished by means of crimped permanent splices, by the use of a terminal lug from which the tongue has been cut off, or by soldering together the broken strands, and applying potting compound. Breaks in large wire (AN size No. 12 and larger) are repaired by means of terminal lugs bolted together.

Splicing broken wires with permanent splice. See Figure 12-1 (next page). When splicing wires by means of permanent splices observe the following procedures.

> **NOTE:** *Make sure that only aluminum splices are used when splicing aluminum wires.*

Left. Inner insulating sleeves should be just long enough to completely cover the permanent splice. The outer sleeve must be long enough to extend beyond the two grounding sheath connectors.

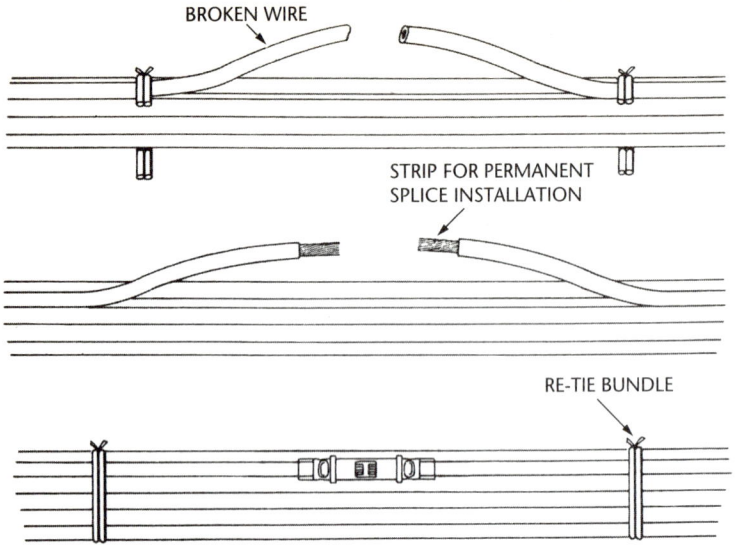

Figure 12-1. Permanent splice repair of wire

1. Cut ties and work the broken wire to the outside of the bundle.
2. Pull sufficient slack from the wire run toward the break so that there will be no strain on the splice.
3. Trim the wire as close to the break as possible so that all strands will be of equal length.
4. Clean the wire for a distance of at least one inch from the break with Stoddard's solvent. This will insure the removal of foreign particles and debris to provide a good insulating surface.
5. Slide a piece of shrinkable tubing slightly larger in diameter than the OD of the splice being used over one end of the severed wire. If shrinkable tubing is not available, a piece of flexible transparent tubing can be substituted.
6. Install the splice as described, in Chapter 4, page 4-18, and re-tie spliced wire into bundle.

Splicing with terminal lug barrel. When a permanent splice is not available, the barrel of a terminal lug can be used.

1. Select a terminal lug with a barrel large enough to accommodate both wires.
2. Cut off the terminal lug tongue.
3. Prepare the wires as described in previous procedure.
4. Insert the wires from opposite ends of the barrel so that each wire protrudes thru the barrel 1/32 inch.
5. Crimp the barrel in the center following the procedures of Chapter 4. See Figure 12-2.
6. If shrinkable tubing is used, follow the procedure below. If flexible transparent tubing is used, slide the sleeving down over the connection so that it extends about 1/2 inch past each end of the crimped barrel and then tie it with nylon cord at each end.

Use of heat-shrinkable tubing. Polyethylene tubing, which is shrunk to the desired size by the application of dry heat, may be used to protect single wires or wire groups where they break out from wire bundles or harnesses. The installation procedure is as follows:

1. Select tubing of an ID that can be slipped easily over the wire or wire group.
2. Use a hot-air gun, hair dryer or other suitable method as a heat source. Hold the heat source four to five inches away from the wire, and apply a heat of 275° F to 300° F for approximately 30 seconds. Rotate the wire while applying the heat, so that the heat is evenly distributed.
3. Remove the heat as soon as the tubing forms to the shape of the wire, and allow to cool for approximately 30 seconds before handling.

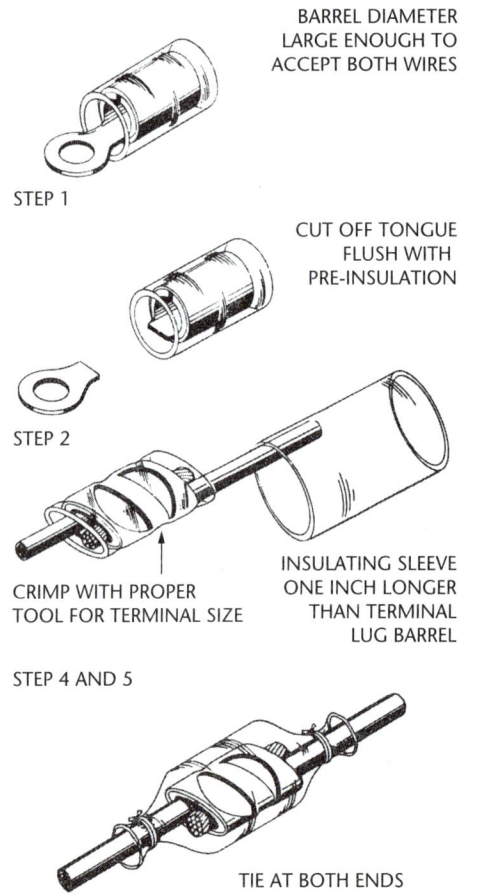

Figure 12-2. Terminal lug barrel repair of wire

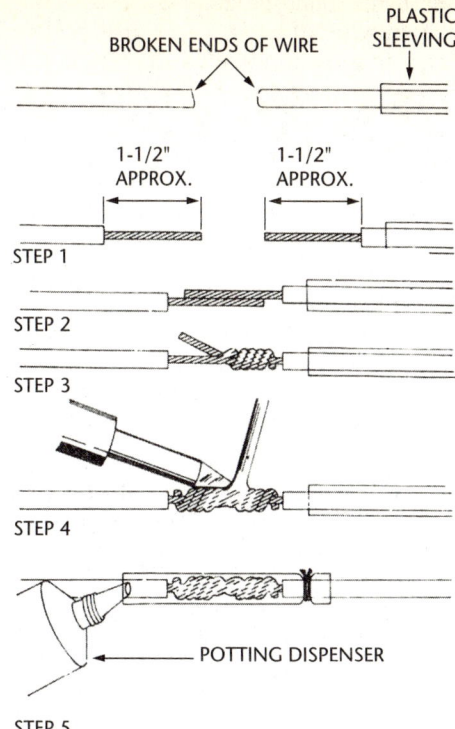

Figure 12-3. Repairing broken wire by soldering and potting

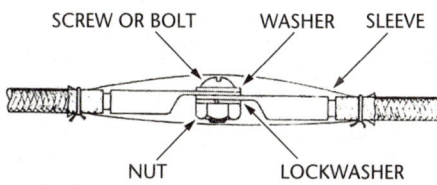

Figure 12-4. Bolted terminal lug repair of large wire

CAUTION: *Do not apply heat higher than 300°F as this may damage the wire. Do not continue to apply heat after the tubing has shrunk onto the wire; further application of heat will not cause it to shrink further.*

4. If the tubing does not shrink onto the wire in approximately 30 seconds, then the tubing selected is probably too large. Select the next smallest size, and repeat the procedure.

Splicing with solder and potting compound. When neither a permanent splice nor a terminal lug is available repair a broken wire as follows (see Figure 12-3):

1. Install a piece of plastic sleeving about 3 inches long, and of the proper diameter to fit loosely over the insulation, on one piece of broken wire.

2. Strip approximately 1-1/2 inches from each broken end of wire.

3. Lay the stripped ends side by side and twist one wire around the other with approximately four turns.

4. Twist free end of second wire around first wire with approximately four turns. Solder wire turns together, using 60/40 tin-lead rosin core solder.

5. When solder is cool, draw sleeve over soldered wires and tie at one end. If potting compound is available, fill sleeve with material prepared in accordance with Chapter 8, and tie securely.

6. Allow potting compound to set without touching for 4 hours. Full cure and electrical characteristics are achieved in 24 hours.

Splicing large wire with terminal lugs. Trim the broken ends of the wire, and install an insulating sleeve over one end of the wire. Strip wire, and crimp an insulated terminal lug of the proper size to each wire end, following procedures described in Chapter 4. Bolt the terminal lugs together as shown in Figure 12-4. Slide the insulating sleeve over the connection and tie securely to the wire at both ends.

Repairing damaged wire insulation. If the wire insulation is damaged but the wire itself is not damaged, repair the insulation in either of the following ways:

- Dip the damaged portion of the wire insulation into a container of potting compound. Instructions for mixing potting compound are given in Chapter 8. Allow potting compound to dry in air (70°-75°F) for 4 hours before touching. Full cure and electrical characteristics are achieved in 24 hours.

- If potting compound is not available, repair damaged wire insulation by using a transparent sleeve of flexible tubing 1-1/2 times the outside diameter of the wire and 2 inches longer than the damaged portion of the insulation. This sleeving is split lengthwise and wrapped 1-1/2 times around the wire at the damaged section. Tie with nylon braid at each end and at one inch intervals over the entire length. See Figure 12-5.

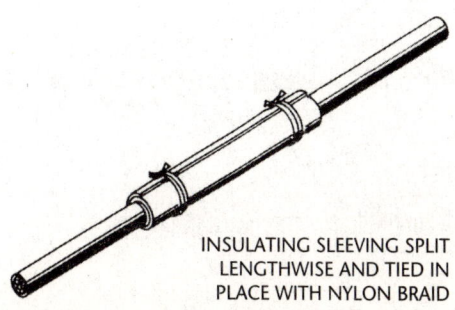

Figure 12-5. Insulation repair with sleeving

12-4 | Emergency Repairs

Repairing shielded cable. When shielded cable is severed it can be repaired in the following manner (see Figure 12-6):

1. Select a grounding sheath according to instructions in Chapter 1, page 1-19, steps 1 and 2.

2. Prepare the severed ends of the cable for application of a grounding sheath connector as described in Chapter 1, page 1-19, steps 3 and 4.

3. Slide two insulating sleeves, either shrinkable or flexible transparent tubing, and the inner one just large enough to pass over the grounding sheath connector, and the outer one large enough to accommodate the inner insulating sleeve and the grounding lead. The inner insulating sleeve should be just long enough to completely cover the permanent splice. The outer sleeve must be long enough to extend beyond the two grounding sheath connectors as shown in Figure 12-6.

4. Attach a grounding sheath connector to one end of the severed wire per Chapter 1, page 1-19, steps 5 through 10. The grounding wire should be long enough to span the repair.

5. Install a grounding sheath connector on the other side of the break. Do not crimp this yet.

6. Use a permanent splice to join the severed inner conductor, or use the barrel of a terminal lug when a permanent splice is not available. See page 12-2.

7. Slide inner insulating sleeve into position as shown in Figure 12-6. If shrinkable tubing is used apply heat.

8. Push the free end of the grounding wire from step 3, into the uncrimped grounding sheath connector. Crimp securely.

9. Slide outer insulating sleeve into place. If shrinkable tubing is used proceed as in step 7. If flexible tubing is used tie both ends with nylon braid as shown.

An alternate method of repairing shielded cable is to be used if grounding sheath connector, as described in the previous procedure, is not available. The alternate method shown in Figure 12-7 is as follows:

1. Prepare the severed ends of the cable for pigtail method of shield termination as described in Chapter 1, page 1-22, steps 1 through 4.

2. Use pre-insulated splice connector to join inner conductors as described in Chapter 4, page 4-18.

3. Use two splice connectors to add short length of insulated wire as extension to complete shield connection.

Repairing damaged shielding. When the shielding braid of shielded cable has been damaged, cut the cable sharp and square, and repair as described on page 12-4.

CAUTION: *Do not attempt to repair damaged shielding braid by covering with tape, as it is not possible to seal off severed ends, and these may puncture the wire insulation.*

Repairing coaxial cable. Do not attempt to patch up broken or damaged coaxial cable. If possible, replace the entire cable. If this is not possible, install a matching plug and jack of the proper size and type at the broken or damaged part, using the procedures described in Chapter 3.

CAUTION: *As every extra connection in a coaxial cable means a loss in efficiency, replace repaired coaxial cables at the earliest possible time.*

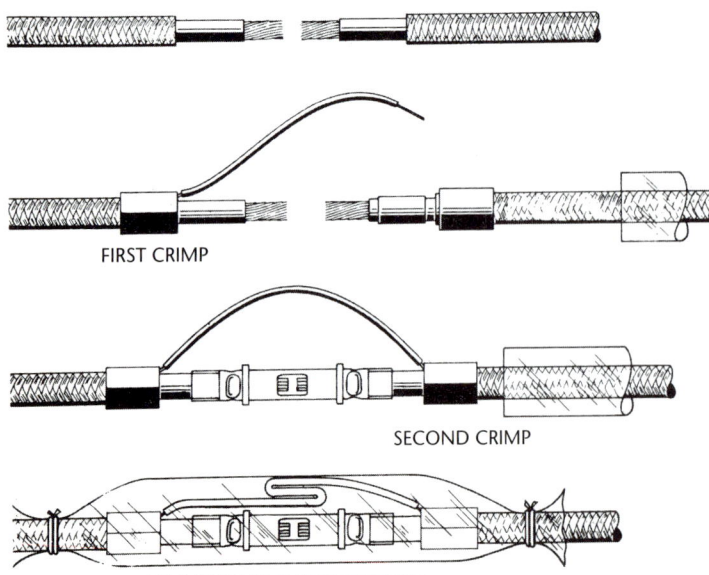

Figure 12-6. Repair of shielded wire

Section 2
Repairing Damaged MS Connectors

Repairing unpotted connectors. Defective MS connectors, which have broken pins, can be temporarily repaired in the following manner:

- Where it is possible to get at both halves of the connector one of the spare wires provided may be used by splicing the wire from the damaged or broken pin to the spare wire following the procedures of Chapter 4 on pages 4-18 through 4-20. This procedure must be followed for the wire leading to both halves of the connector. The unit must then be marked that this repair has been done. Replace both altered connectors at the next major overhaul of the aircraft.

Repairing potted connector. Potted connectors are equipped with spare wires on all spare pins. If a pin becomes defective the repair is made by cutting the wire leading to the defective pin and using a permanent splice (as previously described) to join the wire to a spare wire. The mating connector must also be so modified.

> **CAUTION:** *Tag both connector halves with complete information on the modification. Replace both connector halves at the earliest opportunity.*

If a spare wire is not available, it is possible to replace pins in potted resilient connectors that do not have metallic back shells. The following procedure should be carefully followed:

1. Cut away the potting compound (sealant) with a thin knife blade or scalpel. Use long nose pliers to pull the sealant while cutting. Be careful not to cut into the wire insulation.
2. Carefully scrape away sealant from defective pin.
3. Use a small (pencil) soldering iron or a soldering gun to unsolder the wire lead from pin.
4. Use long nose pliers to pull pin out of resilient insert.
5. Solder wire to new pin and push pin into insert from rear.
6. Pour new potting compound into area of repair and air cure at room temperature for 24 hours. The new compound will seal satisfactorily to the old compound remaining in the connector.

Occasionally a wire will fail inside the potted area of a connector. When the connector has a back shell, slide a thin knife blade around the outside edge of the sealant and unscrew the shell. This may take considerable force depending on how tightly the sealant adheres to the shell. Follow the same steps as described above to reach the soldered conjunction. Do not remove the pin, but solder a new wire to the contact and repot the connector.

Replacing terminal board covers. When a terminal board cover is lost or damaged so as to be unusable, cover the board temporarily with a piece of large vinyl tubing, split lengthwise, and tie securely around the terminal board. This procedure is described in Chapter 9, page 9-8.

> **CAUTION:** *All repairs described in this chapter are temporary, for emergency use only. Replace all temporary repairs as soon as possible with permanent repairs.*

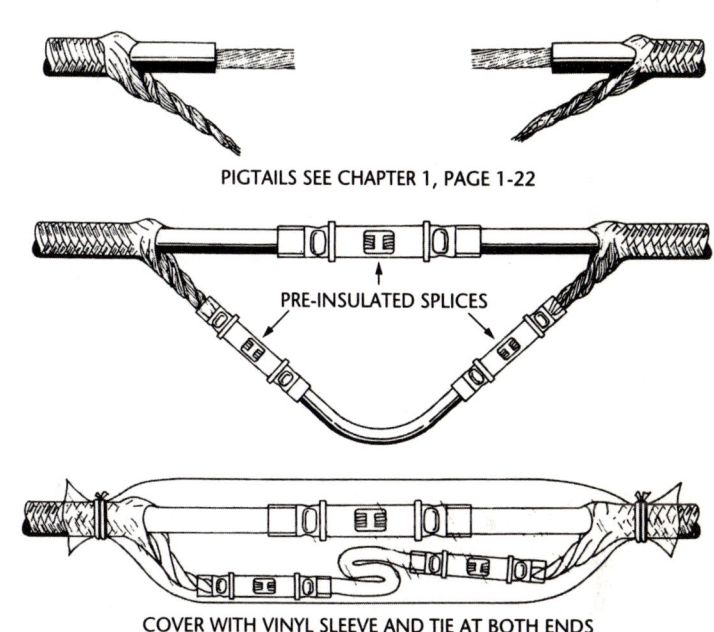

Figure 12-7. **Alternate repair of shielded wire**

Appendix A: Reference Specifications & Drawings

AN3427	Crimping tool, electric cable terminal, hand
AN5537	Connector assembly-thermocouple lead
AN5538	Terminal-thermocouple lead soldering
AN5539	Terminal-thermocouple, brass
AN5542	Terminal-thermocouple
AN5548	Terminal-lug, thermocouple, chromel and alumel
AN735	Clamp, loop type, bonding
AND10380	Fitting installations, standard AN conduit
AND10406	Thermocouple leads-iron-constantan, installation of
AND 10449	Circuit breaker installation
AND10460	Indentors and nests-copper electric terminal, power operated press
EMB 14-55	Aircraft electrical aluminum wire and terminals, installation practices for
LLL-R-626	Rosin, gum; rosin, wood; rosin, tall oil
MIL STD 429	Printed circuit terms and definitions
MIL-A-6091	Alcohol, ethyl, specially denatured, aircraft
MIL-B-5087	Bonding; electrical, aircraft
MIL-C-1140	Glass fiber; yarn, cordage, sleeving, cloth and tape
MIL-C-17	Cables, radio frequency, coaxial
MIL-C-21565	Clamp, loop, plastic, wire support
MIL-C-23216	Contacts, crimp type, electric connector, general specification for
MIL-C-23329	Connectors, coaxial, RF, series BNC, TNC, N and C
MIL-C-25038	Cable, electrical, aircraft, high temperature and fire resistant
MIL-C-25516	Connectors, electrical, miniature, shielded or unshielded, environment resisting type
MIL-C-26482	Connectors, electric, circular, miniature, quick disconnect
MIL-C-26500	Connectors, general purpose, electrical, miniature, circular, environment resisting, 200°C ambient
MIL-C-26636	Contacts, crimp type, for electrical connectors
MIL-C-3607	Connectors, coaxial, RF, pulse series
MIL-C-3608	Connectors, coaxial, RF, series BNC
MIL-C-3643	Connectors, coaxial, RF, series HN
MIL-C-5015	Connectors, electrical, AN type
MIL-C-5649	Cord, cotton, braided, prewaxed
MIL-C-6136	Conduit, flexible, shielded aluminum alloy
MIL-C-7078	Cable, power, electrical, 600 volts, shielded
MIL-C-71	Connectors, N, for radio frequency cables
MIL-C-7931	Conduit, electrical, flexible, radio frequency shielding
MIL-D-6998	Dichloromethane, technical
MIL-E-7080	Electrical equipment; installation of, aircraft, general specification
MIL-F-15160	Fuses; instrument, power and telephone
MIL-F-19207	Fuseholders, extractor post type, blown fuse indicating and non-indicating
MIL-F-20329	Flux, soldering, rosin base, general purpose
MIL-F-21608	Ferrule, shield grounding, insulated, crimp style
MIL-F-5372	Fuse, enclosed link, aircraft, 400-cycle AC
MIL-F-5373	Fuseholders, block type, aircraft
MIL-HDBK-216	RF Transmission Lines and Fittings
MIL-I-18057	Insulation sleeving, electrical, flexible, glass fiber, silicone rubber treated
MIL-I-23053	Insulation sleeving, electrical, flexible, heat shrinkable
MIL-I-3158	Insulation tape, electrical glass-fiber (resin filled); and cord, fibrous glass
MIL-I-3190	Insulation, electrical, sleeving, flexible, treated
MIL-I-631	Insulation, electrical, synthetic resin composition, non-rigid
MIL-I-7444	Insulation sleeving, electrical, flexible

MIL-I-7798	Insulation tape, electrical, pressure sensitive adhesive, plastic
MIL-I-8660	Insulating and sealing compound, electrical
MIL-L-18276	Lighting, aircraft interior; installation of
MIL-L-6363	Lamps, incandescent, aviation service, general requirements for
MIL-L-6723	Lights, aircraft, general specification for
MIL-L-6730	Lighting equipment, exterior, installation of
MIL-M-3171	Magnesium alloys; processes for corrosion, prevention of
MIL-M-4528	Marking machine, wire and plastic tubing, identification
MIL-P-6889	Primer, zinc-chromate, for aircraft use
MIL-S-23190	Strap, cable, adjustable, plastic
MIL-S-572	Cords, yarns and monofilaments, organic synthetic fiber
MIL-S-6872	Soldering process, general specification for
MIL-S-8516	Sealing compound, synthetic rubber electric connectors and electric systems, accelerator required
MIL-T-22520	Tool, crimp type, for contacts of electrical connectors
MIL-T-5679	Thermocouple leads, iron and constantan, chromel and alumel, and copper and constantan, installation of
MIL-T-6094	Thinner; dope and lacquer (cellulose nitrate)
MIL-T-7099	Terminals, lug and splice, crimp style aluminum, for aluminum aircraft cable
MIL-T-713	Twine and tape, lacing and tying, for use in electrical and electronic equipment
MIL-T-7928	Terminals, lug and splice, crimp style, copper
MIL-W-16878	Wire, electrical, insulated, high temperature
MIL-W-22759	Wire, electric, fluorocarbon insulated, copper
MIL-W-27300	Wire, electrical, polytetrafluoroethylene insulated, copper, 600 volt
MIL-W-5086	Wire, electrical, insulated aircraft
MIL-W-5088	Wiring aircraft, installation of
MIL-W-5845	Wire, electrical, iron and constantan, thermocouple
MIL-W-5846	Wire, electrical, chromel and/or alumel, thermocouple
MIL-W-5908	Wire, electrical, copper and constantan, thermocouple
MIL-W-7072	Wire, electrical, 600 volts, aluminum
MIL-W-7139	Wire, electrical, copper insulated, 600 volts, 400 degrees F
MIL-W-8777	Wire, electrical, copper, 600 volts, 150 degrees centigrade
MS17821	Strap, cable, adjustable, self-clinching, plastic
MS17822	Strap, cable, identification, adjustable, self-clinching, plastic
MS17823	Tool, hand, for adjustable plastic cable straps, thin type
MS17824	Tool, hand, for adjustable plastic cable straps, thick type
MS18091	Fuseholder cartridge, extractor post style FHN46G
MS18092	Fuseholder cartridge, extractor post Style FHN47G
MS20659	Terminals, lug, crimp style, copper uninsulated, class I
MS20995	Wire, lock
MS21265	Grommet, plastic, split
MS21919	Clamp, cushioned, support, loop type, aircraft
MS23002	Dies, MS25441 crimping tool, for use with MS25036, size 8 through 4/0 terminals
MS23003	Gages for MS23002 crimping dies
MS24000	Fuse holder, block type, 1-, 2-, and 3-Pole, 1-to 30-Ampere, Aircraft
MS24001	Fuse Holder, Block Type, 1-, 2-, and 3-Pole, 35 to 60-Ampere, Aircraft
MS24256	Tool-contact connector, assembly and disassembly
MS25036	Terminal, lug, crimp style, copper insulated, class I
MS25037	Crimping tool, hand, for copper insulated terminal
MS25123	Terminal board, electrical, assembly of
MS25181	Splice, electric, permanent, crimp style, copper, insulated, class I
MS25183	Connector, plug, electric for potting
MS25274	Cap, wire end (class I)
MS25281	Clamp, loop, plastic, wire support
MS25311	Ferrule, shield grounding, one piece, insulated
MS25312	Tool, crimping, hand, for insulated shield-grounding ferrule
MS25316	Gauges, for MS25312 crimping tool

MS25340	Gages, for MS25037 crimping tool, electric terminal
MS25435	Terminal-lug, crimp style, straight type for aluminum aircraft wire
MS25436	Terminal-lug, crimp style, 90° upright type, for aluminum aircraft wire
MS25437	Terminal-lug, crimp style, left angle type, for aluminum aircraft wire
MS25438	Terminal-lug, crimp style, right angle type, for aluminum aircraft wire
MS25439	Splice, permanent, crimp style, 2-way type, for aluminum aircraft wire
MS25441	Tool, crimping, electric, hydraulic, for wire terminals sizes 8 thru 4/0
MS25442	Dies, for MS25441 electric, hydraulic crimping tool
MS25472	Gages, for MS25442 Crimping Dies
MS25474	Fuseholder, cartridge, extractor post, style FHN28G
MS25494	Tool, manual, hydraulic, for crimping electric wire terminals
MS26572	Fuseholder cartridge, extractor post
MS27212	Terminal board assembly, molded-in stud, electrical
MS2S189	Terminal, lug, flag type, crimp style, copper
MS3103	Connectors-receptacle, electric, for potting
MS3191	Tool, hand, crimp, class i-for contacts of electric connectors
MS3196	Gage, inspection, for contact crimping tool
MS33540	Safety wiring, general practices for
MS33560	Installation of thermocouple leads in chromel and alumel 8 ohm system and copper and constantan 22 ohm system
MS33599	Thermocouple leads-iron-constantan, 2 ohm and 8 ohm system, installation of
MS35083	Jumper assemblies, bonding and current return
MS35489	Grommet, rubber, hot oil and coolant resisting
MS35490	Grommet, rubber, split, general purpose
Navy/BuWeps	
EMC No. 89-55	Electric connector sealing
O-F-499	Flux, brazing, silver alloy, low melting point
QQ-S-561	Solder, silver
QQ-S-571	Solder, lead alloy, tin lead alloy and tin alloy
QQ-T-25	Tape, electrical wire, flexible insulating sleeving, marking machine, (foil, wire identification marking)
VV-P-236	Petrolatum, technical
W-L-111	Lamp, incandescent, (electric, miniature, tungsten filament)

Index

A

air connector 2-4
 MS3100 2-4
 MS3101 2-4
 MS3102 2-4
 MS3106 2-4
 MS3107 2-4
 MS3108 2-4
aluminum wire 1-6
Amphenol connectors 2-28
 Cannon MS class A 2-28
 class E and class R 2-30
 MS class B 2-29
 MS class C 2-30
AMP power tools 4-4
AN-MS connectors 2-1
 classes 2-5
  class A 2-5
  class B 2-5
  class C 2-5
  class E 2-5
  class K 2-5
  class R 2-5
AN connectors 2-1
ARINC 429 10-21
ARINC 629 10-21
asbestos 1-5
assembly of wires to connectors 2-19
 contacts 2-19
  crimp-type 2-19
  solder cup 2-19
 preparation 2-20
 solders 2-19
 wire types 2-19

B

Bendix-Scintilla connectors 2-34
 class A 2-34
 class C 2-35
 class R 2-36
 fireproof 2-37
bend radii 10-3
BNC series connectors 3-5
 captivated contact version 3-6
  attaching to coaxial cable 3-7
 improved version 3-5
  Attaching to coaxial cable 3-6
bonding & grounding 6-1
 cylindrical surfaces 6-6
 hardware 6-3
 methods of 6-5
 preparation of surfaces 6-4
 refinishing 6-7
 testing 6-7
Burndy power tools 4-4
busbar 9-1, 10-10
 cleaning when making connections 10-10
 connecting terminal lugs to 10-10
 connecting two terminals to same point on 10-11
 hardware for connection to 10-10
 insulation 9-2
 mounting hardware 9-2
 plated aluminum and copper 9-2
 preparation of 9-1
 protection of 9-3, 10-11
 repairing damaged plating 9-2
 unplated aluminum alloy 9-1

C

cable 1-1
cable clamp 2-25, 10-7
 installation of 10-7
  to tubular structure 10-7
 MS connector 2-13
  MS3057 2-25
  MS3057A 2-26
  MS3057B 2-26, 2-27
cable identification code 1-6
Cannon connectors 2-38
 class R 2-40
 DPD connectors 2-19
 fireproof 2-40
 MS class B 2-38
 MS Class C 2-38
 MS class E 2-39
circuit assembly 7-6
circuit components 7-7, 7-10
 replacement of defective 7-10
circuit function letter 1-7
coaxial cable 1-9
 repair of 12-4
combing wires 10-2
conductor 7-10
 repair of major defects 7-10
 repair of minor defects 7-9
conductor interfacial connections 7-10
 repair of 7-10
conductor pattern 7-9
 repair of 7-9
conduit 9-3, 10-12
 bending limitations 9-3
 bonding or grounding 9-4
 capacity 10-12
 drainage 9-3
 feeding wires into 10-12
 installation of 9-3
 spacing of supports 9-3

    supporting hardware 9-3
        installation of 9-3
    supporting wires at end of 10-13

connector 10-13
    assembly to receptacles 10-13
    coding of 10-13
    disassembly from receptacles 10-13
    installation of 10-13
    installing conduit on 10-14
    mounting 10-14
    special precautions 10-14

connector sealing 8-1

contact tools 2-41
    tool inspection 2-41

continuity test 2-53

copper wire and cable 1-16

crimp-type contacts 2-13

crimping 2-44
    eyesight 2-45
    MS3191-3 2-44
    procedure 2-44
        AMP power tools 4-5
        Burndy power tools 4-5
        MS25037 standard hand tool 4-4
        MS series power tools 4-11
        T & B power tools 4-6

crimping tool, hand 2-41
    AN3427 standard hand tools (Burndy Tool #MY-28) 4-10
    MS25037 4-3
    MS3191-1 2-41
    MS3191-3 2-41

crimping tool, large copper terminal lugs 4-7

crimping tool, power 4-3
    inspection and adjustment 4-4

C series connectors 3-8
    attaching to coaxial cable 3-8
    C connectors 3-8
    SC connectors 3-8

cutting aluminum wire 1-6

D

dead-ending 1-23
    alternate methods 1-23
    grounding sheath connector 1-23
    shielded cable 1-23
    tape wrap 1-23

dielectric insert 2-3

dip soldering 7-3. See also **soldering**

dip-tinning 1-17

disassembly of connectors 2-15
    AN-MS class E connectors 2-16
    Cannon DPD connectors 2-19
    MS connector 2-17
        class R 2-17
        miniature 2-17
        removable crimp-type contacts 2-18

    removal of back shells 2-15
    removal of contacts 2-15

E

electrical hardware 9-1
installation of 9-1
electrically protected wiring 10-1
electrically unprotected wiring 10-1
electrical resistance soldering 2-21
extraction tools 2-43

F

FEP - (fluorinated ethylene propylene) 1-5
fireproof connectors, environment resisting 2-15
fire resistance 1-5
flux 7-2
fuse 11-3
    precautions 11-4
fuseholder 11-4

G

glass braid 1-5
grommet 10-8
    installing 10-8
ground letter 1-7

H

hand stripper 1-16
hard solder 7-2
HN series connectors 3-9
    captivated contact version 3-9
        attaching to coaxial cable 3-11
    improved version 3-9
        attaching to coaxial cable 3-10
hook-up wire 1-5
hot blade stripper 1-15

I

identification sleeves, installing 1-12
identifying wire and cable 1-6
insertion tools 2-43

insulated wire 1-1

interconnecting wire 1-5

J

jackets 1-5

junction box 9-4, 10-14
- *bonding or grounding* 9-5
- *covers* 9-4
- *drainage of* 9-4
- *identification* 9-5
- *installation of* 9-4
- *insulation* 9-4
- *lacing or tying in* 10-14
- *mounting hardware* 9-4
- *support inside* 10-14
- *vapor-tight* 9-5
- *wire entry holes*
  - preparation of 9-4

L

lacing 10-15
- *branch-offs* 10-16
- *double cord* 10-16
- *precautions* 10-15
- *single cord* 10-15

lamp 11-1
- *abbreviations* 11-1
- *precautions* 11-3
- *types* 11-1

lead mounted components 7-10, 7-11
- *replacement of* 7-10

Liquidometer S62 and S63 series connectors 3-24

lock wire 10-19

lock wiring 10-20
- *AN-MS connector* 10-20
- *connector to structure* 10-20
- *electrical connector* 10-20

lug mounted components 7-11
- *replacement of* 7-11

M

marking machines, wire identification
- *procedure* 1-10
- *set-up* 1-10

miniature rectangular connectors 2-14, 2-49

MS connectors 2-1
- *cable clamps* 2-13
- *contacts* 2-12
- *marking* 2-7
- *miniature* 2-5
  - classes 2-7
    - class E 2-7
    - class F 2-7
    - class FL 2-7
    - class J 2-7
    - class P 2-7
  - crimp contacts 2-6
  - solder contacts 2-5
- *potting connectors* 2-13
- *repairing* 12-4
  - potted connector 12-5
  - unpotted connector 12-4

multiconductor cable 1-8

N

non-metallic conduit
- *use of heat-shrinkable tubing* 12-2

non-standard connectors with removable crimp-type contacts 2-46
- *Amphenol #69 series* 2-46
- *Bendix #10-214000 series* 2-46
- *Bendix CE series* 2-46
- *Cannon EXA series* 2-46

nylon 1-5

N series connectors 3-11
- *captivated contact version* 3-12
  - Attaching to coaxial cable 3-13
- *improved version* 3-12
  - attaching to coaxial cable 3-12

O

open wiring 10-1

P

phase letter 1-7

pin-and-socket contacts 2-3

plated aluminum and copper busbar 9-2. *See also* **busbar**
- *preparation of* 9-2

plug 2-3

polyimide 1-5

potting compound 8-1. *See also* **sealing compound**

potting connectors, MS 2-13

power rotary stripper 1-15

printed circuit assemblies 7-6
- *desoldering components* 7-7
- *repair of* 7-6
  - precautions 7-6

protection of electrical connectors 2-52

protective devices 9-5
- *identification* 9-7
- *mounting*
  - circuit breakers 9-6

Index

    into a blind hole 9-6
    into tapped hole or nut plate 9-6
    relays 9-6
    with through bolts or screws 9-6
  mounting hardware 9-5
  protection 9-6
  protective coating for electrical connections 9-6
  special precautions 9-6

pulse series connectors 3-15
  ceramic insert version 3-15
    Attaching to coaxial cable 3-15
  rubber insert version 3-15
    attaching to coaxial cable 3-17

PVC - (polyvinyl chloride) 1-5

R

R, S and T circuits 1-7

radio, radar & special electronic circuits 1-7

receptacle 2-3

rectangular connectors, miniature 2-14

rectangular shell connectors 2-14

resistance soldering 7-2. *See also* **soldering**

RF connectors 3-1, 3-2
  BNC series 3-2
  C series 3-2
  HN series 3-2
  MB miniature connector series 3-19
    attaching to coaxial cable 3-20
  N series 3-2
  Pulse series 3-2
  SC series 3-2
  subminiature RF connectors 3-21
    attaching to coaxial cable 3-21
    crimping procedure for 3-22
  TNC series 3-2

S

SC connectors 3-8

sealing compound 8-1. *See also* **potting compound**
  dispensers for 8-4
  high temperature 8-1
    preparation of 8-4
  precautions 8-2
  preparation of 8-2
    hand mixing 8-2
    machine mixing 8-3
    mechanical mixing 8-2
  preparation of TFE-insulated wire 8-4
  storage of mixed 8-4
  storage of unmixed 8-4

seal wire 10-19

seal wiring 10-21
  emergency devices 10-21

self-clinching cable strap 10-18
  installing 10-18

shear wire 10-19

shear wiring 10-20
  split shell assemblies 10-20

shield and multiple connections 2-49

shielded cable 1-23
  repair of 12-4
  terminating 1-19

shield termination 1-19
  grounding ferrule method 1-19
  grounding sheath connector method 1-20
  pigtail method 1-23
    alternate pigtail method 1-23

silicone rubber 1-5

sleeve stamping 1-11
  large sleeving 1-12
  machine 1-12

soft solder 7-2

soldering 2-19, 7-1
  coaxial cable to RF connectors 3-4
  electrical resistance soldering 2-21
  heat application methods 7-2
  holding connectors 2-23
  precautions 7-3, 7-8
    amount of solder 7-5
    choice of soldering tip 7-5
    cleanliness 7-4
    heat application time 7-5
    heating capacity 7-4
    pre-tinning 7-4
    selection of flux and solder 7-4
  procedure 2-20
  procedures 7-3, 7-8
    application of heat and solder 7-5
    base laminate repairs 7-9
    cleaning 7-5
    removal of protective sealing (conformal) coating 7-8
    repair of conductor interfacial connections 7-10
    repair of conductor pattern 7-9
    repair of major conductor defects 7-10
    repair of minor conductor defects 7-9
    replacement of defective circuit components 7-10
    replacement of lead mounted components 7-10
    replacement of lug mounted components 7-11
  sequence 2-24
  soft solder-60/40 tin-lead 2-19
  soft solder-lead-silver 2-19
  torch soldering 2-22
  types of
    dip 7-3
    resistance 7-2
    torch 7-3
  typical operations 7-2

soldering iron 7-2, 7-5
  holder 7-5
  maintenance 7-4
  preparing 7-3
  protection against overheating 7-5
  selection of 7-4

soldering iron tinning 1-19

soldering methods 7-1

solder joint 7-1, 7-6
  acceptable 7-6
  cooling 7-5
  inspecting 7-6

    securing 7-5
    unacceptable 7-6
solid shell angle plug
    wiring 10-21
splicing 4-18, 12-1
    aluminum 4-20
    large copper wires 4-19
    large wire with terminal lugs 12-3
    multi-splicing 4-20
    permanent splice 12-1
    small copper wires 4-18
    solder and potting compound 12-3
    terminal lug barrel 12-2
stripping wire and cable 1-13
    aluminum wire 1-13
    copper wire 1-13
    hand stripper 1-16
    hot blade stripper 1-15
    knife stripping 1-16
    power rotary stripper 1-15
    stripping jacket on shielded cable 1-19
subminiature connectors 2-13

T

terminal 10-10
    aluminum 10-10
        installation torques for 10-10
    large copper 10-10
        installation torques for 10-10
terminal boards 9-7, 10-9
    alternate method of attachment 9-7
    attaching busbar to 9-7
    hardware for wiring 10-9
    identification 9-7
    installation of 9-7
    insulation 9-7
    method of attachment 9-7
    mounting hardware 9-7
    protection of 9-7
terminal board cover
    replacing 12-5
terminal lugs 4-12, 10-9
    aluminum 4-14
    and splices 4-1
    connecting to equipment 10-12
    high temperature 4-12
    large copper wires 4-6
    pre-insulated 4-2
test leads 2-53
TFE - fluorocarbon (polyretrafluoroethylene) 1-5
    heat-setting identification marking 1-10
    wire, special instructions for marking 1-10
thermocouple 5-1
    soldering 5-5
        hard soldering 5-5
        resistance soldering 5-7
        silver solder 5-5
        soft-soldering 5-8
        torch soldering 5-6

    wire 1-9
        installation 5-10
    wire leads 5-1
    wire preparation 5-4
tinning copper wire and cable 1-16
    dip-tinning 1-17
    soldering iron tinning 1-19
TNC series connectors 3-18
    attaching to coaxial cable 3-18
tool inspection 2-41
torch soldering 2-22, 7-3. See also **soldering**
twisting wires 10-2
tying 10-15
    making 10-16
    precautions 10-15
    securing with pressure-sensitive tape 10-17
    self-clinching cable straps 10-18
    sleeves to wire groups or wire bundles 10-17
    temporary 10-17
    wire groups into wire bundles 10-17

U

unit number 1-7
unplated aluminum alloy busbar 9-1. See also **busbar**
    preparation of 9-1

V

vapor-tight junction box 9-5. See also **junction box**

W

wire. See **insulated wire**
    repair of 12-1
wire and cable preparation 1-13
    stripping 1-13
wire bundle 1-12, 10-1, 10-15
    identity of groups within 10-2
    size of 10-2
    spliced connection in 10-3
wire group 10-1, 10-15
wire harness 1-12, 10-1
wire identification code 1-6
    circuit function letter 1-7
    ground letter 1-7
    phase letter 1-7
    thermocouple letter 1-7
    unit number 1-7
    wire number 1-7
    wire segment letter 1-7
    wire size number 1-7

wire insulation
- *repair of* 12-3

wire marking 1-7
- *location of sleeve* 1-7
- *objectives* 1-7
- *spacing* 1-7

wire number 1-7

wire segment letter 1-7

wire separation 10-2
- *from plumbing lines* 10-6

wire size number 1-7

wiring installation 10-1
- *in conduit* 10-12
- *in junction boxes* 10-14
- *precautions* 10-3
- *protection* 10-5
- *routing* 10-3
- *slack* 10-3